CONSTRUCTION LAW

AUSTRALIA
LBC Information Services
Sydney

CANADA and USA
Carswell
Toronto

NEW ZEALAND
Brooker's
Auckland

SINGAPORE AND MALAYSIA
Sweet & Maxwell Asia
Singapore and Kuala Lumpur

CONSTRUCTION LAW

Law and Practice relating to the
Construction Industry

By
JOHN UFF

**Q.C., Ph.D., B.Sc.(Eng.)
F.R.Eng., F.I.C.E., F.C.I.Arb.**

*Chartered Engineer, Barrister
Recorder of the Crown Court
Bencher of Gray's Inn
Nash Professor of Engineering Law,
King's College, London*

SEVENTH EDITION

SWEET & MAXWELL

1999

First Edition 1974
Second Edition 1978
Third Edition 1981
Fourth Edition 1985
Second Impression 1989
Fifth Edition 1991
Second Impression 1994
Sixth Edition 1996

Sweet & Maxwell Limited of
100 Avenue Road, Swiss Cottage, London, NW3 3PF.
Phototypeset by LBJ Typesetting Ltd
of Kingsclere
Printed in England by Clays Ltd, St Ives plc.

A catalogue record for this book is
available from the British Library.

ISBN 0–421–667 206

*This edition is dedicated to Diana
who was my first law teacher*

PREFACE

This seventh edition marks the twenty-fifth anniversary of the publication of the first slim edition of *Construction Law*, then priced at £1.30. It is a suitable occasion to recall the circumstances in which that first edition came to be written and published, recording events that would not normally see the light of day.

The first edition was substantially drafted while working as Donald Keating's pupil (as it happened, his last) and during the first two years in practice at 11 King's Bench Walk. There were, in the early 1970s, few texts on the subject. *Hudson* and *Keating* were the standard fare, with *Abrahamson on Engineering Law* and one or two texts dealing with architects and surveyors. *Emden* had a small corner of the market, but my researchers revealed that the text had been largely borrowed in 1932, no doubt by the mysterious W. E. Watson, from the first edition of *Halsbury's Laws of England*, originally written by Alfred Hudson himself. My conclusion was that there was certainly room for a new book and my intention was to write a text which would be of interest primarily to engineers and architects.

The text was originally offered to Butterworths, without success. Sweet & Maxwell responded more positively. My original title was "Law for Engineers and Architects" but Sweet & Maxwell's Managing Editor, the formidable Radmila Barnicot saw this as a potential candidate for the Concise College Text Series, provided that the text would satisfy the syllabus of a number of building and civil engineering courses. Another requirement was to avoid footnotes and to include all the material within the text. During the negotiations the title came up for consideration. On April 23, 1971 Miss Barnicot wrote:

> "Secondly, I am not very happy about the title which gives the impression of excluding builders, and I think something like *Building and Construction Law* would be better. This would fit in better with the titles of the series in which we would put your book, the Concise College Texts, which are as short as we can make them."

The reply of May 2, 1971 makes interesting reading.

> "With regard to the title of the book, I have not given this much thought and have no strong views. I am sure your suggestion is an improvement on the original. The proper title should be something like *Law Relating to Building and Construction* since building law is really the specialised part which applies only to building. However, if this is too long, perhaps we could set a new trend with *Construction Law*".

Thus, the new title and a new subject came into being. The manuscript, bearing the strong imprint of Radmila's firm direction was another 18 months in the completion, followed by a further delay waiting for the new fifth edition of the ICE Conditions. Publication was a much more leisurely affair in the 1970s. The work finally appeared in hard and soft covers in 1994, substantially in the form it now takes but with various chapters which are now split into two, so that the original 11 chapters has grown to 16.

Among the serious changes which have occurred since 1974, perhaps the most far reaching has been the systematic (if not overwhelming) availability of construction law decisions from the Technology and Construction (formerly Official Referee) Courts. This revolution was pioneered by my old friend the late John Parris with his *Building Law Reports*, first published in 1976. Only those in practice at the time will recall the almost cloak and dagger atmosphere that accompanied the publication of typed manuscripts of Official Referee judgments, intended only for private circulation among the bar and specialist solicitors. Parris was openly accused of not playing the game. Fortunately, he persevered and the subject became transformed, supported today by a vast range of texts, reports and journals, was well as the twin institutions: the Society of Construction Law and the Centre of Construction Law & Management, King's College, London, with both of which I have been proud to have a long association.

This new edition incorporates the latest clutch of leading construction cases, among which the final denouement of *Crouch* is of particular note. Standard Forms continue to proliferate with substantial changes to the JCT Form and complete revision to GC/Works/1 and FIDIC. Additions to the script include coverage of PFI, tree root damage and foreign corporations. The 1996 Arbitration Act, covered in anticipation in the sixth edition, produced a few surprises at the enactment stage, including the abandonment of the courts' discretionary power over stay. The CIMA Rules are a positive product of the 1996 Act. The new CPR has also required extensive changes to Chapter 2. Of greatest moment, however, is the final bringing into force on May 1, 1998 of the Housing Grants, Construction & Regeneration Act 1996, Part II, together with its controversial Scheme for Construction. This has led to major changes in practice and in the Standard Forms. The full impact of this sea-change must await another edition.

May 24, 1999   Keating Chambers
                                                  10 Essex Street
                                               London WC2R 3AA

ACKNOWLEDGMENTS

Grateful acknowledgment is made to the following publishers for permission to quote from their works:

BUTTERWORTHS: All England Law Reports and Construction Law Reports

THE INCORPORATED COUNCIL OF LAW REPORTING FOR ENGLAND AND WALES: Law Reports and Weekly Law Reports

LLOYD'S OF LONDON PRESS LIMITED: Lloyd's Reports

LONGMAN GROUP U.K. LIMITED: Solicitors' Journal and Building Law Reports

TOLLEY PUBLISHING: Local Government Reports

RIBA PUBLICATIONS: JCT Standard Form

THOMAS TELFORD PUBLISHING: ICE Conditions of Contract

While every care has been take to establish and acknowledge copyright, and contact the copyright owners, the publishers tender their apologies for any accidental infringement. They would be pleased to come to a suitable arrangement with the rightful owners in each case.

CONTENTS

TABLE OF CASES

TABLE OF STATUTES

CHAPTER 1

CONSTRUCTION AND THE LEGAL SYSTEM

"CONSTRUCTION law" is neither a legal term of art nor a technical one. It is used to cover the whole field of law which directly affects the construction industry. Construction law is not concerned only with law. It has been found, for example, that efficient and workable construction contracts require that the needs of the construction process should be taken into account by applying the principles of management. Construction law is, thus, an inter-active subject in which both lawyers and construction professionals, including managers, have an essential part to play.

The term "building contracts" is often encountered in text books. This refers to those areas of law which govern the interpretation of these contracts. Construction law thus includes building contracts but extends further into the construction process itself and the part which the law has to play in the efficiency of the process.

Most of the book is concerned with substantive law, that is, law which lays down rights and duties of individual parties. The concern of most people is to keep well within the bounds of the law and to avoid becoming involved in legal action. Substantive law is all that concerns them. But the nature of the construction process seems inevitably to lead parties frequently to the brink of legal action, including arbitration, and often beyond.

Many individuals in the construction industry will at some time in their careers become professionally involved in litigation or arbitration. And there can be few who will not have contact with claims, which may be a prelude to a formal dispute. This book extends beyond substantive law, to give some account of the systems of dispute resolution which may be resorted to.

THE NATURE OF LAW

Different types of law create different kinds of rights and duties. Some law applies only between individuals; if A behaves in a way which causes loss to B, then the civil law may allow B to recover his loss from A. Much of the law of contract and tort applies in this way, B's legal remedy being the only sanction against A. In other types of law the State may be involved; if A drives his motor car negligently he may be liable not only for B's resulting loss (through injury to him or his property) but also to criminal prosecution. Some areas of law concern only the rights or duties of the individual against the State; for example planning law or tax law. Most of the principles discussed in this work relate to civil law.

The factor which connects various types of law is that they all exist in order to allow persons to regulate their lives with reasonable certainty and in a manner which society presently considers to be just. The law is often accused of unwarranted delay, and of being old fashioned and too complicated. All these things are at times true. But law is made by and for human beings, who today lead lives of unprecedented complexity. The fact that the law will, if called upon, provide a reasoned solution to any type of human problem is in no small degree remarkable.

Law and technology

Many of those involved in construction law have a technical background. It may therefore be informative to examine the nature of law through the ways in which law differs from technology.

First, in technical fields there are always some problems which, for the time being perhaps, cannot be solved. Until the invention of large-diameter bored piles engineers considered that tall buildings could not be built in London, because of the compressible clay subsoil. In law such a situation never arises. No matter how complex the facts of a case are or how confused the law, a judge must always decide which of the parties wins. This does not mean that the judge is always right. There may be an appeal against his decision; but the same problem then confronts a higher court.

The way in which the courts solve a difficult problem may be illustrated by a judicial example. In the case of *S.C.M. v. Whittall*[1] the Court of Appeal had to decide whether the plaintiff's loss was excluded by a rule of law. The precise extent of the rule was a matter of uncertainty. In giving judgment, Lord Denning said:

> "Where is the line to be drawn? Lawyers are continually asking that question. But the judges are never defeated by it. We may not be able to draw the line with precision, but we can always say on which side of it any particular case falls."

---

[1] [1971] 1 Q.B. 337.

Secondly, in engineering and building practice approximation and simplification play a large part: small light structures may be safely designed using approximate methods, while on large structures many more factors have to be taken into account. In law there is (or should be) no such scale effect. The law applied to a claim for £1,000 is the same as that applied to a claim for £10 million. This perhaps explains the old adage that a lawyer will never give a simple answer to a simple question. Simplification of the facts of a case does not simplify the law involved, and one may need to know every factual detail before even a guess can be made at the probable legal result. There are many examples in the law reports of leading cases, often going to the House of Lords, where the sum in issue between the parties was very modest.

Thirdly, while technology proceeds upon logical induction and deduction, the development of the common law is different. General principles of law may be arrived at by induction. But when a judge seeks to extend an established principle of law, he may do it in a way which does not amount to logical deduction. The common law in its practical application has been said to be more the embodiment of common sense than reason. The point was expressed judicially in the *S.C.M.* case above:

"In the varied web of human affairs the law must abstract some consequences as relevant not perhaps on the grounds of pure logic, but simply for practical reasons."

Fourthly, although technology advances with increasing rapidity not everyone uses the latest theories and methods. Indeed, one could today build a perfectly satisfactory bridge using a design by Brunel. A modern bridge would simply be much cheaper. In law the situation is quite different. Legal rights and duties depend only upon what the law says at the relevant time.

A change in statute law can mean that a man may do an act on one day with impunity but on the next at his peril. Common law, or case precedent, is rather different. In theory the judge states what the law is and has always been, so that a restatement applies retrospectively. The practical effect as to the future is, however, the same as a change in the law. Thus, when considering the law on some point, the very latest sources must be consulted and often likely future changes may need to be considered. It is by no means unknown for a court to come to a wrong decision because a recent change in statute law or recent case on the point was not brought to its attention.

CATEGORIES OF LAW

English law may be categorised in a number of ways. It may be divided into substantive law and procedure. Substantive law refers to all the branches of law which define a person's rights and duties, such as contract, tort and crime. The substantive law determines in a particular case what facts must be proved to achieve a certain result, such as to establish that a binding contract was concluded. This may depend, for example, upon whether the defendant communicated his alternative conditions of contract to the plaintiff before starting the work in question (see Chapter 6).

Procedure deals with the often complex rules by which the process of law is set in motion to enforce some substantive right or remedy. Procedure properly arises only when there is resort to legal action, but nevertheless it can be as important in practice as the substantive law. In the context of construction disputes, procedure includes arbitration and other dispute resolution processes, which all have their own procedures.

A further division of law is into common law and statute law; that is, into judge-made law and legislation. Then there is another type of division between common law (used in a rather different sense) and equity, which is a distinction based upon the two great independent roots of English law. These latter two divisions are discussed separately in the following sections because they are fundamentally concerned with the sources of English law. While considering each of the divisions it should be borne in mind that they are not mutually exclusive. Statute law deals with procedure and substantive law; judge-made law comprises equity as well as common law, and so on.

Another division is between private law and public law, sometimes called administrative law. Private law relates to rights exercisable by and against individuals (including corporations and other legal entities). Public law, however, applies in the domain of powers and duties exercisable by public bodies, which may affect the rights or expectations of individuals. The way in which public law rights may be enforced is significantly different to private rights, and consists essentially of applying to the courts for orders to review the actions of the public body in question. The enforcement of public law rights is dealt with below. Yet another division exists between so-called domestic and international law. And international law itself falls into two distinct categories: private international law, which applies to disputes having an international character but concerning individuals or other legal entities; and public international law, which applies primarily between states. Private international law comprises the rules applied in a particular country to resolve conflict between the domestic law of that country and of other countries whose laws may affect a dispute. International law is discussed below.

The common law

English law is based upon the common law system. The common law means literally the law which was applied in common over all parts of the realm. It was created in the twelfth and thirteenth centuries by the King's judges and has been developed and handed down to the present day. The essential feature of the common law which distinguishes it from other systems, is that it is based entirely on evolving precedent, with no written principles from which the precedent stems. The common law has thus been found a wonderfully flexible instrument, capable of rapid adaption to wholly new circumstances without obvious strain. This is in contrast to problems which frequently occur under statutes, which may be found inadequate to cover some new situation. Likewise under the Civil Codes which constitute the laws of many foreign countries, it may be found that the code cannot be made to cover a new case, and the judge is unable to do anything other than to apply the existing code. The common law is fundamentally different in that English (and Commonwealth) judges are creating law when they give judgments in cases which progressively add to, develop and sometimes amend the law in accordance with the needs of the community. Judges under the common law system thus play a more significant constitutional role than under the civil law of other countries.

One effect of the spread of the English-speaking peoples from the sixteenth century onwards, was that they took with them their laws. As a result, by the nineteenth century there was established literally throughout the world, a greater common law, subject to the effect of statute law in any particular state. Within the British Empire, and later the Commonwealth, this law was maintained as a truly common system by the establishment of the Privy Council (composed largely of members of the judicial committee of the House of Lords) as the final court of appeal. This still applies in a limited number of countries although the major commonwealth countries have long since set up their own courts of final appeal, such as the High Court of Australia. This means that the law in such countries tends inevitably to diverge from English law with the passage of time. The United States, which has developed its own law for 200 years, has adopted some notable differences from English law, such as stricter liability under the law of tort. English courts will, however, take note of and draw guidance from decisions from other common law countries and, to an often greater extent, English decisions are regularly followed or adopted in many different countries. The common law can now be regarded as having more sources than only English law.

In theory, the common law is not written down. It is stated each time a judgment is given at the end of a case, when the judge gives reasons for the legal principles embodied in his decision. In practice, the common law is found in the reports of judgments, and the law on any topic is to be discovered by reading those cases which turn on related facts. In some areas there will be only one or two cases, but in others there will be many dozens of cases, perhaps going back several centuries. There have been a number of attempts to "codify" the common law, that is, to write down in a statute the effect of the common law as it stands, with the object of making the law more accessible. This was done with considerable success at the end of the nineteenth century in a number of important commercial areas, including the of sale of goods (see Chapter 7). In the twentieth century, codification and the general reform of areas of old and unsatisfactory law have generally been entrusted to the Law Commission. This is a statutory body which prepares reports, carries out consultations and makes proposals for amending legislation to codify and clarify the common law. It is then a matter for Parliament to accept their recommendations. One of the latest Law Commission Bills to be placed before Parliament concerns rights of third parties in contract (see page 161).

Law reports

Not every case before the courts makes new law. It is only cases which involve a point of law which can do so; that is, when the law is uncertain in its application to the particular facts. Many cases depend simply upon conflicting versions of the facts. In the construction field, points of law are often involved, not least because the effect of complex conditions of contract is often unclear. Further, over the past 20 years, construction cases have become a major generator of new commercial law, having brought before the courts a range of new legal issues ranging from the application of the tort of negligence to defective buildings to various principles of the law of contract, such as rights of third parties.

Of the cases which do lay down new law, the important ones are published in reports. A law report contains, verbatim, the essential parts of the judge's decision. Subsequently, academic writers of textbooks and articles, and judges in later cases, will comment upon the judgments and mould them into propositions of law. This is an essential process to make the law manageable. But when a point of law arises the lawyer must go back to the reports to discover the words in which the judges have previously stated the law.

Given the importance of judgments it is curious that, in England, there has not, in modern times been an official organisation to produce law reports. In mediaeval times official court "rolls" were maintained (in Latin) which are still the subject of legal research. From the sixteenth to the nineteenth century, law reports were produced by private individuals, the quality of reports not always being very high. From about 1870 there has been a semi-official series called "The Law Reports," which are fully authoritative. There are also today a substantial number of commercially produced series. Two well-known series which contain cases of general interest are the *Weekly Law Reports* (W.L.R.) and the *All England Law Reports* (All E.R.). Building cases in the Law Reports will appear in *Queen's Bench reports* (Q.B.) or *Chancery reports* (Ch.) or in *Appeal Cases* (A.C.). Commercial cases, including many important decisions in the field of arbitration and contract, appear in *Lloyd's Reports* (Lloyd's Rep.). The abbreviation in brackets is the citation for a particular series (the year, volume and page also being given).

In addition to these general series, a new series was introduced in 1976 to deal exclusively with cases affecting the construction industry. *Building Law Reports* (B.L.R.) do not follow the traditional pattern of publishing reports as they appear, but include earlier decisions not previously reported or not readily available. Some volumes cover particular topics. A further series more recently introduced is the *Construction Law Reports* (Con L.R.). Frequently, cases will be reported in more than one series, so that a leading case may appear in both the Law Reports and in one of the commercial series of reports and also in the specialist reports such as B.L.R. The main difference in these reports is in their format. Most appellate cases in the Law Reports appear with a note of the argument of counsel whereas in B.L.R. the reports appear with a commentary by the editors.

If a case fails to get into the reports it may still be relied upon in court. This is done usually by obtaining a transcript of the judgment. Some unreported cases and the more important

Commonwealth decisions may be found noted in text books such as *Hudson's Building and Engineering Contracts*, and *Keating on Building Contracts*. A notable disadvantage of case law is that the law can change or develop only when a suitable case comes before the court. Thus, if a decision is given which is considered to be erroneous it remains law until a case comes before that or a higher court in which it can be reconsidered. Furthermore, the judges are concerned solely with stating what the common law is, and not what it should be. Both these disadvantages are overcome by augmenting the common law through the separate system of statute law.

Statute law

While the judges declare and apply the common law, Parliament in its legislative capacity passes enactments to change the law. Since the seventeenth century Parliament has had supreme authority so that in theory it can make or unmake any law. The passing of a Bill through Parliament and the argument at different stages in its passage can be followed in the media. The end result is an officially printed document which states, in the words chosen by Parliament, the law on some topic or group of topics.

Once enacted and in force, the words of the statute are themselves law. But there naturally arise situations where the words call for interpretation, and this is done by the courts. The declarations of the judges on the interpretation of statutes thus become a sub-branch of the common law, with which the statutes must be read to ascertain their meaning. A notable example of this process was the Arbitration Act 1979, s.1, which, as drafted appeared to be intended to facilitate appeals. The House of Lords, however, interpreted the section in *The Nema*[2] so as to restrict the right of appeal to exceptional cases. This piece of judicial interpretation or development is now codified in the new Arbitration Act 1996 (see Chapter 3). In the interpretation of statutory material the courts are not limited to the words chosen by Parliament. In *Pepper v. Hart*,[3] the House of Lords confirmed that parliamentary debates and ministerial statements could be considered in order to determine the intention of the legislature.

Delegated legislation

In addition to statute law proper there has, principally during this century, grown up a great body of delegated or sub-legislation.

---

[2] *BTP Tioxide v. Pioneer Shipping* [1982] A.C. 724.
[3] [1993] A.C. 593.

This is written not by Parliament itself, but by some other body or official to whom Parliament has given authority. The sub-legislators range from ministers of the government to statutory bodies such as local authorities. This delegated legislation goes under such names as rules, regulations or by-laws. It takes effect as though it were contained in the parent Act, which sets out the delegated power. As an example, the Building Act 1984 contains authority under which the Building Regulations are made. A great deal of day-to-day activity in industry, including the construction industry, is covered by delegated legislation. In general this lays down more stringent and specific duties than those which are to be found in the common law.

Equity

In the division between common law and equity, each branch comprises both judge-made law (found in case reports) and statute law. The difference arises because before 1873, when the systems began to be jointly administered, there were two separate legal systems which operated in different courts. Equity was applied in the old Court of Chancery, located in Lincoln's Inn. Charles Dickens found much to criticise in the courts of early nineteenth century England. He reserved the most biting condemnation for the interminable delays of Chancery.[4] The delays in the Chancery Division are today no more than in the rest of the High Court.

The differences between law and equity are still of importance. One essential distinction is that a common law remedy is a right, whereas a remedy in equity is, theoretically, discretionary. It depends on the justice of the cause. The distinction may be illustrated by the consequences of a breach of contract. The common law remedy is damages. These will be awarded, however unjustly the plaintiff has acted, and whether or not damages will make good the loss suffered. Alternatively, in equity the plaintiff can ask for the remedy of specific performance, that is, that the defendant be compelled to fulfil his obligation. But this will be available only under certain conditions, inter alia, that the plaintiff has acted fairly, that he has not delayed in seeking his remedy and that damages would not adequately compensate him.

Where to find the law

From the foregoing it may be said that the law proper is to be found only in law reports and in statutes, regulations and the like.

---

[4] *Jarndyce v. Jarndyce*, Charles Dickens, *Bleak House*.

However, textbooks and articles by academics, practitioners and judges play an important part in stating the law. Their role is not only to present the source material in a convenient form, but also to analyse, comment and speculate upon gaps in the law. In some fields, of which building and engineering contracts are a good example, what is said in the established textbooks can often be a valuable guide to the court in deciding a point of law. Sometimes the court cites with approval a particular passage from a textbook as being an accurate statement of the law. The status of any book depends on the standing of its author and current editor. As a general rule, it is said that the courts pay less attention to the views of an author while he is still alive.

Legal textbooks are of several different kinds. Many are written by individuals on particular topics. Some are published in series, intended to create an encyclopaedia of law. The best known of these is *Halsbury's Laws of England*, first published in 31 volumes between 1907 and 1919. The current fourth edition is in 52 volumes and is periodically up-dated. The companion work, *Halsbury's Statutes of England*, deals with legislation. Each topic in these works is written or edited by one or more specialist contributor. *Halsbury* is often cited in court as a convenient summary of the law.

THE COURTS

There are a number of different courts in which civil actions may be tried. A case will be heard at first instance in the High Court or in the County Court. Both the High Court and the County Court are to be found in different locations throughout the country. Appeals may then be brought to the Court of Appeal and finally to the House of Lords. The appeal courts generally sit only in London.

Courts of first instance

The High Court with its judges has three divisions: the Queen's Bench, the Chancery and the Family Division. Although each division administers the common law and equity and could theoretically deal with any matter, in practice a particular case will be assigned to one division. Matters concerning the construction industry come usually before the Queen's Bench Division, but occasionally before the Chancery Division. The Queen's Bench

Division deals with most common law work, such as contract and tort. The Chancery Division deals with contracts relating to land, company and partnership disputes, copyright and intellectual property.

Within the Queen's Bench Division of the High Court, there are two particular divisions where construction cases may be found. The first is the Commercial Court where mercantile, banking, insurance and shipping cases are tried, before High Court judges assigned from the Queen's Bench Division. The second was for over a century known as the Official Referee's Court. In 1998 it was re-named the Technology and Construction Court (TCC). Here, matters specifically relating to the construction industry are tried by circuit judges appointed to deal with TCC business. In each case, an action may be started in the Commercial Court or in the TCC; or it may be started in the Queen's Bench Division and later transferred.

For smaller civil cases trial may take place in the County Court by a circuit judge. Every district in the country has its local County Court. In the past, the County Courts have had a modest limit on their monetary jurisdiction. This has now gone and in theory the County Courts have unlimited jurisdiction. In practice, larger cases are started in the High Court, but may be transferred for trial. Government policy in recent years has been towards decentralisation with more civil work being tried in the County Courts. There are a number of provincial TCC judges who will try TCC cases locally; and in London there has been established a "business court" at County Court level and a County Court patents court. The latest development has been the merger, in 1999, of the rules of procedure applied in the High Court and County Court, the new rules being known simply as Civil Procedure Rules (see Chapter 2).

Appeals

After the hearing of a case at first instance either party may consider an appeal. From the High Court or the County Court there may be an appeal to the Court of Appeal. Changes to the rules introduced in 1998 mean that all appeals, whether on fact or law require leave, either from the trial judge or from the Court of Appeal. This applies equally to TCC cases. An important difference between an appeal and the original trial or action, is that the appeal will usually be concerned solely with argument. New evidence is admitted very rarely in the Court of Appeal. On an appeal, the court will give its own decision on matters of law. In matters of fact, however, while the Court of Appeal will review the written record of evidence given, it will also attach weight to the trial judge's assessment of the witnesses.

After an appeal to the Court of Appeal, a further appeal may be available to the House of Lords. Leave is sparingly given. House of Lords appeals are almost always on an important point of law. Where both the High Court and the Court of Appeal are bound by previous decisions it is possible to "leapfrog" by appealing direct from the High Court to the House of Lords. Ordinarily, however, the House of Lords requires issues to have been fully considered by the Court of Appeal and will not usually entertain a ground of appeal not raised in the court below.

Every court must apply statute law. However, with case law, courts are generally bound only by the decisions of higher courts, and to an extent by their own decisions. Despite some views to the contrary, the Court of Appeal is bound by its own decisions (*stare decisis*). The same generally applies within a particular division of the High Court but TCC judges, who technically rank as circuit judges, do not strictly bind each other and have been known to reach conflicting decisions on the same issue. Where a court is bound by a previous decision, it may nevertheless be avoided if it is possible to "distinguish" the previous case from the one under consideration, or to confine its application to the particular facts of that case. This was done in the well-known case of *Junior Books v. Veitchi*,[5] a decision of the House of Lords which was generally considered wrong but never overruled.

The law as stated by the House of Lords is generally regarded as fixed and binding on all other courts. Nevertheless, in 1966, the House of Lords itself decided that it could depart from its own previous decisions. In the important case of *Murphy v. Brentwood District Council*[6] the House of Lords consisting, uniquely in modern times, of seven members including the Lord Chancellor, decided to depart from the long established authority of *Anns v. London Borough of Merton*[7] and in doing so effectively reversed two decades of litigation. The House of Lords took another unprecedented step when it decided, in 1998, to set aside the decision of a different division of the House in the *Pinochet* case.[8] Ordinarily, decisions of the House of Lords are altered only by statute.

---

[5] [1983] A.C. 520.
[6] [1991] 1 A.C. 398.
[7] [1978] A.C. 728.
[8] *R v. Bow Street Stipendiary Magistrate, ex p. Pinochet No. 2* [1999] 2 W.L.R. 272 and see *Dimes v. Grand Junction Canal* below.

PUBLIC LAW

Public law, sometimes called administrative law, concerns the exercise of powers and duties by public bodies, usually arising under statute. In some cases this may give rise to a direct right of action in civil law against the public body. For example, a claim against a local authority alleging negligence in the enforcement of Building Regulations relates to such public law duties or powers. Where statutes do not provide expressly whether or not an individual who suffers damage may bring an action, it is necessary for the courts to construe the statute. The case of *Anns v. London Borough of Merton*[9] contains an analysis of whether the Building Regulations and their governing statutes create a right for individuals to sue. In that case the answer was affirmative but the question remains far from settled.

A different aspect of public law is the right of an individual who is affected by the exercise of such a power or duty to seek an order from the courts restraining or controlling the way in which the public body acts. This involves a distinct form of civil procedure known as judicial review. The law and procedure have become transformed in recent years. Formerly, an aggrieved individual had the right to ask the court to make one of three types of "prerogative order": (1) certiorari, to quash the decision of a public body or tribunal; (2) prohibition, to prohibit a public body or tribunal from acting in a particular manner; and (3) mandamus, to compel the public body or tribunal to act in accordance with its duty. These specific remedies have become enlarged by new statutory provisions and rules of court which now permit an individual to apply for the more general remedies of injunction or declaration, and on an application for judicial review the court may award damages to the applicant.[10]

Judicial review applies to an almost unlimited range of matters, including decisions of government ministers and local authorities, public bodies, inferior courts and tribunals, and covers all types of law, both civil and criminal, including, for example, the application of prison regulations. Perhaps the most numerous are claims relating to immigration decisions and to actions of local authorities. The procedure now laid down is simple. An applicant must first obtain leave of the court and this requires only the filing of certain papers and particulars. The decision to grant or refuse

---

[9] *ibid.*
[10] RSC, Ord. 53.

leave is now generally made by a single judge without a hearing, on a review of the papers. Where leave is refused, the application may be renewed in open court. The application for leave is *ex parte*, and if granted, a hearing then takes place with the body or authority which is the subject of the complaint appearing and being represented. The rules require applications to be made promptly and in any event normally within three months of the relevant event.

The principles of law which the courts apply on application for judicial review are common law principles, developed by the courts themselves, principally in a series of cases following the introduction of the new procedure. On an application for judicial review, the court is not concerned with deciding whether it agrees with the decision or action of the relevant authority, nor is the process an appeal. The court is concerned only with restraining the wrongful exercise of public law powers and duties, and the grounds on which the court will intervene are limited. The principal grounds are the following:

(1) Want or excess of jurisdiction, which may include error of law.

(2) Irrationality, which is colloquially referred to as the "Wednesbury" principle following the leading case of *Associated Provincial Picturehouses v. Wednesbury Corporation*[11] In this case the local authority granted a licence for cinema performances on a Sunday on the condition that no children under 15 years of age should be admitted. The owners challenged the decision as an unreasonable exercise of discretion. Lord Greene M.R. said:

> "It is clear that the local authority are entrusted by Parliament with a decision on a matter which the knowledge and experience of that authority can best be trusted to deal with. The subject-matter with which the condition deals is one relevant for its consideration. They have considered it and come to a decision upon it. It is true to say that if a decision on a competent matter is so unreasonable that no reasonable authority could ever come to it, then the courts can interfere. . . . It is not what the court considers unreasonable, a different thing altogether."

The principle is that the court will intervene if the decision is such that no authority properly directing itself on the relevant law and acting reasonably could have reached it.

---

[11] [1948] 1 K.B. 223.

(3) Procedural impropriety, which covers failure by a body to observe its own procedural rules. The main area of application here is in breach of the rules of natural justice. These rules broadly require public bodies to act fairly in the particular circumstances. For example, a person liable to be dismissed from a public office must be given a hearing, and must be notified of the allegations against him. A particular requirement of natural justice is that the person exercising a power or giving the decision must not have an interest in it. This was the subject of the celebrated case of *Dimes v. Grand Junction Canal*[12] where, in the course of a long dispute between the company and an adjoining landowner, the Lord Chancellor gave a decision, after which it was found that he was a substantial shareholder in the company. The House of Lords subsequently expressed their views on the matter, Lord Campbell saying:

"No one can suppose that Lord Cottenham could be, in the remotest degree, influenced by the interest that he had in this concern; but, my Lords, it is of the last importance that the maxim that no man is to be a judge in his own cause should be held sacred. And that is not to be confined to a cause in which he is a party, but applies to a cause in which he has an interest. Since I have had the honour to be Chief Justice of the Court of Queen's Bench, we have again and again set aside proceedings in inferior tribunals because an individual, who had an interest in a case, took a part in the decision. And it will have a most salutary influence on these tribunals when it is known that this High Court of last resort, in a case in which the Lord Chancellor of England had an interest, considered that his decree on that account a decree not according to law, and was set aside. This will be a lesson to all inferior tribunals to take care not only that in their decrees they are not influenced by their personal interest, but to avoid the appearance of labouring under such an influence."

This case was recently applied in the *Pinochet* case (see above) where the House of Lords set aside their own judgment, holding that bias was not limited to financial interest but covered the situation of a judge who was a director of a charity controlled by a company which had an interest in the case before the court.

There are many situations in which judicial review might be appropriate in the context of construction contracts, for example, decisions of local authorities regarding their tender lists. Such

---

[12] (1852) 3 H.L. Cas. 759.

decisions are open to judicial review and may be set aside if appropriate grounds are established. For example, in *R. v. London Borough of Enfield*[13] a decision was set aside because the Borough had not complied with the appropriate procedural rules. In such a case, the decision about the tender list is that of the authority and the court will not intervene except on the "Wednesbury" principle (see above).

EUROPEAN COMMUNITY LAW

In addition to Common Law and Statute Law, the European Economic Community has, since 1973, formed a third independent and increasingly major source of law applying throughout the United Kingdom. The foundation of the Community was and remains the Treaty of Rome, signed by the original six members in 1957. Membership of the Community has grown since then and continues to do so. Development in the 1980s was towards the Single European Market, for which 1992 was set as the target. This involved many radical harmonisation proposals, which included several related directly to construction. Also in 1992, at Maastricht, a new Treaty on European Union was signed and subsequently ratified (with some hesitation) by all Member States. This made certain amendments to the Treaty of Rome and also gave the EEC the new name of the European Community (E.C.).

The European treaties form a "framework" of measures, expressed as broad aims to be achieved, which are intended to be filled in by detailed measures. Areas for detailed legislation include: competition law and public procurement, health and safety, environment and consumer protection, and the more general harmonisation of the laws of Member States to the extent required for the functioning of the common market.

The EEC is unique in having achieved more in terms of interstate integration than any comparable organisation in history. It operates in some ways as an international body and in other ways as a federal government. The principal institutions through which it operates are:

(1) The European Parliament, which is now a directly elected body which exercises defined powers, but falling far short of a full legislative assembly;

---

[13] (1989) 46 B.L.R. 1.

(2) The Council of Ministers, which is a fluctuating body of Ministers from individual Member States who meet when required and generally represent the interests of their own Governments. Presidency of the Council circulates among the Member States. When the Ministers are heads of state, the Council is referred to as the Council of Europe;

(3) The European Commission, which is the equivalent of the European Civil Service, headed by permanent Commissioners who, although drawn from the Member States, should represent the interests of the Community, unlike the members of the Council. Every Member State contributes one Commissioner and the larger states contribute two;

(4) The European Court of Justice, which comprises judges appointed by each Member State and has the function of interpreting and applying Community law.

An early example of the effect of the E.C. occurred in *Bulmer v. Bollinger S.A.*[14] which became known as the "Champagne" case. The defendant French company claimed that the use of the word champagne to describe an English beverage contravened community law. The court was asked to refer the issue to the European Court of Justice under Article 177 of the Treaty of Rome. Lord Denning took the opportunity to describe the effect of the new law:

"The first and fundamental point is that the Treaty concerns only those matters which have a European element, that is to say, matters which affect people or property in the nine countries of the common market besides ourselves. The Treaty does not touch any of the matters which concern solely England and the people in it. These are still governed by English law. They are not affected by the Treaty. But when we come to matters with a European element, the Treaty is like an incoming tide. It flows into the estuaries and up the rivers. It cannot be held back. Parliament has decreed that the Treaty is henceforward to be part of our law. It is equal in force to any statute. . . . In future, in transactions which cross the frontiers, we must no longer speak and think of English law as something on its own. We must speak and think of the community law, of community rights and obligations, and we must give effect to them."

Lord Denning's judgment now requires qualification in two respects. First, community law is not simply of "equal force" to English statute or other law. It is clear that it must take precedence and any domestic rule running contrary to European law must give

---

[14] [1974] Ch. 401.

way to European law. Secondly, matters subject to Community law are not simply those where other members of the European Community are involved. Increasing areas of English domestic law and procedure are now governed or influenced by Community law. An important example of this is English domestic health and safety law relating to the design and construction of buildings (see Chapter 16).

European legislation emanates from the Council and the European Commission, principally in the form either of a regulation or a directive. Regulations have direct binding force on all Member States and comprise fully detailed measures. Conversely, directives specify the result to be achieved and are intended to be acted upon through individual legislation enacted in each Member State. Thus, in terms of English law, European regulations comprise another category of delegated legislation which takes effect in England by virtue of section 2 of the European Communities Act 1972, which provides that rights and obligations created under the European treaties "are without further enactment to be given legal effect". In the case of directives, however, they must be enacted by the United Kingdom Parliament in the form of a domestic Act. The fact that the United Kingdom Parliament has effectively bound itself by treaty to pass such enactments gives rise to the contention that Parliament has surrendered its power to Brussels. In legal terms, such powers were subordinated to the European legislative bodies upon accession to the Community, subject to the representation of the United Kingdom interests with the European Council and commission.

In later chapters reference is made to the principal elements of European Community law where they affect the construction industry, both here and within Europe.

INTERNATIONAL CASES

International law has taken on new dimensions in the past two decades, in response to increasing international trade. But despite a number of individual initiatives, there has been no notable movement towards harmonisation of national laws. There have, however, grown up a number of international bodies which play an increasing role in international law. As regards the United Kingdom, the most important is the European Community whose constitution and applicable legislation is dealt with elsewhere in this book.[15] Despite the breadth of influence of European Com-

---

[15] See Chaps 6 and 16.

munity law, contracts between community members remain essentially governed by the law of individual Member States. Inter-state E.C. litigation remains rooted in the courts of one or other of the Member States, the jurisdiction of the European Court being limited to enforcement of, and where necessary giving declarations on European Community law. Of more practical significance is the rise of international arbitration, particularly within the European Community. As discussed in Chapter 3, international arbitration involves no conflict of jurisdiction and also permits much greater scope in terms of the applicable law.[16] Certain international transactions are subject to the rules of international institutions which make provision for matters of jurisdiction and applicable law. Foremost among these is ICSID[17] which applies primarily to loan agreements made with a state or state corporation, but also has potentially wider applications. International commercial law is a constantly developing subject.

International litigation requires the application of a particular set of legal rules referred to as the conflict of laws. These rules are the principal subject of this section. Whenever a dispute exists between private parties which has a foreign element, two preliminary questions must be answered before any court can determine the matter. First, does the national court in question have jurisdiction and should that court determine the dispute? Secondly, what law should the court apply? A third question which may arise after the dispute has been determined is how the judgment is to be enforced. The first and second questions are covered here. Enforcement internationally is dealt with in Chapter 2. Most commercial international disputes are likely to involve the law of contract or tort and these are dealt with below. Outside the field, international disputes may arise in many areas such as family and succession law. Questions of jurisdiction and enforcement as well as choice of law are part of the domestic law and therefore differences will exist between one state and another.

Jurisdiction and procedure

The English courts normally assume jurisdiction to hear actions in contract and tort in three cases:

(1) if the defendant is served with a writ while present in England; a foreign company is regarded as being present if it carries on business here;

---

[16] Through recognition of Equity clauses and, to some limited extent, *lex mercatoria*.
[17] International Centre for Settlement of Investment Disputes.

(2) if a defendant submits to the jurisdiction, for example, by bringing an action in the English courts;

(3) the English courts may give leave for a writ to be served abroad so that the action can proceed, if necessary, in the absence of the defendant. The principal grounds on which leave to serve abroad may be given are:

   (i) that the defendant is normally resident in England;

  (ii) that the dispute arises from a tort committed in England; or

 (iii) that the dispute arises from a contract which was made or broken in England, or agreedto be governed by English law.[18]

Where a foreign defendant is brought before the English courts he may apply for the proceedings to be stayed in favour of some foreign court on the ground of *forum non conveniens*. On such an application, or on the application for leave to serve the writ abroad, the court will consider all matters including convenience to the parties and witnesses and the cost of the proceedings. In regard to disputes involving a party domiciled elsewhere in the European Community (and in some other European states) the question of jurisdiction is governed by the Civil Jurisdiction and Judgments Act 1982[19] which entitles the plaintiff to bring proceedings as of right against a foreign party. This provides an alternative basis of jurisdiction broadly analogous to those under which the English courts will grant leave for service abroad. The jurisdiction is, however, mandatory and cannot be avoided on the ground of inconvenience.

If an English court accepts jurisdiction over a case, it will proceed to trial in the same way as a case with no foreign element, and generally English rules of procedure and evidence are applied. If it is an issue in the case, the court will decide which country's law is to be applied. If the applicable law is English law, the judge will treat the case as a domestic one and decide upon the law himself. If the law to be applied is a foreign law, this is treated as a question of fact, the relevant provisions of that law being proved to the court by expert legal witnesses. In the absence of proof to the contrary the English courts assume foreign law to be the same as English.

---

[18] RSC, Ord. 11, r. 1(1).
[19] Extended by the Civil Jurisdiction and Judgments Act 1991.

Applicable law in contract

In the case of an international dispute arising in contract, the law to be applied depends upon the nature of the dispute. Most aspects are governed by one law referred to as the "applicable law", or sometimes as the "proper law" of the contract. This law can be chosen by the parties, but in the absence of express choice the court will determine it as the law of the country having the closest connection with the contract. This is governed by statute[20] which largely enacts the common law rule. The Act does not apply to arbitration. The choice of law may depend on many factors including the country in which the contract was made, where it was performed and the place and currency of the payment. No one factor is conclusive. The applicable law, when identified, will determine such matters as whether a binding contract has been made, how the contract is to be construed, what is the effect of a misrepresentation, and whether an exclusion clause is valid. If a contract is illegal by the applicable law it is unenforceable in England. There are, however, some topics which are governed by a different law. A transfer of land is governed generally by the law of the place where the land is situated. Further, arbitration procedure is normally governed by the law of the venue or seat of the proceedings (see Chapter 3).

The International Conditions of Contract for civil engineering work (FIDIC) provide for the national law governing the contract to be specified. Before agreeing to a foreign law a party should ascertain the effects of that law, for instance, on whether clauses in the contract for his benefit will be enforceable. In the past it was frequent for English law to be chosen as the governing law. It is now much more common for the national law of the employer to be chosen, usually being also where the works are sited. This can pose problems where the country in question is part of the developing world. It may be found that principles of law assumed by the draftsmen of the contract do not exist in any developed form.

Applicable law in tort

If a tortious act is committed in England by, or to, a foreign party, any action brought here is tried as an ordinary English domestic case applying English law. The conflict of law arises only

---

[20] Contracts (Applicable Law) Act 1990.

where a torts is committed abroad. Despite acceptance in other common law countries, the concept of a "proper law" of tort is not accepted in England. The applicable law in a foreign tort action is a compromise between the law of the place of commission and English law. In general, to found a suit in England, a tort committed abroad must be actionable both under English law and also at the place of commission. Such matters as the amount of damages claimable may then be determined by English law, while the law of the place of commission may affect the defences available to the defendant. An incidental problem may be to decide where a tort is committed, for example, where an act in one country causes damage in another. The English rule is generally that a tort is committed where the wrongful act takes place.

CHAPTER 2

DISPUTE RESOLUTION—LITIGATION, ADR AND ADJUDICATION

DISPUTES in the construction industry may be resolved by a wide variety of means. This chapter deals first with court practice in terms of procedure, and with the rules of evidence which apply to civil litigation. Chapter 3 covers Arbitration, which is used in construction contracts as the major means of final dispute resolution. Between these two formal systems lies a variety of consensual procedures generally referred to as Alternative Dispute Resolution (ADR). These procedures are generally free from formal rules but can still generate difficult legal issues. In addition, since 1998, construction contracts falling within a statutory definition are subject to mandatory provisions for rapid and temporarily binding "adjudication". This procedure takes precedence over any other form of dispute resolution.

COURT PROCEDURE

Procedure is a general term which covers the various steps necessary to turn a legal right into a satisfied judgment of the court. While procedure is properly the concern of lawyers, it can have a far-reaching effect on the course of an action, and the trial is often profoundly affected by the procedural steps which precede it. The pre-trial proceedings will usually extend over months, or even years. Appeals, enforcement of a judgment, and the assessment and enforcement of costs orders, may prolong the matter further after the trial. Procedure covers all these different stages.

Procedure in the civil courts is governed by statutory rules. Those applying in the High Court were formerly known as the Rules of the Supreme Court. They are now replaced by Civil

Procedure Rules (CPR) which apply throughout the civil courts, including the county courts. The rules themselves are contained in separate "Parts", which were formerly known as Orders. The CPR implement recommendations of the Woolf report[1] and seek to improve the accessibility, speed and efficiency of civil court procedure.

Basic steps in court

The steps involved in a civil action in the Queen's Bench Division, where most cases concerning the construction industry will be brought, are as follows. The action is begun by issuing and serving a claim form (formerly known as a writ). The defendant must file a defence or an acknowledgement of service. There may be a counterclaim from the defendant and reply from the claimant, in which the issues are defined. Disclosure and inspection of documents follows. The solicitors or others preparing the case will seek to collect the evidence which will be needed to prove the case or to discredit the opposing case. This will include both oral and documentary evidence. The culmination of this process is the trial itself which results in a judgment. If there is no appeal, the matter is concluded by enforcement of the judgment and orders for costs. The principal steps are enlarged upon below.

Very few actions proceed in such an apparently straightforward manner. At every stage there are alternative courses, and in fact the great majority of court actions (well in excess of 90 per cent) are disposed of before reaching trial. The preparatory stages between the issue of the claim form and trial are referred to as interlocutory proceedings. Where any decision may be given by the court at an interlocutory stage it is usually given by a Master of the court. He is an official who exercises most of the powers of a judge including giving judgments and other decisions in advance of the full trial. In a High Court action, the Master will make most of the orders required during the interlocutory stage of the case. This means that the judge in a Queen's Bench action often has no knowledge of a case until shortly before the trial. In cases before the Technology and Construction Court or in the Commercial Court, interlocutory orders are made by the trial judge, which has the advantage of ensuring familiarity with the issues before trial. The same advantage applies in an arbitration.

---

[1] Access to Justice, 1994.

Starting proceedings

A typical action is begun by issuing a claim form, which places the matter on the official record. Particulars of the claim must either be contained within the claim form or served on the defendant within 14 days (CPR, Part 7). The Civil Procedure Rules are accompanied by detailed practice directions applying to each part of the rules. Since procedure in the High Court and County Court is now merged, the CPR practice direction provides for the appropriate court in which claims should be started. Claims are to be brought in the High Court only if they exceed £15,000 (or £50,000 in the case of personal injuries) or on other grounds Other rules provide for transfer between the courts. An important innovation by the new rules is the requirement for a claim form and particulars to be accompanied by a "statement of truth" (CPR, Part 22) which may lead to proceedings being brought in the case of a false statement made without an honest belief in its truth. A copy of the claim form must be served on the defendant, either by delivering it to him personally, or by other means, such as service on his solicitor. The general rule is that the defendant must be made aware of the proceedings against him. But there is an important exception in respect of limited companies,[2] which may be served by leaving the writ at the registered office, or sending it there by post. A claim form must normally be served within the period of its initial validity. For many years this was 12 months, giving effectively one year's extension to the period of limitation. The period of validity is now four months from the date of issue. The court has power sparingly to extend the validity of a claim form. After he has been served with a claim the defendant must file either an admission, a defence or an acknowledgement of service. The rules provide for the defendant to file with his defence a counterclaim against the claimant. A counterclaim may be served at a later date, but then requires the court's permission (CPR Part 20).

Joinder of claims and parties

A particular advantage of litigation is the ease with which other parties may be joined in an action, unlike arbitration proceedings which are ordinarily limited as between the two parties to the arbitration agreement (see Chapter 3). Any number of plaintiffs

---

[2] Companies Act 1985, s. 725.

who have similar interests in the subject matter of the litigation may join together in a claim. Alternatively, they may issue separate claims which may be consolidated, that is, treated as a single action. Claims may be brought against two or more defendants. The general rule is that joinder is available where all the relevant claims can be conveniently disposed of in the same proceedings. The defendant may bring in another party as "sub-defendant" to the claim against him, and then that party may similarly bring in other parties. These were formerly known as third and fourth parties respectively. All such subsidiary claims are now known as "Part 20 claims" under the CPR. Under this part of the rules, provision is also made for any party to bring counterclaims or claims for contribution for indemnity against any other party. The court then has an overriding discretion to decide, as part of the case management, which claims should be heard together and which severed and dealt with as separate proceedings.

Case management

At the heart of the new Civil Procedure Rules are extensive powers which the court is required to exercise in the interests of efficiency and expedition. Thus, at the outset of the proceedings, after the defendant has filed a defence, the court will inquire into the nature of the proceedings by issue of a questionnaire. The court has an initial power, whether on the application of a party or on its own initiative, to stay the proceedings for one month while the parties explore settlement. Thereafter the court will allocate the action (and may re-allocate if it becomes necessary) to one of the three "tracks" on the following basis:

(1) Small claims track—is appropriate for claims not exceeding £5,000.

(2) The fast track—is the normal track for other claims up to £15,000 where the trial is likely to last no more than one day and where expert evidence is limited to one expert per party in any field and no more than two expert fields.

(3) The multi-track—is appropriate to any other claim.

In the context of civil litigation of any substance, the multi-track will, therefore, segregate off smaller and simpler claims, the bulk of substantial claims remaining as part of the "multi-track" (CPR, Part 29). In relation to such claims, the new case management rules require the court to set up a timetable for steps leading up to the trial or to fix a case management conference for pre-trial review. The court is also required to fix the trial date or the period in which the trial is to take place as soon as practicable. These provisions are subject to the overriding case management powers of the court including general powers to decide which issues need full investigation and trial, and the order in which issues are to be resolved (CPR, r. 1.4). There is also power to strike out a statement of case where there has been a failure to comply with the rules and to the enter judgments in consequence (CPR, Part 3). The stated objective of these procedural forms is to involve the court in a more pro-active role with a view to resolving the real issues between the parties quickly and efficiently.

Pleadings

The object of pleadings is to define the areas of dispute between the parties before the action comes to trial. A party will not normally be allowed to raise a matter at the trial unless he has pleaded it. Leave may be given to amend pleadings even during the trial, but this will invariably involve payment of the costs thrown away by the amendment; and an amendment may be refused where the other party will be prejudiced beyond the incurring of additional costs. The new CPR contain provisions as to the content of pleadings (CPR, Part 16). The claim should contain a concise statement of the facts relied upon, but not the evidence by which they will be proved; and matters of law should not normally be pleaded. These rules, however, are often difficult to satisfy and it is not uncommon in construction cases to find both matters of law and evidence included in pleadings. A useful rubric both for pleaders and judges or arbitrators is that any matter which the opposing party needs to answer positively should be pleaded.

Global claims and particulars

In construction cases it is often appropriate to plead facts in great detail. This is sometimes the result of voluntary "particulars" delivered with a claim document. More often, however, there will be a dispute over the extent to which the party asserting a claim must give detailed particulars in advance of the hearing. This has given rise to a particular type of procedural dispute concerning "global" or "rolled up" claims. The issues often concern the extent to which the claiming party must specify in his pleading the causal connection alleged to exist between the causes of action relied upon and the damages or other relief (such as extension of time) claimed. In *Crosby v. Portland UDC*[3] Donaldson J. upheld the award of an arbitrator on a global claim in the following terms:

> "Since, however, the extent of the extra cost incurred depends upon an extremely complex interaction between the consequences of the various denials, suspensions and variations, it may well be difficult or even impossible to make an accurate apportionment of the total extra cost between the several causative elements. An artificial apportionment could of course have been made: but why (the contractor asks) should the arbitrator make an apportionment which has no basis in reality? I can see no answer to this question . . . provided (the arbitrator) ensures that there is no duplication, I can see no reason why he should not recognise the realities of the situation and make individual awards in respect of those parts of individual items of the claim which can be dealt with in isolation and a supplementary award in respect of the remainder of these claims as a composite whole."

This issue was taken up in a later Hong Kong case before the Privy Council. In *Wharf Properties v. Eric Cumine*[4] it was said that the *Crosby* case had:

> "no bearing upon the obligation of a plaintiff to plead his case with such particularity as is sufficient to alert the opposite party to the case which is going to be made against him at the trial. (The defendants) are concerned at this stage . . . with the specification of the factual consequences of the breaches pleaded in terms of periods of delay. The failure even to attempt to specify any discernible nexus between the wrong alleged and the consequent delay provides, to use Mr Thomas' phrase "no agenda" for trial."

The Privy Council upheld the decision that the claim be struck out as embarrassing the fair trial of the action or as an abuse of the process of the court. As a result of this decision, attempts are usually made to provide seemingly adequate particulars of the causation alleged between the individual grounds of claim and the damages or financial consequences alleged. This issue does, however, leave a number of matters still within the discretion of a tribunal dealing with such a claim. How far should the party asserting a claim be pressed to give particulars which it may be artificial or even impossible to give with precision? And where full particulars are not given, should the party bringing the claim be permitted to call detailed evidence or to "unroll" the claim at trial?

---

[3] (1967) 5 B.L.R. 121.
[4] (1991) 52 B.L.R. 1.

These are matters of particular relevance to arbitrators, since the power to strike out an embarrassing pleading is available only in court.[5]

Pleading by schedule

A common feature of construction and other technical litigation, as well as arbitration, is the use of schedules to plead particulars. This covers both the pleading of facts and of damages or other quantum particulars. The point of a schedule is that it allows otherwise indigestible prose to be split into short entries under common headings. It also puts the case of two or more parties on the same issue or item on the same sheet of paper. It is important to devise the most appropriate form for the schedule, but once set up, they can be of great use in collecting together details in the most convenient form. Schedules were first used in official referees building defects cases and were named "Scott Schedules" after a former official referee. In modern use, schedules lend themselves well to production by electronic word processors which can be used to produce conveniently formatted tables or spread sheets without need for retyping.

Counter-claim and set-off

A defendant, in addition to serving a defence to the claim made against him, may serve a counter-claim against the party bringing the original claim. The counter-claim need not relate to the subject matter of the original claim. Typically in construction litigation, a contractor's claim for payment will be met with a counter-claim for damages for delay; or it may be for damages for defects in work other than that for which payment is being claimed. An important question in relation to a counter-claim is whether it ranks merely as a separate cross-action or whether the defendant may rely on it as a defence to the original claim. A counter-claim which operates as a defence is called a set-off.

A cross-claim need not arise out of the same transaction to rank as a set-off; but there must be a sufficiently close connection with the original claim. In *Hanak v. Green*[6] a builder who was sued for defective work was held entitled to set-off a greater sum found due upon his counter-claim for payment and damages. Sellers L.J. said:

"Some counter-claims might be quite incompatible with a plaintiff's claim, in no way connected with it and wholly unsuitable to be used

---

[5] CPR, r. 3.4 (2).
[6] [1958] 2 Q.B. 9.

as a set-off. But the present class of action involving building or repairs, extras and incidental work so often leads to cross-claims for bad or unfinished work, delay or other breaches of contract, that a set-off would normally prove just and convenient, and in practice, I should have thought, has often been applied, as indeed it was in the referee's report. It would serve to reduce litigation and the consequent costs. I would not be astute to restrict the right but rather to develop it and discourage litigation when no or little monetary benefit ensues on balance. It cannot, as I see it, make any difference which side commences proceedings in which cross-claims arise. If there is a set-off at all each claim goes against the other and either extinguishes it or reduces it."

This type of set-off is referred to as "equitable" set-off. In *Gilbert-Ash v. Modern Engineering*[7] Lord Diplock described a different form of set-off:

"The principle is that when the buyer of the goods or the person for whom the work has been done is sued by the seller or contractor for the price 'it is competent for the defendant . . . not to set-off, by a proceeding in the nature of a cross-action, the amount of damages which he sustained by a breach of the contract, but simply to defend himself by showing how much less the subject matter of the action was worth by reason of the breach of contract'. . .[8] This is a remedy which the common law provides for breaches of warranty and contracts for sale of goods and for work and labour. It is restricted to contracts of these types. It is available as of right to a party to such a contract. It does not lie within the discretion of the courts to withhold it. It is independent of the doctrine of 'equitable set-off' developed by the court of Chancery to afford similar relief in appropriate cases to parties to other types of contract. . ."

Thus, a defendant may rely on a "common law" set-off to reduce or extinguish the value of the goods or services for which payment is claimed in the action. Alternatively, he may bring a cross-action and rely on equitable set-off provided that the claim and cross-claim are sufficiently related. Where a claim is extinguished by set-off this may have an important effect on the right to recover costs, as in *Hanak v. Green* where the defendant, who recovered more than the plaintiff, was awarded the costs. Another similar but distinct remedy occasionally relied on is abatement.[9]

Where a contract is subject to the Housing Grants, etc., Act 1996, the right to withhold payment after the final date for payment is conditional upon notice being given as to the amount to

---

[7] [1974] A.C. 689.
[8] *Mondel v. Steel* (1841) 1 B.L.R. 108.
[9] See *Hutchinson v. Harris* (1978) 10 B.L.R. 19.

be withheld and the grounds for withholding payment. Notice must be given within a prescribed period: in the absence of agreement, the Scheme for Construction Contracts provides for notice not later than seven days before the final date for payment, which is to be 17 days from the date that payment becomes due.[10]

Damages and other relief

The claim, and any counterclaim, must expressly state the remedy sought. In contract and tort actions the remedy is usually damages, that is, the payment of a sum of money in compensation. But there are other remedies, which may be appropriate in different circumstances, such as an injunction, or specific performance, or rectification of a contract. Damages are sometimes categorised as general or special. General damages are claimed where the plaintiff has suffered loss which cannot be calculated in terms of money, for example, damages for pain and suffering. General damages must be assessed by the judge and no specific sum is claimed in the pleading. Special damages are those which can be calculated in money as an actual or prospective loss. In construction cases it is rare to find claims for damages which cannot be calculated. The difficulty is usually in terms of how the damages, for example for disruption of the progress of construction works, should be calculated.[11] There is no rule as to how a claim for damages should be calculated. There are often alternative approaches, for example, between claiming wasted expenditure or loss of anticipated profit. It is a matter for the plaintiff (or counter-claimant) to specify the claim he wishes to pursue and there is nothing to prevent alternative damages claims being pleaded. The plaintiff must specify which is the primary claim, but may rely on a second or third alternative in the event the preferred claim does not succeed.

"Liquidated" damages, in the context of a building contract, refers to specified sums payable in defined circumstances, particularly for delay in completion. They must be claimed by the employer as a specific remedy. The contractor, in such a case, may claim for an extension of time and for the consequent return of liquidated damages deducted.

Where the parties wish simply to establish their legal rights without claiming monetary relief, it is often appropriate to claim a "declaration". This will be appropriate, for example, where the

---

[10] Scheme for Construction Contracts, Pt II, paras 8, 10.
[11] See "global" claims above.

parties are in dispute as to the meaning of some term of the contract or as to whether some action taken by one party was contractually justified.

Subsequent claims

It sometimes happens that a construction project leads to more than one set of proceedings being brought between the same parties, whether in the form of two separate arbitrations, two sets of court proceedings or some combination. A problem can arise where a decision is given in earlier proceedings which bears on matters in issue in subsequent proceedings. The same principle can apply where different claims in the same proceedings are dealt with by way of an interim award or judgment on preliminary issues.

The broad principle which applies in all these circumstances is that a plaintiff may not bring a subsequent claim which involves re-opening a matter already decided. This principle is referred to as *issue estoppel* or *res judicata*. Nor may a plaintiff bring a claim which seeks some relief which was or should have been included in the claim already decided. The application is often far from simple. In *Conquer v. Boot*[12] the defendant builder had contracted to build a bungalow for the plaintiff, who brought an action for breach of contract to complete in a good and workmanlike manner. After recovering damages in this action, the plaintiff then brought another action in identical terms but alleging failure to build with proper materials. The Divisional Court held the plaintiff not entitled to bring the second action. Talbot J. held:

> "The contract is an entire contract. No claim for payment could have been made by the defendant unless and until he had finished the bungalow. There is one contract and one promise to be performed at one time, although no doubt the defendant may have failed to perform it in one or in many respects. There may of course be many promises in one contract, the breach of each of which is a separate cause of action . . . here there is but one promise, to complete the bungalow."

Another consequence is that damages or relief arising from any cause of action must be claimed once and for all. It follows that any claim for damages whether in tort or for breach of contract, must claim for all future anticipated loss, which will be assessed at the date of the hearing. For example, in the case of *Batty v. Metropolitan Realisations*[13] the court awarded damages in respect of a

---

[12] [1928] 2 K.B. 336.
[13] [1978] Q.B. 554.

house which was deemed not fit for habitation because it had been built at the top of a potentially unstable slope. The plaintiff recovered for the anticipated loss even though it had not collapsed and, apparently, did not subsequently collapse. There is no mechanism whereby either plaintiff or defendant can ask for damages, once assessed, to be reassessed in the light of subsequent facts once a judgment becomes final. One of the few exceptions is in the case of claims for damages due to withdrawal of support where the rule is that the plaintiff may recover only the damage actually suffered, even though other damage may be imminent.[14] A further exception exists in respect of personal injury cases where the court can, in particular circumstances, defer final judgment until the extent of the plaintiff's injury is known.

When bringing claims under a construction contract, consideration must be given to whether different causes of action exist which allow separate claims to be brought. In many cases, the terms of the contract will permit such separate claims, but the plaintiff can recover his loss once only. If, therefore, the separate claims are simply alternative ways of recovering the same loss, they must be pleaded as alternatives. The plaintiff must also exercise caution in ensuring that all damages or remedies arising from the causes of action relied on are claimed for, otherwise they may be lost.

Judgments without trial

There are several instances in which the court may decide some issues in a case or may make orders, which have the effect of terminating the case, without waiting for the trial. First, there are a number of circumstances in which one party may invoke the power of the court to terminate or strike out the case of the other party who is in default. For example, where the defendant fails to file a defence to the claim,[15] judgment may be entered. Such "default" judgment may be set aside by the court but many simple actions for the recovery of debt are in fact concluded in this manner. Where one party is ordered to do something, for example, to serve particulars or to give discovery, the court may in an appropriate case order that the claim or defence of that party be struck out in default. Where a plaintiff (or a defendant who has a counterclaim) fails to take any action to bring the claim to trial and the other party suffers prejudice, the court has an inherent power to strike out that claim for want of prosecution, which is backed up by rulesof court.[16]

---

[14] *Darley Main Colliery v. Mitchell* (1886) 11 App. Cas. 127.
[15] CPR, r. 15.3.
[16] CPR, r. 3.4 (2).

Summary judgment

The plaintiff may apply to the court for judgment on his claim (or the defendant on his counterclaim) on the ground that there is no sufficient defence. There are two sections of the CPR which may be relied on. Under CPR, Part 24, the plaintiff may apply for summary judgment on the claim or some particular part of it, on the ground that the defendant has no real prospect of successfully defending the claim or issue. If the defendant fails to satisfy the court that there is an issue which ought to be tried, the plaintiff will be entitled to immediate judgment on the claim or part of the claim in question. Part 24 (formerly RSC, Ord. 14) now applies equally to the defendant, who may similarly apply to the court for judgment on the ground that the claimant has no real prospect of succeeding on the claim or issue.

Frequently, the only dispute on an application by a plaintiff for summary for judgment will be whether the defendant can establish a credible counterclaim which he is entitled to set-off against the sum otherwise due. In a construction dispute, where a valid certificate has been issued for payments to the contractor, the employer may seek to counter an application for summary judgment by setting up a counterclaim, for example, on the ground of delay or defects in the work done. One drawback of the rules, from the point of view of the plaintiff, is the need to show no defence to a particular claim. It frequently happens that the plaintiff has a series of claims many of which are very likely to be successful, where it is difficult or impossible to identify one particular claim which is bound to succeed. In these circumstances, the plaintiff may apply for an interim payment under CPR, Part 25. This empowers the court to order an interim payment if it is satisfied that, if the cliam went to trial, the claimant would obtain judgment for a substantial amount of money against the other party.[17] This permits the judge to look more broadly at the merits of the claim and the defence and allows judgment to be given for the plaintiff of such amount as the court thinks just, not exceeding a reasonable proportion of the likely amount of the final judgment after taking into account any cross-claim.

The procedures under CPR, Parts 24 and 25 are intended to facilitate "cash flow" on a commercially sensible basis to avoid or deter litigation whose primary purpose is to avoid payment. This is mirrored in the adjudication procedures contained in, for example,

---

[17] CPR, r. 25.7 (1) (c).

the JCT nominated sub-contract form[18] and in the recommendations contained in the Latham report for the wider use of
adjudication.[19]

Disclosure of documents

After the close of pleadings, if an action is to continue, there
follows the work of preparing for trial. An important step is
disclosure and inspection of documents, when each side must
disclose to the other all documents which are relevant to the
matter in dispute. The Civil Procedure Rules require that a party
disclose documents which are or have been in his custody or power.
This includes documents which have been destroyed or may be in
the physical custody of some other person. Such documents must
be listed and described even though they cannot be produced.
Documents may be withheld on the ground of privilege. This
covers letters between the party and his solicitor, and documents
which came into existence as a result of the dispute, such as
experts' reports. The fact that documents were intended to be
private or confidential does not permit a party to withhold them.
Documents which come to light on discovery may have a profound
effect upon the course of an action. Occasionally one party may fail
to be sufficiently diligent in searching for documents in his
possession. The rules of court allow orders to be made for the list
of documents to be verified by affidavit, and for any further specific
documents or classes of document not previously disclosed to be
produced.[20]

One of the inherent problems created by the rules of discovery is
the potentially huge burden of documentation which must be
considered and copied in major commercial disputes, particularly
those relating to construction projects. This may involve examining
hundreds of files to extract relevant documents for discovery and
copying. The courts have in the past given a wide interpretation to
the concept of "relevance". The current rules, however, generally
require only "standard disclosure" which include the following:

(a) the documents on which a party relies;

(b) the documents which:

    (i) adversely affect his own case;

---

[18] NSC/C, cl. 4.30.
[19] *Constructing the Team* (1994), paras 9.11–9.14.
[20] CPR, r. 31.12.

(ii) adversely affect another party's case; or

(iii) support another party's case.

It is also provided that the court may order, or the parties agree (in writing) that disclosure be dispensed with or limited.

There have been various attempts to reduce the burden of discovery and the consequent costs of litigation. In arbitration, the power of the court to order disclosure was removed from the Arbitration Act in 1990 so that the extent of disclosure is now a matter for the discretion of the arbitrator or agreement of the parties. Civil law countries, that is most countries outside the Commonwealth and the United State of America, have no tradition of discovery nor of conducting litigation by seeking to undermine the case of the adversary. In civil law jurisdictions, the parties are generally not obliged to produce documents which are against their interest and the documents produced are usually limited to those relied on to support the claim together with specified documents which may be ordered to be produced. These differences of approach explain why both litigation and arbitration under the civil law systems is much quicker (and cheaper) than under the common law systems. None of the proposed civil justice reforms, including the Woolf report, have gone so far as to propose such fundamental changes.

As part of its inherent power to make orders for the production and preservation of documents, the court may make an order empowering the plaintiff to enter the defendant's premises to search for and seize material documents and articles. This is known as an Anton Piller Order[21] and is now supported by statutory authority[22] and covered by CPR, Part 24. Such an order may be made in circumstances where there is a real possibility that the defendant might destroy the material. The court will, if necessary, sit *in camera*, so that the defendant has no notice of what is intended. The plaintiff must make full disclosure of all material facts. The order usually requires the defendant to permit the plaintiff's representatives to enter, search for and remove to safe custody relevant documents or other evidence. Orders have been made in cases involving patent infringement and video piracy.

When each side has given discovery it is necessary to collect them into an "agreed bundle" for use at the trial. This usually takes the form of lever arch files in which the documents produced by each party are collated chronologically and numbered for ease

---

[21] *Anton Piller K.G. v. Manufacturing Processes* [1976] Ch. 55.
[22] Civil Procedure Act 1997, s. 7.

of reference. The plaintiff has the primary task of preparing the bundle and supplying copies for the court, but it is not unknown for the parties to produce separate bundles. Usually in a substantial case there will be several different categories of documents, for example, correspondence, contract documents, minutes of meetings and perhaps also valuations, certificates and reports. The objective should be to make the documents as accessible as possible in the interest of saving costs.

The trial

English trial procedure is based on the adversary system. The court has no duty and very little power itself to investigate the issues. It is limited to making decisions on the cases presented by the parties. A typical trial starts with the plaintiff's representative opening the case by outlining the issues involved. If there is a counterclaim, he will open the plaintiff's case on this also. In the course of opening, the pleadings and other major documents will be referred to. There will usually be a written summary of the relevant documents. In many cases, either by order or simply to assist the court, a written "skeleton" of the plaintiff's case will be prepared and circulated in advance, and the defendant may do likewise. In appeals this is expressly required by the rules.[23] Time limits are increasingly coming to be used in litigation at all levels and it may be that the plaintiff's opening has to be delivered within a limited period. This will make it essential for a written "skeleton" and other written notes to be used.

After the opening, the plaintiff must call his witnesses, first witnesses of fact and then experts, to give their evidence on the matters in dispute. At the end of his evidence the plaintiff's case is "closed" and the defendant's representative then opens his case and calls evidence for the defence. It is important for the plaintiff to ensure that he has called all necessary evidence before closing, since he will not ordinarily have an opportunity of calling further evidence, should it appear that some issue has not been proved. After the close of all the evidence, there are the closing speeches and submissions on the law, first from the defendant's and then from the plaintiff's representative.

During the course of the proceeding the parties' representatives and the judge must keep a note of the proceedings and particularly the evidence, sufficient to aid the submissions to be made at the

---

[23] *Practice Direction* [1990] 1 W.L.R. 794.

end of the trial, and, in the case of the judge in order to prepare his judgment. Traditionally, this is done by taking, as rapidly as possible, a long-hand note, often in the familiar blue "counsel" notebooks. In larger cases, the parties may agree to provide a transcript at their own expense. Currently the most advanced technology in this area is "live note", which is received by the parties and the judge on lap-top screens as the proceedings take place and can be marked up by individual recipients. The court keeps a tape recording of all proceedings in open court, but these are only transcribed and made available to the parties in the event of an appeal.

Representation and duty of advocates

In most substantial civil litigation each party will instruct a solicitor who will, at the appropriate time, instruct counsel. Junior counsel will be instructed to draw pleadings and to appear in interlocutory hearings. If the case is sufficiently weighty, the solicitor may also instruct leading counsel (Q.C.). He may be instructed during the preparatory states or sometimes only for the trial. In very large and important cases more than one Q.C. and more than one junior counsel may be instructed. The economics of litigation, however, mean that some parties cannot or do not wish to afford "full" legal representation and it is increasingly common even in large and complex cases, to find parties wishing to represent themselves. The rule is that any individual is entitled to represent himself in any court of the land. He is also entitled to be accompanied by an adviser. Some courts, particularly the county court, tend to be more indulgent and allow the adviser to speak on behalf of the litigant in person. The appearance of lawyers in any court is subject to rules as to rights of audience. Until quite recently, barristers had exclusive rights of audience in the High Court, while solicitors had a general right of audience in the county courts and right to appear on interlocutory matters in the TCC and Commercial Court. The position has, however, recently undergone, and is continuing to undergo, substantial change. Solicitors who are admitted as "solicitor advocates" now have rights of appearance in the High Court and some have been appointed Q.C.s. It is likely that further development will lead to solicitors acquiring general rights of audience in the near future.

Many of the larger construction companies operate their own legal departments, staffed by solicitors or employed barristers. They are entitled to do the same work as a private solicitor, including instructing counsel. It should be noted that in the case of arbitration, in the United Kingdom at least, any person, whether legally qualified or not, has free right of audience. The choice of representation in arbitration and to some extent in litigation, is therefore governed by economics, including the importance of the case.

In addressing the court it is the duty of an advocate to put forward all the relevant facts and law, not just those favourable to his client, although he will endeavour to present the matters in the most favourable light. This is particularly important when legally qualified advocates appear before a lay arbitrator. It is informative to note the relevant rule contained in the Code of Professional Conduct by which barristers are bound. This provides:

> "a practising barrister has an overriding duty to the Court to ensure in the public interest that the proper and efficient administration of justice is achieved: he must assist the Court in the administration of justice and must not deceive or knowingly or recklessly mislead the Court."

Judges, referees and court experts

The great majority of actions in the Queen's Bench Division of the High Court take place before a single judge who decides all matters of fact and law. Civil jury cases are extremely rare and are practically confined to actions in defamation. Cases which involve prolonged investigation into technical matters, such as building disputes, are mostly tried in the Technology and Construction Court (TCC). The judges in these courts were, before the Courts Act 1971, known as Official Referees. Now they are designated circuit judges appointed to deal with TCC business. But the original term continues to be recognised. TCC judges have practically the same powers as a High Court judge. They sit in London in a special court building in Fetter Lane. There are designated TCC judges in Bristol, Manchester and in other provincial centres; and the London judges will also sit on circuit when appropriate.

In addition to the judge, court rules provide for the possibility of appointing an assessor in appropriate cases.[24] Following the Woolf Report, more attention has been given to the use of a single joint expert, rather than separate experts appointed by each party. Accordingly, the new Civil Procedure Rules contain express power by which the court may direct that evidence on a particular issue is to be given by one expert only. These rules provide that, where the

---

[24] CPR, r. 35.15, Supreme Court Act 1981, s. 70.

parties cannot agree on a single expert, the court may direct the manner in which the expert is to be selected. Further powers allow the court to give directions about payment of the experts' fees and any inspection, examination or experiment which the expert wishes to carry out.[25] Such procedures are already well known in arbitration and it remains to be seen how frequently the courts (particularly judges of the TCC) will exercise these powers in contrast to the traditional approach or relying on party-appointed experts.

TCC procedure

There are a number of distinct features of TCC procedure now enshrined in a Practice Direction under the CPR,[26] but which have evolved over a number of years. The practice of exchanging experts' reports, now axiomatic under the court rules and applicable to all types of action, originated in the Official Referee's Courts. This has been found to be of great use in informing each party of the case to be advanced by the other side, and in facilitating the narrowing of issues. Orders for experts to meet and discuss their differences on a without prejudice basis similarly originated with Official Referees and are now governed by formal rules.[27]

TCC judges have routinely ordered the holding of a pre-trial summons or meeting at which the parties, represented by counsel, discuss the form which the trial should take. These procedures, again, are now formally included in the CPR Practice Direction, which requires the holding of both a case management conference and pre-trial review. The former Official Referees also led the way in use of computerised trial procedures. As well as widespread use of laptop computers, VDU screens have been introduced for the trial of large actions. The first "paperless" trial was conducted before one of the Official Referees using a fully developed electronic case management system with all documents stored on a retrieval system, the proceedings being recorded on a "live note" transcript system. This was provided as the case proceeded, both to those sitting in court and to lawyers in nearby offices with a directly wired link to the proceedings. Such procedures are unusual but illustrate the types of procedure which may become more commonplace in future. Parallel reforms and developments have also been taking place in the field of arbitration which is described in the next chapter.

---

[25] CPR, r. 35.7, 35.8.
[26] Pt 49.
[27] CPR, r. 35.12.

JUDGMENT AND ENFORCEMENT

After the close of the hearing the judge must come to his decision on the facts (if there is no jury) and on the law. He gives his decision in the form of a reasoned judgment. This is sometimes delivered *ex tempore* at the end of the case, but more usually reserved to a later date. Judgments in cases of any substance are usually typed and provided to the parties' representatives shortly before the judgment is to be given. In most cases, these written judgments will not actually be read out, so that the court proceeds immediately to consider the final orders which ought to be made in terms of ordering payment, interest and costs.

Interest

A further matter to be dealt with in the judgment is interest. Particularly in inflationary times with the long delays of litigation, lost interest on the sums in issue can be very significant. The court is empowered to give interest under section 35A of the Supreme Court Act 1981 on any part of a claim for debt or damages which is either included in a judgment or which is paid before judgment. Thus, even if the defendant pays all or part of the sum claimed, after issue of proceedings but before trial, the court may still award interest on the sum so paid. The power is limited to awarding simple interest. The amount awarded should normally represent a realistic rate for the time during which the successful party has been wrongfully deprived of the sum awarded. This often results in an award of around one per cent above bank rate over the period outstanding. An alternative to the recovery of statutory interest is to claim interest as damages. In *Wadsworth v. Lydell*[28] the defendant had failed to pay an agreed sum which the plaintiff required to finance the purchase of a property. The plaintiff raised the necessary sum by borrowing on mortgage and claimed the interest payments from the defendant as special damages. The Court of Appeal held that the sum was recoverable, and the House of Lords have approved the decision. Accordingly, by pleading the actual outlay of interest incurred, it is possible to recover the actual sum lost. As a further alternative, some forms of contract provide expressly for the payment of interest on overdue certificates.[29]

Where a contract provides for the reimbursement of cost or loss there may be a right, expressly or by necessary inference, to include

---

[28] [1981] 1 W.L.R. 598.
[29] See ICE form, cl. 60(7), Chap. 13.

interest or "financing charges" in the amounts payable. The Court of Appeal so held in *Minter v. W.H.T.S.O.*,[30] where the formula "direct loss and/or expense" under the JCT form of contract was held to include interest. This case should, in principle, apply to other forms of contract which provide for the recovery of claims based on actual loss or cost, including the ICE Conditions of Contract.

Costs and offers of settlement

Prima facie a successful party to litigation is entitled to an order for payment of his costs by the loser, who must also pay his own costs. Because of the very high costs of litigation, it is important to examine the circumstances in which a successful party may not recover a full order for costs. First, it is clear from the authorities that the award of costs is always in the discretion of the court or tribunal and the court may reduce the proportion of costs recovered if the conduct of the successful party so warrants. For example, if a claim has succeeded, the costs awarded may be reduced if the claim was nevertheless exaggerated, or if the plaintiff has wasted time or if elements of the claim were unsuccessful. In appropriate cases the court may go further and order the successful party to pay costs to the unsuccessful party. The relevant law was summarised in the case *Re Elgindata (No. 2)*[31]:

> "(i) Costs are in the discretion of the Court, (ii) they should follow the event except where it appears to the Court that in the circumstances of the case some other order should be made. (iii) the general rule does not cease to apply simply because the successful party raises issues or makes allegations on which he fails, but where that has caused a significant increase in the length or cost of the proceedings he may be deprived of the whole or a part of his costs, (iv) where the successful party raises issues or makes allegations improperly or unreasonably, the Court may not only deprive him of his costs but may order him to pay the whole or a part of the unsuccessful party's costs."

The award of costs becomes less clear when, in addition to (or in lieu of) a defence, the defendant has a counterclaim. If both claim and counterclaim succeed then each party is prima facie entitled to costs on his claim. This is so, even where the defence operates as a set-off. The court, in awarding costs, will look at the issues which

---

[30] (1980) 13 B.L.R. 1.
[31] [1992] 1 W.L.R. 1207.

had to be litigated, not at the sums in dispute. In order to save the laborious process of taxing items of cost as between claim and counter claim, the court will often make a global order whereby one side is to pay a proportion of the costs of the other side, and also to pay his own costs. However, it must be said that the award of costs is not an exact science. The sums involved can often be very large, even in relation to the substantial sums often in dispute in construction cases.

Where a party obtains an order for payment of its costs, the amount has to be ascertained by the process of assessment by the taxing master who is an official of the High Court, unless the parties are able to agree the costs payable. There are two bases on which costs can be awarded, the "standard basis" or the "indemnity basis". When costs are assessed, the taxing master must ascertain first whether the sums claimed for each element are reasonable. Secondly, he must decide whether it was reasonable to incur the cost in question. Where there is doubt on the standard basis of taxation the issue is resolved in favour of the paying party, whereas on an indemnity basis, it is resolved in favour of the receiving party. The net result is that taxation on the standard basis results in recovery of not more than about two thirds of the actual outlay.

Where the defendant considers he is likely to be found liable in some degree he may obtain protection against liability for costs, both his own costs and those of the plaintiff by making an offer of settlement. The rules allow the defendant to make a "payment into court"[32] of the sum offered. The plaintiff will be notified of the payment in, and may within a limited period of time accept the money in settlement of his claim, together with his costs on the claim. If the plaintiff chooses not to accept the payment in and fails to obtain judgment for more than that sum, he will normally be ordered to pay the defendant's costs and his own after the date of notification of the payment, even though he has won the action. The judge must not be told of the payment in until he has determined how much the plaintiff is to recover. A plaintiff may similarly pay money into court in respect of a counterclaim. The calculation of how much to pay into court and the decision whether or not to accept it in settlement may require very careful considera-tion in view of the large sums for costs which may be at risk. There is a similar means of protection available without making use of the statutory process simply by making an offer of settlement stated to be "without prejudice save as to costs". This is known as a

---

[32] CPR, Pt 36.

Calderbank[33] offer and will be considered in the discretion of the court in the same way as a payment. A Calderbank offer is useful where a payment cannot be made under the court rules, for example by one of several defendants to a claim, who might make such an offer to the co-defendants.

Enforcement

The final stage in the action is enforcement of the judgment. If the judgment debtor does not pay, there are a number of methods available to the judgment creditor by which he may obtain at least some payment. The most important of these are: seizure of the debtor's goods; charging the debtor's land; appointment of a receiver over the debtor's business; or obtaining an order that a debt owed to the debtor be paid to the judgment creditor instead. Where the debtor is a limited company there may be an application to the court for winding up. However, if the threat does not produce payment, winding up will not improve the position of the judgment creditor, who will rank equally with other unsecured creditors.

It is a fundamental principle of English law that a plaintiff bringing proceedings takes his chance as to whether there will be assets against which to enforce a judgment. However, in recent years the courts have evolved an important procedural device which, while it does not improve the plaintiff's position, prevents the defendant from worsening it. This is the "Mareva" injunction,[34] which prohibits the party against whom it is directed from disposing of or otherwise dealing with assets within the jurisdiction. Initially, this form of relief was granted against foreign defendants who might remove their assets from England, but developments in case law and now statutory backing[35] allow such injunctions to be granted against any party and in respect of assets within the jurisdiction or abroad. The plaintiff must show that he has a good arguable case against the defendant, and that there is a real risk that a judgment will be unsatisfied because the defendant will dispose of his assets in advance unless restrained from doing so.

Applications for Mareva injunctions are now frequent, and the courts take a strict attitude. As with other forms of injunction, an order may be granted *ex parte*, if necessary on very short notice; but the matter will be reconsidered on an *inter partes* hearing. The

---

[33] *Calderbank v. Calderbank* [1976] Fam. 93.
[34] *Mareva Compania v. International Bulk Carriers* [1975] 2 Lloyd's Rep. 509.
[35] CPR, r. 25.1 (1) (f).

plaintiff must make full disclosure of all material facts. A Mareva injunction does not give the plaintiff any preferential right over the assets restrained. However, a defendant will sometimes offer to put up security in lieu of the injunction, so as to permit him to use the assets in question. A Mareva injunction may be obtained in aid of arbitration proceedings.

Foreign judgments

A judgment may need to be enforced in a country other than that in which it was given. In every country enforcement depends solely on that country's internal laws.[36] For enforcement in England, the courts require to be satisfied that the foreign court had proper jurisdiction, that the judgment is final and for a fixed sum, and that it was properly obtained. If these conditions are satisfied, a foreign judgment may be enforced in England under various reciprocal statutory arrangements. These allow the foreign judgment to be registered and enforced as an English judgment. They similarly allow English judgments to be enforced abroad. Particular arrangements which apply to enforcement of judgments throughout Europe (not limited to the European Community) are contained in the Civil Jurisdiction and Judgments Act 1982. A judgment from a country with which there is no statutory arrangement is treated in England as a simple contract debt, which may be enforced by suing in the English courts. Enforcement of an English judgment abroad, without the aid of reciprocal arrangements, depends on the internal law of the country where enforcement is sought. Enforcement of arbitration awards in different countries is dealt with in Chapter 3.

EVIDENCE AND WITNESSES

In a civil action the relevant facts must be proved on a balance of probabilities, unlike criminal cases where proof beyond reasonable doubt is required. The burden of proving a fact usually lies upon the party asserting the fact. When deciding how much evidence must be adduced, it must be considered that a judge, unlike an arbitrator cannot draw upon his own knowledge, except in very obvious matters, and therefore every fact relied on must be proved.

---

[36] *Adams v. Cape Industries* [1990] Ch. 433.

In practice some of the facts will usually be admitted, for example in the pleadings, or in open correspondence. If one party refuses to admit some fact which, while likely to be true, would be expensive to prove formally (for example, that the resident engineer signed hundreds of day-work sheets appearing to bear his signature), the other party may serve a "notice to admit facts".[37] If not then admitted, the court may order the first party to pay the costs of proof, whoever wins the action.

Admissibility

A major part of the law of evidence is concerned with admissibility. The principal application in construction disputes is in relation to "hearsay" evidence. Normally a witness should relate only what he himself perceived. Alleged facts which another person has related to him are hearsay. The admission of hearsay evidence is now governed by the Civil Evidence Act 1995, under which a party proposing to adduce hearsay must give notice and provide particulars to the other party, subject to rules of court. The provisions governing hearsay may be excluded by agreement, or waived. In long construction cases, it is often prudent to take statements from elderly or infirm witnesses so that, should they be unfit to attend the trial, their evidence may be admitted in accordance with the Act.

A witness of fact may not normally give his opinion; although an expert may do so, and is usually brought in to give his professional opinion on the facts he is given. This question can create difficulty in construction and other technical cases, where the factual witnesses may wish, as part of their evidence, to give opinions on the matters in issue. Such difficulty can be resolved by a mutual acceptance that technically qualified witnesses may proffer their opinions, where relevant to their evidence. Alternatively, it may be necessary to seek to admit the relevant witnesses as experts (see below).

Giving evidence

The law of evidence also deals with the way in which testimony may be given. The normal practice, which still applies in some civil actions and in criminal cases, is that evidence must be given orally and from memory. This process is slow and costly, and in many

---

[37] CPR, r. 32.18.

civil actions (and particularly in arbitration) the practice of exchanging witness statements and of taking the contents of such statements as read is frequently adopted. This is now enshrined in the Rules of Court,[38] which apply to proceedings in the TCC and Commercial Court. There are various procedures for dealing with contentious parts of statements. In the Commercial Court, the practice is to require particular parts of the statement objected to by the opposing party to be adduced by conventional question and answer. In arbitration, particularly in international cases, there is a strong trend towards putting all evidence in writing and dispensing with oral questioning as far as possible.

Oral evidence must normally be adduced by question and answer. Evidence given for the party who calls the witness is known as examination-in-chief. The advocate adducing such evidence must not ask "leading" questions, *i.e.* those which suggest the answer. This is not merely to avoid unfair advantage: leading questions in-chief may succeed in giving quite misleading evidence, because witnesses often try to be helpful. After a witness has given his evidence in-chief (whether orally or simply by putting in a signed statement) he must be tendered for cross-examination by the opposing party or his advocate. Cross-examination is not restricted to the matters on which the witness has given evidence in-chief. A witness may be questioned on any matter relevant to the case, including his truthfulness. Cross-examination is a fundamental part of the English adversarial system, and many lawyers will have experience of the process having a material or even dramatic effect in exposing the truth. Cross-examination is the more important when written statements are used, because there is a constant temptation for witnesses to include material which is not properly within his knowledge.

A party may compel the attendance at the trial of any person whom he wishes to give evidence or produce documents. Attendance is enforced by serving a witness summons.[39] In arbitration proceedings the High Court has power to issue a witness summons.[40] There is no right or property in witnesses. Any person may be called by either side. The potential witness is, however, entitled to refuse to give a statement in advance. A party may even compel his opponent to give evidence. But it is unwise to call a person who may be hostile because he cannot normally be cross-examined by the party calling him.

---

[38] CPR, Pt 32.
[39] CPR, r. 34.2.
[40] Arbitration Act 1996, s. 43.

Expert evidence

Evidence which is proposed to be given by an expert is admissible only subject to court rules. These require permission to be obtained for the calling of such evidence, which will be conditional upon the exchange of reports between the parties in advance.[41]

There is no precise definition of expert evidence. Its function depends upon the tribunal before which it is to be adduced. In the High Court, expert evidence is necessary to explain technical features of a case. Conversely, in arbitration, expert evidence should be unnecessary where the arbitrator is appropriately qualified. Despite this, it is usual for such evidence to be introduced. A common practice before the TCC is to invite the court to adjourn to read all the expert evidence, which is then either taken as read or introduced briefly by the witness, before he is cross-examined. In arbitration, it is often appropriate for the arbitrator himself to ask questions of the expert.

Until the report of an expert is exchanged, his views and opinions remain privileged. If a party obtains an unfavourable opinion from one expert, he may go to another and rely exclusively upon his opinion, should it be more favourable. A party who takes this course, however, runs a risk that the identity of the first expert may be discovered by the opposition, who may then compel him to give evidence. Parties are usually well advised to accept the first opinion they are given. An expert should always give an independent and unbiased opinion on the issues. It is, however, quite proper for the expert, both in his report and in his evidence to emphasise any technical points in his client's favour. Technical issues are often arguable in just the same way as legal issues are.

Experts' costs

An important distinction between an expert and a witness of fact, is that the former is entitled to be paid a proper professional fee, which may be recovered on taxation. This sometimes leads to dispute about whether a particular witness is an expert. In the case of *James Longley v. S.W. Regional Health Authority*[42] the claimant contractor in an arbitration sought to include a substantial sum in the bill for costs in respect of the fees of an unqualified "claims consultant." The respondent objected that the consultant was not qualified to give expert evidence, and that his evidence was not

---

[41] CPR, Pt 35.
[42] (1983) 25 B.L.R. 56.

admissible. On a review of taxation in the High Court, it was held that such evidence was admissible, and an expert might be appropriately qualified by skill and experience. Where a successful party in an arbitration is represented by a claims adviser rather than a solicitor, it has been held that the costs of the adviser are recoverable on taxation.[43]

Alternative Dispute Resolution

The term ADR has been in circulation for some years, having been imported from the United States of America. There are different views as to what it includes. The Woolf report includes both arbitration and the "ombudsmen" system; but the more general view is that it refers to the various forms of ad hoc procedure which are all (unlike litigation and arbitration) consensual and unsupported by any coercive or directive powers of the court, save to the extent of enforcing what the parties have agreed.

ADR procedures tend to be relatively informal, but their range is very wide. At one extreme, the process usually described as mediation can consist of little more than settlement negotiation through an intermediary, from which either party can withdraw at any stage. At the opposite extreme are processes which may be referred to as contractual adjudication, which have a formal structure and a decision which may become binding. There is considerable scope for confusion in the use of the terms and it is safer to define what is meant. A general distinction which needs to be drawn is between processes which (pursuant to the agreement of the parties) are mandatory, and those which are voluntary. A mandatory procedure operates as a condition precedent to the pursuit of any further remedy, for example by arbitration or litigation. Such procedures will generally be enforced by the courts, as in the *Channel Tunnel* case. Here the employers had sought an injunction from the English court, despite the existence of an elaborate dispute resolution procedure involving an initial reference to a panel of three independent experts. Lord Mustill, holding that the agreed procedure should be enforced, said:

> "Having made this choice I believe that it is in accordance, not only with the presumption exemplified in the English cases cited above that those who make agreements for the resolution of disputes must

---

[43] *Piper Double Glazing v. D.C. Contracts* (1992) 31 Con. L.R. 149.

show good reasons for departing from them, but also with the interests of the orderly regulation of international commerce, that having promised to take their complaints to the experts and if necessary to the arbitrators, that is where the (employers) should go. The fact that the (employers) now find their chosen method too slow to suit their purpose, is to my way of thinking, quite beside the point."[44]

The same principle applies whenever a customised dispute resolution procedure is provided in a contract and expressed in mandatory terms, and equally when the parties enter into an ad hoc agreement to pursue a particular procedure. However, if the parties have merely agreed to embark upon a process of non-binding negotiation, then the court would be unlikely to enforce it.

The Woolf report stopped short of recommending court-annexed ADR but did recommend that parties to litigation should be required, at the pre-trial stage, to state whether they have discussed ADR. This is in contrast to the position in the United States of America, where the courts are more pro-active in promoting ADR. The following section reviews briefly the major forms of ADR, excluding arbitration which is the subject of Chapter 3.

Conciliation and mediation

These terms are often used interchangeably but they are two essentially different processes. Mediation, as the name implies, involves a neutral mediator finding middle ground between the position of the parties with the aim of achieving a negotiated solution acceptable to all parties. The role of the mediator includes separate and private negotiation with each party in order to discover, by a process of accelerated settlement discussions, at what figure (or on what terms) each party will settle. The actual settlement is achieved by a legally enforceable contract setting out the terms agreed.

The essence of mediation is that the mediator does not publicly express his view on the case. Indeed, his function may be to concentrate on the commercial aspects of the dispute. In contrast, a conciliator may be empowered to or required to express his provisional view on the merits of the case. The ICE Conciliation Procedure (1994), referred to in clause 66 of the ICE Conditions of Contract, requires the conciliator initially to act in a mediating role, in the sense discussed above. If an agreed settlement is not

---

[44] *Channel Tunnel Group v. Balfour Beatty* [1993] A.C. 334.

achieved, the conciliator gives a recommendation which may become binding upon the parties if a notice to refer to arbitration is not given.[45]

Mini-trials

This refers to a form of aided settlement in which each side presents a summary of its case, in trial mode and using advocates and experts, before a tribunal composed a senior representative of each side and a neutral chairman. The objective is to demonstrate directly the strengths and weaknesses of the respective cases to those in a position of responsibility so that they may seek to negotiate an informed settlement, with the aid of the neutral. The process is not inexpensive and necessarily involves preparation and the employment of professional advocates in order that the case is seen in its best light.

ADJUDICATION

The standard forms of contract have, for some years, contained provisions for adjudication.[46] Although there is no definition of the term (and this continues to be the case under the new Act), adjudication was recommended for construction contracts in the Latham Report[47] and has now become mandatory in certain construction contracts by the Housing Grants, Construction and Regeneration Act 1996. "Statutory" adjudication, which is dealt with below, is not to be confused with the original contractual process, which is dependant upon the particular terms and status agreed between the parties. "Contractual" adjudication may, however, comply with the requirements of the Housing Grants, etc., Act so that the two processes may overlap. As examples of adjudication under contract, the traditional decision of the Engineer under the ICE Conditions complies with what is generally understood by the term; and under the Engineering & Construction Contract (ECC), the equivalent process is called adjudication, in this case being carried out by an independent third party.

The important distinction between adjudication and other forms of ADR lies in the fact that adjudication is intended to result in

---

[45] ICE Conditions, cl. 66 (5).
[46] See generally Mark McGaw, "Adjudicators, Experts and Keeping out of Court" (1992) 8 Const. L.J. 332.
[47] *Constructing the Team* (1994).

one party being compelled, at least temporarily, to submit to the decision, for example, by paying money to the other party. Consequently, the status of an adjudicator's decision can give rise to dispute. In *Cameron v. Mowlem*[48] the plaintiff sub-contractor requested the appointment of an adjudicator under Form DOM/1 to determine a dispute over payments due. Without giving reasons, the adjudicator determined that the sum of £52,800 should be paid, but Mowlem resisted payment relying on a right of set-off, in respect of which the required notice had been given. The Court of Appeal held, first, that the adjudicator's award could not be enforced as an arbitration award; and secondly that the adjudicator's award did not preclude the contractor from exercising a contractual right of set-off, and that the adjudicator's powers did not include determination of sums due under the terms of the sub-contract. The case illustrates the need for clarity in determining the status of an adjudicator's decision.

Can the decision of an adjudicator be challenged on its merits, either on fact or law? The *Cameron v. Mowlem* case confirmed that the decision did not rank as an award for the purpose of challenge in accordance with the Arbitration Act. There is a line of authority dealing with the circumstances in which the decision of an independent expert might be called into question, for example where reasons for the decision are given which are demonstrably wrong. It has now been held by the Court of Appeal, however, that the decision of an expert, which would include an adjudicator, cannot be challenged unless it can be shown that the expert has departed from the instructions given to him in a material respect.[49]

Statutory adjudication

Following the Latham Report of 1994 the Department of the Environment issued consultation papers intended to explore ways of legislating to achieve the reforms recommended. This proved somewhat controversial,[50] but a Bill was introduced dealing with a limited number of the Latham recommendations which was passed into law as Part II of the Housing Grants, Construction and Regeneration Act 1996. The Act deals with three construction-related matters. First, there is an elaborate definition of the term "construction contract", to which the substantive provisions apply

---

[48] (1990) 52 B.L.R. 24.
[49] *Jones v. Sherwood Services* [1992] 1 W.L.R. 277.
[50] See "Contemporary Issues in Construction Law", *Construction Contract Reform: A Plea for Sanity*, Vol. 2 (John Uff Q.C. ed.) Construction Law Press, 1997.

(or do not apply if the definition is not satisfied). Secondly, there are measures providing for the compulsory availability of adjudication and its consequences. Thirdly, there are important provisions dealing with the right to payment under a construction contract, as defined. The first two matters are dealt with in this section and the third in Chapter 9.

"Construction contract" is defined in sections 104 and 105 of the Act as meaning an agreement for carrying out construction operations (as defined), including sub-contracted work and (importantly) architectural design or surveying work or advice on building, engineering, decoration or landscape.[51] The term "construction operations" is widely defined but (significantly) excludes a long list of construction operations such as extraction of oil, gas or minerals, installation of plant for nuclear processing, power generation or water or effluent treatment, bulk storage of chemicals, oil, gas, steel, or food or drink. Also, excluded is the manufacture or delivery of components, materials, plant and machinery unless the contract also provides for installation.[52] The Act applies only to construction operations in England and Wales (and Scotland), whatever the applicable law of the contract. Accordingly, the Act will apply to an Italian sub-contractor supplying marble if it is also to be installed by the sub-contractor; but it will not apply to a United Kingdom supplier who merely delivers components for heating and ventilation, drainage or fire protection, etc. It should be noted that this is the first attempt to "ringfence" the construction industry for the purpose of special legislation. The success of the new measures remain to be seen.

Adjudication is dealt with in section 108 of the Act which provides as follows:

"(1) A party to a construction contract has the right to refer a dispute arising under the contract for adjudication under a procedure complying with this section.
For this purpose 'dispute' includes any difference.

(2) The contract shall—

(a) enable a party to give notice at any time of his intention to refer a dispute to adjudication;

(b) provide a timetable with the object of securing the appointment of the adjudicator and referral of the dispute to him within 7 days of such notice;

(c) require the adjudicator to reach a decision within 28 days of referral or such longer period as is agreed by the parties after the dispute has been referred;

---

[51] ss. 104 (1), (2).
[52] ss. 105 (2).

(d) allow the adjudicator to extend the period of 28 days by up to 14 days, with the consent of the party by whom the dispute was referred;

(e) impose a duty on the adjudicator to act impartially; and

(f) enable the adjudicator to take the initiative in ascertaining the facts and the law.

(3) The contract shall provide that the decision of the adjudicator is binding until the dispute is finally determined by legal proceedings, by arbitration (if the contract provides for arbitration or the parties otherwise agree to arbitration) or by agreement. The parties may agree to accept the decision of the adjudicator as finally determining the dispute.

(4) The contract shall also provide that the adjudicator is not liable for anything done or omitted in the discharge or purported discharge of his functions as adjudicator unless the act or omission is in bad faith, and that any employee or agent of the adjudicator is similarly protected from liability.

(5) If the contract does not comply with the requirements of subsections (1) to (4), the adjudication provisions of the Scheme for Construction Contracts apply."

Of particular note is the requirement that a party must be enabled to give notice "at any time". This provision, at a stroke, removes the traditional authority of the Engineer or the Architect to render decisions which could be challenged only by a subsequent process of arbitration, often after completion of the contract. The new measure entitles the contractor (or even the Employer) to require immediate adjudication on any matter of difference. This measure also has a profound effect on the procedure under clause 66 of the ICE Conditions whereby a dispute is to be referred first to the Engineer. No such provision may now hold up the right to adjudication. The consequential amendments to clause 66 are dealt with in Chapter 13.

The other provision of particular note is that clause 108 is not restricted to payment disputes, but includes disputes relating to time, quality and any other matter capable of giving rise to a difference between the parties. Bearing in mind that all sub-contracts must also contain such provisions, the possibility of over-lapping and conflicting adjudication decisions is immediately apparent (see below). A further notable feature, to which attention was drawn during the parliamentary debate, is that section 108 (4) provides only for immunity in contract, which will be apt to bind only the immediate parties, and not any third party who may suffer damage as a result of the adjudicator's decision. This is to be contrasted with section 29 of the Arbitration Act 1996 (see Chapter 3).

Procedure for adjudication

Clause 108 operates by requiring either a conforming contractual adjudication scheme, or in default the statutory Scheme for Construction Contracts is to apply. The Scheme (which proved highly controversial in its drafting) was published shortly before the primary legislation came into force, on May 1, 1998.[53] The Scheme contains a much more detailed procedure for the giving of notice, the appointment, the adjudication procedure, the decision and its enforcement. As regards overlapping disputes (for example where the same issue arises under the main contract and one or more sub-contract) the Scheme provides only that, where the dispute in question is the same or substantially the same as one which is already the subject of an adjudication decision, the adjudicator "must resign".[54] The parties may, of course, agree some other course, and the issue may be dealt with by way of a contractual scheme. As regards enforcement, the adjudication decision takes effect as a contractual debt, since the procedure itself is expressly required to form part of the construction contract. Little consideration appears to have been given to the problems of enforcement which, in the event, have proved to be a major difficulty. In principle, there is nothing to prevent the party against whom an adjudication decision is to be enforced claiming a right of equitable set-off (see above). This may be excluded by the terms of the contract, but to do so would elevate the adjudication decision to a preferred status that may not be intended. This is a matter for agreement between the parties. In terms of the mechanism of enforcement, the existence of an arbitration clause creates a further problem, in that the court procedure for summary judgment will not now be available in the light of recent authority.[55] Various solutions are possible. The standard forms of contract now contain a special provision in the arbitration clause entitling a party who seeks enforcement of an adjudication decision to abrogate the arbitration clause, thereby permitting enforcement through court action. Secondly, the statutory Scheme enables the adjudicator to order that his decision is complied with peremptorily, in which case, by the application of an amended version of section 42 of the Arbitration Act 1996 (Enforcement of Peremptory Order) the court is given jurisdiction to enforce the decision.[56] Some of the

---

[53] Scheme for Construction Contracts (England and Wales) Regulations 1998 (S.I. 1998 No. 649).
[54] Scheme for Construction Contracts, para. 9 (2).
[55] *Halki Shipping v. Sopex* [1997] 1 W.L.R. 1268: see Chap. 3 above.
[56] Scheme for Construction Contracts, paras 23–24.

potential difficulties of enforcement were considered in the recent TCC decision in *Macob v. Morrison Construction*.[57]

The statutory Scheme, which was introduced by the new Labour Government in 1998 (the Act having been introduced by the previous Conservative Government), deals with immunity of the adjudicator in paragraph 26, which closely follows section 29 of the Arbitration Act 1996, providing as follows:

> "26 The adjudicator shall not be liable for anything done or omitted in the discharge or purported discharge of his functions as adjudicator unless the act of omission is in bad faith and any employee or agent of the adjudicator shall be similarly protected from liability."

However, while section 29 of the Arbitration Act creates an immunity which will be effective against third parties, the Scheme, where it applies, appears to take effect as a matter of contract[58] which will not, therefore, bind third parties. Additionally, the Scheme will not apply where there is a conforming contractual procedure. Again, it is unclear whether the statutory Scheme can apply in part, for example, where immunity is not dealt with in the contractual procedure.

A number of contractual adjudication schemes have been published complying with the requirements of section 108, such as that published by CEDR.[59] Any of these procedures may be incorporated into the contract to provide a conforming scheme. The effect of this is to avoid the application of the Scheme for Construction Contracts, so that the parties should ensure that any provision of the Scheme they may wish to apply is included in the contractual procedure.

The success or otherwise of these statutory measures remains to be established. Among the major drawbacks of the Scheme may be mentioned the following:

(i) the process is temporary and necessarily inexact. In most cases there will not be adequate opportunity to consider detailed arguments within the stipulated period of 28 days. This may be seen to operate unfairly against the Respondent, where the Claimant has had unlimited time to prepare his claim;

(ii) the overall objective of adjudication is to avoid "cashflow" being held up by the inability to challenge decisions of the

---

[57] [1999] B.L.R. 93.
[58] HGCRA, s. 114(4).
[59] Centre for Dispute Resolution.

Engineer or Architect. The extent to which this is achieved, however, is entirely dependant on the ability to obtain and enforce decisions in timely manner. The technical complexity of the provisions militate against this being achieved;

(iii) where an adjudication decision is given and enforced, there is no requirement for security for repayment, in the event of the decision being reversed. The subsequent insolvency of the receiving party will, therefore, create a loss, for which recovery may ultimately be sought from the adjudicator, subject to the immunity provisions. It appears questionable whether appropriate indemnity insurance will be available;

(iv) the Scheme undermines the status and authority of the Engineer and Architect operating under traditional forms of contract, and their ability adequately to preserve the interests of the Employer. This aspect of the Scheme has given rise to much adverse comment from the professional institutions.

The extent to which statutory adjudication effectively replaces other forms of dispute resolution, including arbitration, in the construction industry remains to be seen.

CHAPTER 3

DISPUTE RESOLUTION—ARBITRATION

THE term "arbitration" has several meanings. In popular usage, it denotes the placing of a dispute before a third party to obtain a fair or equitable decision, based on discretion rather than on fixed rules. In industrial law, it refers to a process of conciliation, where attempts are made to find a formula acceptable to two parties in disagreement.[1] In commercial law, arbitration has acquired a more definite and fixed meaning, as a process, subject to statutory support by which formal disputes may be determined in a binding manner by a tribunal of the parties' own choosing. It is in the third sense that arbitration has become widely adopted for the resolution of disputes under construction contracts. It is the principal alternative to determination of disputes in court. But, as will be seen, there are cases where arbitration is an exclusive remedy and, even where there is a theoretical choice of tribunal, one party may be able to compel a reference to arbitration.

STARTING AN ARBITRATION

Three things are required before there can be an arbitration. First, there must be a dispute. This requires one party to make a claim or assertion and the other party to deny it. Thus, there can be no dispute about a claim which has not previously been put forward, or which has not been rejected. Secondly, there must be an agreement to arbitrate. Thirdly, there must be a submission of the dispute to arbitration.

In construction contracts, the agreement to arbitrate is often included as one of the clauses of a standard form of contract, such

---

[1] Under the Advisory Conciliation and Arbitration Service (ACAS) set up by the Employment Protection Act 1975.

as article 5 and clause 41 of the JCT form and clause 66 of the ICE form. In such clauses the parties agree to submit specified future disputes to arbitration. There may also be an agreement to arbitrate made after the dispute has arisen, usually referred to as an ad hoc agreement.

In either case there must be a submission (sometimes called a reference) of specified disputes to arbitration, by one party serving notice to refer on the other. No particular form is required for a submission but it is an important step, as it usually constitutes the commencement of the arbitration for the purpose of limitation, and is thus equivalent to the issuing of a claim form in court. Subject to agreement of the parties, arbitral proceedings are commenced when one party gives notice initiating whatever step is required in accordance with the arbitration agreement, for example requesting the President of the ICE to appoint an arbitrator.[2] No particular form is required for an arbitration agreement, whether made in advance or after a dispute has arisen. If the agreement is in writing, which is invariably so in the case of a construction contract, the arbitration will be governed by the Arbitration Act 1996. An arbitration can exist at common law outside the Act and could, in theory, be enforced in the same way as any other private dispute resolution procedure agreed between the parties.[3]

A reference to arbitration is deemed to be to a single arbitrator unless some other number is agreed.[4] Many commercial arbitrations (such as shipping and commodity disputes) employ three arbitrators who either sit as a court of two arbitrators (one appointed by each side) with an umpire to settle any disagreement; or alternatively as three arbitrators, one acting as chairman. The difference is that an umpire is required to act only where the two appointed arbitrators disagree, upon which he takes over as sole arbitrator.[5] The umpire may or may not attend the hearing with the arbitrators, as the parties may agree. A chairman, on the other hand, acts throughout as one of the arbitrators, usually on the basis that he will make the decision if there is no majority.[6] Multiple tribunals have never been favoured in building and engineering disputes. The JCT and the ICE forms of contract refer specifically to one arbitrator. The international (FIDIC) conditions contemplate that there may be more than one arbitrator, and tribunals

---

[2] Arbitration Act 1996, s. 14.
[3] See under Alternative Dispute Resolution, Chap. 2.
[4] Arbitration Act 1996, s. 15(3).
[5] s. 21.
[6] s. 20.

consisting of three arbitrators are commonplace under ICC procedure (see below). The function of an umpire is, however, virtually unique to English arbitration law and other laws deriving from it.

The selection and appointment of an arbitrator follows the reference to arbitration. The arbitrator may be named in the arbitration agreement, but it is more usual to find a requirement that the arbitrator be agreed between the parties and in default appointed by a specified or identified person. The JCT and ICE forms provide for appointment, in default of agreement, by the presidents of the RIBA or ICE respectively. Where the parties cannot agree and there is no mechanism for the appointment, the court has power to appoint an arbitrator.[7]

What is arbitration?

Arbitration proper is to be distinguished from various less formal processes met in construction contracts. The essentials of arbitration are that there must be a dispute, which is referred to an independent arbitrator for decision, usually after hearing the parties and receiving any evidence they wish to put forward. His decision is final, subject to the possibility of review of points of law by the courts. There are other analogous processes which do not constitute arbitration. Certifying requires the certifier to act independently and fairly, but there is no dispute and the decision need not be final. Negotiation requires that there be a dispute, but there is no independent decision (there may be no "arbitrator") and no finality. Conciliation is used to reach a settlement, but the parties do not bind themselves to accept the result. Valuation requires that the parties agree to accept an independent opinion, but there is usually no hearing or submission of evidence. Adjudication (see Chapter 2) is the nearest process to arbitration. The result is intended to be binding but the process lacks many of the essential qualities of arbitration and does not fall under the Arbitration Act 1996.

It may be difficult to categorise a particular dispute resolution process. For example, the engineer's decision on a dispute under clause 66 of the ICE conditions (see Chapter 13) must be made independently and may become binding. But there is no duty to hear representations, so the process does not amount to arbitration. Arbitrators (like judges) are generally immune from action by

---

[7] s. 18.

the parties.[8] It was suggested in the case of *Sutcliffe v. Thackrah*[9] that such immunity would not extend beyond a "quasi arbitration" that is, one to which the Arbitration Act did not apply, but which was an arbitration in all but name. Thus, certifiers, valuers and conciliators owe a duty of care to the parties, and may be liable for negligent acts.

Arbitration and the courts

Arbitration is a private alternative to litigation as a means of settling disputes. Inevitably, there are many connections between the two processes. Arbitration must be conducted within and in accordance with the law. The underlying function of the court is to support and enforce the arbitration process through a number of specific powers (see below). Arbitration proceedings are sometimes conducted in a manner closely analogous to court proceedings, for example, with pleadings, disclosure of documents and evidence closely following the Civil Procedure Rules. This is not necessary, however, and it should always be remembered that arbitrators and parties to an arbitration have a very wide discretion as to the way in which they conduct proceedings, the courts themselves being much more closely restricted by their own rules.

Disputes may generally be brought either in court or in arbitration, so that there is a possibility of conflict. Historically, the courts have been jealous of their supremacy, but the modern approach of the courts is to encourage parties to take disputes, particularly technical ones, to arbitration. Where a conflict could arise, the courts take the view that it is a matter for the parties whether they wish to proceed with the resolution of their disputes by arbitration or in court. In the case of *Lloyd v. Wright*[10] the parties commenced an arbitration but the plaintiff subsequently issued a writ repeating the same claims, and the question arose whether this brought the arbitration to an end. Eveleigh L.J. in the Court of Appeal held:

> "The principle that the court will not allow its jurisdiction to be ousted is at the root of the defendant's argument. However, the court does not claim a monopoly in deciding disputes between parties. It does not, of its own initiative, seek to interfere when citizens have recourse to other tribunals. The court exercises its jurisdiction when appealed to. Until then, the court is not conscious of ignominy if an arbitrator decides a question with which the court is competent to

---

[8] s. 29 and see also s. 74.
[9] [1974] A.C. 727.
[10] [1983] Q.B. 1065.

deal. Furthermore, the court will not refuse to allow the subject matter of an action already begun to be referred to arbitration, if the parties so agree. . . . The court, however, will not permit its assistance to be denied to a party who has invoked it except by that party's consent or by its own ruling."

The Arbitration Act 1996

This major piece of new legislation has a long history. The principal arbitration Act up until 1996 was the Arbitration Act 1950 which substantially re-enacted the Act of 1934. Many of the provisions dated back to the Act of 1889. For most of the present century arbitration in London was conducted with little regard to foreign laws or practices. Because of its commercial importance, London attracted a large amount of international work, much of this in the maritime, insurance and commodity fields. A high proportion of arbitrators in all commercial fields, including construction were technically, and not legally qualified. The system operated satisfactorily through the "case stated" procedure,[11] whereby any point of law could readily be brought before the commercial court. This procedure had the advantage of ensuring compliance with the law and, at the same time, aiding the development of English commercial law through the cases. Almost invariably, the underlying contract would be subject to English law, whatever the nationality of the parties.

Major changes began to occur from the 1950s onwards. In 1958 the New York Convention on the enforcement of foreign awards was produced, though only ratified in the United Kingdom by the Arbitration Act 1975. This convention recognised the increasing importance of international trade and of the "internationalisation" of arbitration. By the 1970s the view was held that London was losing its international business to other foreign centres because of the ease with which appeals could be mounted, whether in domestic or international cases. The lack of finality was regarded as commercially unacceptable. This led to the Arbitration Act 1979 which abolished case stated and substituted a qualified right of appeal, dependent upon leave of the court. Although not made clear by the Act, the House of Lords soon laid down that leave should be sparingly given[12] so that in the majority of cases an award would be final. The next and most fundamental development was the launch in 1985 of the UNCITRAL[13] Model Law on inter-

---

[11] Arbitration Act 1950, s. 21
[12] *BTP Tioxide v. Pioneer Shipping (The Nema)* [1982] A.C. 724.
[13] United Nation Commission on International Trade Law.

national commercial arbitration. This was intended to be adopted, with or without amendment, in place of national arbitration laws, to produce a harmonised system for international arbitration law, with the option also of adopting the Model Law for domestic arbitration also. A committee was set up by the DTI, initially chaired by Lord Mustill and subsequently by Lord Steyn and Lord Saville, to consider how England should respond. The first report in 1989 rejected adoption of the Model Law in England and Wales, either to replace existing English domestic law or for international arbitration, but recommended that a new and updated arbitration law be prepared. Subsequently, a DTI committee considering the law of Scotland recommended adoption of the Model Law, which has accordingly been incorporated into Scots law. The DTI committee for England and Wales produced a number of interim reports which led to amendments to the existing Arbitration Acts. Finally, a complete draft Bill was produced in 1994 and substantially revised in 1995. This supersedes with major amendments all the pre-existing Arbitration law. It also codifies the major common law principles of arbitration which had not previously appeared in any of the Acts. Without reversing its earlier decision to reject the Model Law, the DTI committee nevertheless adopted substantial elements of the Model Law, which has been referred to as the most important influence over the new law. The Act makes major changes in the underlying approach to arbitration and will have a fundamental effect on the process. The Act applies to any arbitral proceedings commenced on or after January 31, 1997. The new Act applies in Northern Ireland but not in Scotland. The discussion in the following sections is based exclusively on the new Act. For an account of the pre-1996 arbitration law, reference should be made to Chapter 3 of the fifth edition of this book.

Effect of an arbitration agreement

Since arbitration is a matter of private agreement, the arbitrator's authority depends upon the scope of that agreement. This may be completely general, such as an agreement to refer "any dispute or difference arising under or in connection with the contract"; or it may be limited to specified areas of dispute, for example, an agreement in a lease that disputes as to rent increases are to be settled by arbitration. Building and engineering contracts usually contain wide arbitration clauses, but the clauses in the JCT and ICE forms are expressed to be subject to certain time limits. Also, arbitration under some forms cannot, in respect of most disputes, usually be opened until after completion of the works (see

Chapter 12). This may have repercussions in seeking to challenge adjudication decisions by arbitration.

Before 1984 cases before the courts proceeded on the basis that, where there was an arbitration agreement in the contract, the court would exercise the same powers as the arbitrator, for example by reviewing certificates and extensions of time as necessary. In the case of *Northern RHA v. Derek Crouch*,[14] the Court of Appeal held that the court did not have the same powers and interpreted the words "open up review and revise any certificate" in the JCT arbitration clause as creating a power exclusively to be exercised by an arbitrator. The result was that the court was held to have no jurisdiction over such matters. The case was followed in a series of decisions of the Official Referees (now TCC judges) and in other decisions of the Court of Appeal. The decision in *Crouch* was finally and decisively overruled by the House of Lords in *Beaufort Developments v. Gilbert-Ash NI*,[15] an appeal from the Court of Appeal of Northern Ireland. It was there held that clear and unequivocal words would be needed to deprive a party of recourse to the court and that this was not the effect of the JCT arbitration clause. Lord Hope put the matter this way:

> "If the contract provides that the sole means of establishing the facts is the expression of opinion of an architect's certificate, that provision must be given effect to by the court. But in all other respects, where a party comes to the court in the search of an ordinary remedy under the contract or for a remedy in respect of an alleged breach of it, the court is entitled to examine the facts and to form its own opinion upon them in the light of the evidence. The fact that the architect has formed an opinion on the matter will be part of the evidence. But, as it will not be conclusive evidence, the court can disregard his opinion if it does not agree with it".

Thus, with the exception of a final or other certificate which is expressed to be conclusive evidence in subsequent proceedings, the courts or an arbitrator have parallel jurisdiction to entertain disputes arising from the contract.

Multi-party proceedings

One of the recurrent difficulties of construction industry arbitration is that disputes often involve more than two parties. Ordinarily an arbitration must be limited to the two parties to the contract,

---

[14] [1984] Q.B. 644.
[15] [1999] 1 A.C. 266, 88 B.L.R. 1.

there being no power to join other parties comparable to powers available in court (see Chapter 2). Under the old law (prior to January 1997) it was possible to bring court proceedings against two or more defendants and rely on the discretion of the court to refuse to grant a stay in favour of arbitration. Thus an owner might bring parallel proceedings in respect of defects against the contractor and against the designer, alleging alternative claims. This solution, however, is no longer available under the Arbitration Act 1996 because the court no longer has discretion to refuse a stay (see below).

Multi-party arbitration can take two forms. Consolidation involves treating two or more arbitrations as a single case, to be heard in one set of proceedings and determined by one single award. In effect, consolidated arbitrations merge into a single arbitration in which the arbitrator determines the issues of liability as between all parties. Alternatively, there may be concurrent hearings of two separate arbitrations which remain separate and lead to two separate awards, save only that the awards may be expected to be consistent. It is clear that consolidation could never be brought about without the consent of all parties. It is less clear whether an arbitrator who is appointed in two related disputes has a discretion to direct concurrent hearings. Where separate arbitrators are appointed, the problem does not arise. But it is often the case that an appointing body will appoint the same arbitrator, for example, in a main contract and related sub-contract dispute. The expectation is that there will be consistent findings, but one or more of the parties does not consent to concurrent hearings. The problem is for the arbitrator in such circumstances to adopt a form of procedure which avoids inconsistent findings but which also respects the individual privacy (autonomy) of the parties.

The position is confirmed by a new provision in the Arbitration Act 1996 as follows:

"**35**—(1) The parties are free to agree—

    (a) that the arbitral proceedings shall be consolidated with other arbitral proceedings, or

    (b) that concurrent hearings shall be held,

on such terms as may be agreed

  (2) Unless the parties agree to confer such power on the tribunal, the tribunal has no power to order consolidation of proceedings or concurrent hearings."

This provision does not assist the arbitrator who is appointed in two related disputes involving different parties, where there is no such agreement. In the case *Abu Dhabi v. Eastern Bechtel*,[16] the court was asked to appoint an arbitrator in a main contract and a sub-contract arbitration which were closely related and where the parties had agreed that the arbitrator be appointed by the English court. The court had to weigh up the competing arguments for and against appointing the same arbitrator. The Court of Appeal concluded that they could appoint the same arbitrator on the parties' agreement that there could be an application to replace the arbitrator if one party thought that it was being prejudiced. Lord Denning M.R. expressed the problem as follows:

"The sub-contractors, for instance, might say that the arbitrator's decision in the first arbitration might affect his decision the second arbitration. If he had already formed his view in the first arbitration, they would be prejudiced. It would be most unfair to them: because he would be inclined to hold the same view in the second arbitration. On the other hand, as we have often pointed out, there is often a danger in having two separate arbitrations in a case like this. You might get inconsistent findings if there were two separate arbitrators."

The problem for the common arbitrator, in the absence of agreement for concurrent hearings, is to avoid such prejudice, possibly by allowing a representative from the party not involved in the first arbitration to be present. That, however, involves problems of confidentiality, as does the disclosure of documents from one arbitration in the other. The sanction would be for the aggrieved party in the first arbitration whose privacy is threatened, or the aggrieved party in the second arbitration who considers he is prejudiced, to apply to the court for the arbitrator to be replaced. No such application has been reported in the cases. The solution to these difficulties clearly lies in agreement such as that found in some of the standard forms of contract, such as the JCT 80 forms. These oblige the employer, contractor and sub-contractor to consent to the joining of disputes which are "substantially the same or connected" (See Chapter 12). However, it is often found in practice that while one of the relevant contracts has such an arbitration clause, the other has not. The problem remains a matter of considerable difficulty, which may be solved by the general adoption of appropriate arbitration rules throughout the construction industry (see below).

---

[16] (1992) 21 B.L.R. 117.

Stay of proceedings

If one party to an arbitration agreement starts court proceedings in respect of a matter covered by the agreement he is technically in breach of contract. The courts will not, however, order specific performance. Instead, the party wishing to enforce arbitration may apply for a stay of the court proceedings which, if granted, leaves arbitration as the only remedy. If a stay is not granted the action continues in court.

The court's power to order a stay of proceedings is now contained in sections 9 of the Arbitration Act 1996 as follows:

> "**9** (1) A party to an arbitration agreement against whom legal proceedings are brought (whether by way of claim or counterclaim) in respect of a matter which under the agreement is to be referred to arbitration may (upon notice to the other parties to the proceedings) apply to the court in which the proceedings have been brought, to stay the proceedings so far as they concern that matter.
>
> (2) An application may be made notwithstanding that the matter is to be referred to arbitration only after the exhaustion of other dispute resolution procedures.
>
> (3) An application may not be made by a person before taking the appropriate procedural step (if any) to acknowledge the legal proceeding against him or after he has taken any step in those proceedings to answer the substantive claim.
>
> (4) On an application under this section the court shall grant a stay unless satisfied that the arbitration agreement is null and void, inoperative, or incapable of being performed."

This section enacts the provisions of the New York Convention 1958 by which arbitration clauses are required to be enforced (or recognised) save on the grounds set out. During the previous century, English law had adopted a difference approach, by which the court had a discretion to refuse to grant a stay (and thereby to refuse enforcement of the arbitration agreement). The grounds on which the court might exercise its discretion included the bringing of alternative claims against two or more defendants, where enforcement of the arbitration agreement would result in multiple proceedings.[17] When the United Kingdom acceded to the New York Convention by the Arbitration Act 1975, it was provided that the mandatory recognition of arbitration agreement (with no discretion as to stay), applied to international and not to "domestic" arbitration. The availability of the courts' power to refuse a

---

[17] See Arbitration Act 1950, s. 4(1) and *Taunton-Collins v. Cromie* [1964] 1 W.L.R. 633.

stay of proceedings in domestic arbitration was preserved by section 86 of the Arbitration Act 1996, in respect of which the Act also included a power to repeal in the event that the different domestic regime should be regarded as discriminatory under European law. In fact, before the 1996 Act came into force a decision of the Court of Appeal indicated that the provision would be regarded as discriminatory and section 86 was not brought into effect. Consequently, since January 1997, the English courts have been obliged to grant a stay save where any of the grounds set out in section 9 of the 1996 Act is established. In practice, the most important ground is now the taking of "any step in those proceedings to answer the substantive claim". Cases under the old law indicate that any response to the court proceedings, such as filing a defence or even applying for an adjournment of the proceedings, will be regarded as a "step", depriving the defendant of any further right to enforce the arbitration agreement.

Time-bar clauses

Commercial contracts frequently require arbitration proceedings to be commenced within a limited period which is considerably shorter than the period of limitation. Periods of months or even weeks are not uncommon. Failure to comply with such limits effectively bars any subsequent right of action. Consequently, a provision was inserted into the Arbitration Acts[18] which empowered the courts to extend the time for beginning arbitration proceedings. Construction contracts tended to include a different type of provision which would render a certificate or decision binding unless challenged by giving notice of arbitration within a specified period. Thus, the Engineer's decision under clause 66 of ICE Conditions was to be binding unless challenged by notice of arbitration; and similarly the architect's final certificate under JCT 80, clause 30.9 (see Chapters 12 and 13 below).

In *Crown Estate Commissioners v. Mowlem*[19] the employer sought to rely on the power of the court to extend time for commencing arbitration proceedings in order to avoid the binding effect of a final certificate in accordance with clause 30.9 of JCT 80. The Court of Appeal held that the power to extend the time for arbitration did not empower the court to override the binding effect of the final certificate. The relevant power under the Arbitration Act 1996 is now contained in section 12 as follows:

---

[18] Previously s. 27, Arbitration Act 1950
[19] (1994) 70 B.L.R. 1.

"**12**—(1) Where an arbitration agreement to refer future disputes to
arbitration provides that a claim shall be barred, or the
claimant's right extinguished, unless the claimant takes within
a time fixed by the agreement some step—

(a) to begin arbitral proceedings, or
(b) to begin other dispute resolution procedures which must
be exhausted before arbitral proceedings can be begun

the court may by order extend the time for taking that step.

. . .

(3) The court shall make an order only if satisfied—

(a) that the circumstances are such as were outside the
reasonable contemplation of the parties when they
agreed the provision in question, and that it would be
just to extend the time, or
(b) that the conduct of one party makes it unjust to hold the
other party to the strict terms of the provision in
question."

It is to be noted that the new section includes the words "or the
claimant's right extinguished", which will now permit the court
both to avoid a time-bar and to avoid a binding certificate, where
relief is to be granted, thereby reversing the effect of *Crown
Estates*. The new clause further sets out the grounds upon which
the court may exercise its discretion, which are likely to be
sparingly exercised.

PROCEDURE IN ARBITRATION

Who is in charge of procedure? There has been long debate as to
the proper balance under English law between the powers of the
arbitrator and the ability of the parties (or their representatives) to
impose their own procedure. Previously, the law stated that the
parties should do "all . . . things which during the proceedings on
the reference the arbitrator or umpire may require".[20] However,
this was usually interpreted as being subject to the agreement of
the parties, even if communicated orally during the hearing. The
result was that, while the arbitrator was "master of the proceedings"[21] his mastery was subject at any time to an agreement
between the parties, for example, extending the length of the
hearing.

---

[20] Arbitration Act 1950, s. 12(1).
[21] *Bremer Vulkan v. South India Shipping* [1981] A.C. 909.

The result is that arbitration has often been regarded as slow, costly and out of touch with commercial requirements, with the powers of the arbitrator being very limited. The 1996 Act has responded positively to the debate over both the powers of the arbitrator and the efficiency of the proceedings. Section 1 of the Act now states the founding principles as follows:

"(a) the object of arbitration is to obtain the fair resolution of disputes by an impartial tribunal without unnecessary delay or expense;

(b) the parties should be free to agree how their disputes are resolved, subject only to such safeguards as are necessary in the public interest."

This is supplemented by an important general duty on the arbitrator or tribunal as follows:

"33—(1) The tribunal shall—

(a) act fairly and impartially as between the parties, giving each party a reasonable opportunity of putting his case and dealing with that of his opponent, and

(b) adopt procedures suitable to the circumstances of the particular case, avoiding unnecessary delay or expense, so as to provide a fair means for the resolution of the matters falling to be determined.

(2) The tribunal shall comply with that general duty in conducting the arbitral proceedings, in it decisions on matters of procedure and evidence and in the exercise of all other powers conferred on it."

Further, in setting out the list of procedural and evidential matters requiring decision, it is stated:

"34—(1) It shall be for the tribunal to decide all procedural and evidential matters, subject to the right of the parties to agree any matter."

Section 5 of the Act provides that any such agreement must ordinarily be made in writing. The result is that the arbitrator will be bound by such written agreement (usually comprising the arbitration agreement together with incorporated rules) as exists at the date of his appointment. Subject to this, he has the power "to decide all procedural and evidential matters" unless the parties enter into a further written agreement. In this event, if the further agreement involves fundamental changes or restrictions on the power of the arbitrator, it would be open to him to decline to be bound by the new agreement. This would create an impasse which would require either further agreement as to procedure between the parties and the arbitrator, or the parties' agreement to revoke the arbitrator's appointment[22] or the arbitrator's resignation.[23] While these circumstances are unlikely to arise save in an extreme case, they are a necessary element in establishing, for the first time in English arbitration law, that the arbitrator, subject to the terms of his appointment, is to be in control of the procedure.

Detailed procedural matters

Section 34(2) lists the following matters which are to be decided by the arbitrator in the absence of written agreement:

(a) when and where any part of the proceedings is to be held;

(b) the language or languages to be used in the proceedings and whether translations of any relevant documents are to be supplied;

(c) whether any and if so what form of written statements of claim and defence are to be used, when these should be supplied and the extent to which such statements can be later amended;

(d) whether any and if so which documents or classes of documents should be disclosed between and produced by the parties and at what stage;

(e) whether any and if so what questions should be put to and answered by the respective parties and when and in what form this should be done;

(f) whether to apply strict rules of evidence (or any other rules) as to the admissibility, relevance or weight of any material (oral, written or other) sought to be tendered on any matters of fact or opinion, and the time, manner and form in which such material should be exchanged and presented;

(g) whether and to what extent the tribunal should itself take the initiative in ascertaining the facts and the law;

(h) whether and to what extent there should be oral or written evidence or submissions.

Ordinarily, many of these matters will be the subject of rules incorporated within the arbitration agreement (see below). The 1996 Act also sets out a number of specific powers which the

---

[22] Arbitration Act 1996, s. 23.
[23] s. 25.

arbitrator may exercise unless otherwise agreed by the parties. These are as follows:

(1) power to appoint experts or legal advisors or to appoint assessors to assist on technical matters[24];

(2) power to order security for the costs of the arbitration;

(3) power to give directions in relation to any property which is the subject of the proceedings, including inspection, preservation, taking samples or making tests.[25]

The 1996 Act also contains two powers of great importance to construction industry arbitrations which are stated to be available only if the parties so agree:

(4) a power to order consolidation or concurrent hearings of two related arbitrations in which the same tribunal is appointed[26];

(5) a power to make a provisional order or award.[27]

This latter power is considered further below in relation to the range of decisions available to an arbitrator. All the above powers are supplemented by a general duty of co-operation placed upon the parties themselves as follows:

"**40**—(1) The parties shall do all things necessary for the proper and expeditious conduct of the arbitral proceedings.
   (2) This includes—
      (a) complying without delay with any determination of the tribunal as to procedural or evidential matters, or with any order or directions of the tribunal . . . "

Arbitrators are often requested, in the absence of other applicable procedure, to follow court practice in ordering pleadings, disclosure of documents and other matters. While this may be a useful guide, the 1996 Act makes it clear that these rules do not apply to arbitration. The arbitrator should in every case use his discretion as to what is required. There may be cases where, in lieu of formal pleadings, it is sufficient for the claimant to rely on an existing claim document, and for the respondent to be ordered to submit details of grounds of disagreement. This may save considerable time and cost to the parties. Similarly, there is no requirement for

---

[24] s. 37.
[25] s. 38.
[26] s. 35.
[27] s. 39.

an arbitrator to direct a formal hearing following practice in the High Court. He might, if appropriate, direct the parties to attend on site and explain their dispute to him there, or he might require the parties to deliver documents to him so that he may investigate the dispute before proceeding further; or again, if he forms the view there is some issue of principle upon which the dispute will turn, he might order the parties to deal with this issue at the outset. In all such matters, the arbitrator is required to decide upon the procedure to be adopted, taking account of the express duties contained in the 1996 Act.

Duties of the arbitrator

The express requirement under section 33(1)(a) to act "fairly and impartially as between the parties", although derived from the UNCITRAL Model Law[28] expresses in part the common law duty which applies to arbitrators and other tribunals, to comply with the rules of "natural justice". This is an unwritten concept whose boundary can be traced through the cases in which arbitrators have been accused of what was formerly called "misconduct",[29] but will now be known as "serious irregularity[30] (see below). The arbitrator must be impartial and must act so as to convey a continuing impression of impartiality. This does not mean that an arbitrator can have no connection whatsoever with either of the parties—in the construction industry this would be an impossibility since arbitrators and lawyers alike are regularly engaged in disputes involving major players in the industry. The arbitrator should, however, be satisfied that he is in fact impartial in relation to the dispute in question and should disclose any circumstances which, if known, might create doubts. The arbitrator must also ensure that he possesses any qualifications required by the arbitration agreement and that he is in all ways capable of properly conducting the proceedings. The 1996 Act provides specifically for the removal of an arbitrator on the ground:

> "24—(1)(a) that circumstances exist that give rise to justifiable doubts as to his impartiality;
> (b) that he does not possess the qualifications required by the arbitration agreement;
> (c) that he is physically or mentally incapable of conducting the proceedings or there are justifiable doubts as to his capacity to do so;

---

[28] Art. 18.
[29] Arbitration Act 1950, s. 23.
[30] Arbitration Act 1996, s. 68.

(d) that he has refused or failed—

> (i) properly to conduct the proceedings, or
> (ii) to use all reasonable despatch in conducting the proceedings or making an award,
> and that substantial injustice has been or will be caused to the applicant."

This, and other sections as well, apply subject to application first to any institution vested with relevant powers.

Remission and setting aside

The sanction available under section 24 is limited to removal: the court has no power to remit, for example, with a direction that the arbitrator should henceforth conduct the proceedings in some different manner. This emphasises the limited extent to which the courts exercise any overseeing role in arbitration. In the case of *Damond Lock v. Laing Investments*[31] an arbitrator decided to maintain a hearing date even though one party had produced a large number of relevant documents at a late stage so that the other party would have no proper opportunity to consider them. The court had no jurisdiction to remit the matter[32] and consequently had to choose between allowing the matter to proceed or removing the arbitrator, in this case choosing the latter.

The power of the court to remit to the arbitrator on the ground of serious irregularity is available only once the tribunal has made an award.[33] The court then has power to remit the award, set the award aside or declare it to be of no effect in whole or in part. The court must also be satisfied that serious injustice has or will be caused to the applicant. The individual grounds upon which an application may be made are:

> "**68**—(2)(a) failure by the tribunal to comply with section 33 (general duty of tribunal);
> (b) the tribunal exceeding its power (otherwise than by exceeding its substantive jurisdiction: see section 67);
> (c) failure by the tribunal to conduct the proceedings in accordance with the procedure agreed by the parties;
> (d) failure by the tribunal to deal with all the issues that were put to it;
> (e) any arbitral or other institution or person vested by the parties with powers in relation to the proceedings or the award exceeding its powers;

---

[31] (1992) 60 B.L.R. 112.
[32] Under s. 22 of the Arbitration Act 1950.
[33] The same result was held to apply in the case of an application under s. 22 of the Arbitration Act 1950 in *Three Valleys v. Binnie & Partners* (1990) 52 B.L.R. 42.

(f) uncertainty or ambiguity of the award;
(g) the award being obtained by fraud or the award or the way in which it was procured being contrary to public policy;
(h) failure to comply with the requirements as to the form of the award;
(i) any irregularity in the conduct of the proceedings or in the award which is admitted by the tribunal or by any arbitral or other institution or person vested by the parties with powers in relation to the proceedings or the award."

The above provisions are a compendium of many provisions formerly found in the cases. It is a useful checklist for the arbitrator. Some of the provisions are more obviously directed towards remission (for example, failure to deal with all the issues or ambiguity in the award); while others are likely to give rise to a serious question as to whether remission or removal is most appropriate (for example failure to comply with the agreed procedure). An application under section 68 may be coupled with an alternative application for removal under section 24, for example on the ground of lack of partiality.

How should the arbitrator proceed in practice so as to avoid an application to the court? As well as being in fact impartial the arbitrator should take pains appear impartial at all stages and be seen to treat the parties equally. He should ordinarily act only upon the evidence or other material presented to him in the arbitration. If he wishes to take any step which might take the parties by surprise or be regarded as unconventional, he should inform the parties what he is doing and give such explanation as will demonstrate that he has proper reasons. Another important practical question is how to control the length of the proceedings. In this regard, he has potentially conflicting duties to give the parties a reasonable opportunity to put their cases, but to avoid unnecessary delay or expense.[34] The important word is "reasonable" and this means that the arbitrator may and on occasions must, set a limit to various stages of the proceedings, unless he is bound by an agreement between the parties.

The arbitrator must take a sufficient note of the evidence and argument to enable him to determine the issues and to deliver a reasoned award, if called for. The parties may agree to provide a shorthand note of the proceedings. Alternatively, some arbitrators choose to make their own tape recordings of the hearing. This should not, however, be regarded as a normal requirement; and it should never be regarded as a substitute for following the argument as it goes on.

---

[34] Arbitration Act 1996, s. 33(1)(a) and (b).

An example of an arbitrator falling foul of the general requirements as to conduct is found in the difficult case of *Fox v. P.G. Wellfair*.[35] The respondent builders were in liquidation at the time of arbitration proceedings which were about a block of flats which were alleged to contain numerous defects. The proceedings were continued to obtain the benefit of NHBC insurance, but were effectively undefended so that the plaintiff's evidence was unchallenged. Instead of making an award in accordance with the plaintiff's evidence, the arbitrator substantially reduced the sums claimed. He relied on his own opinion about the defects, but did not disclose this to the claimant, who had no reason to suppose that their evidence was contested. The Court of Appeal held that the arbitrator had committed misconduct by failing to bring his views to the attention of the claimant. The three judgments show a range of views as to what the arbitrator had done wrong. Essentially, the court considered that the claimant ought to have had the opportunity of knowing what was in the arbitrator's mind.

Power of the courts

Traditionally the courts have exercised a wide range of powers in support of the arbitration process.[36] In recent years, however, there has been a policy of progressively reducing these powers to those regarded as essential. The 1996 Act has continued this policy through a number of express provisions which encourage the transfer of powers to the arbitrator or to an arbitral institution. The remaining essential powers of the court fall into the following categories:

(1) appointment and removal of arbitrators;

(2) extension of time and stay of proceedings;

(3) supportive powers during the proceedings;

(4) challenges to and enforcement of awards.

In regard to (1) the courts have always had power to step in when necessary to ensure that appropriate arbitrators are appointed and this continues under the 1996 Act.[37] (2) is dealt with above and (4) below. This section deals with the remaining

---

[35] [1981] 2 Lloyd's Rep. 514.
[36] Arbitration Act 1950, s. 12(4)(5) and (6).
[37] ss. 17, 18.

supportive powers in the 1996 Act, which are without exception subject to agreement of the parties to exclude such powers.

It is to be noted that section 1(c), following the UNCITRAL model law, now provides expressly in relation to Part I of the Act (sections 1 to 84) that "the court should not intervene except as provided by (Part I)". It appears to follow that the court has no remaining inherent powers, for example to grant a declaration or an injunction. It will remain to be seen whether there are any circumstances in which the court will be persuaded to take action beyond that expressly empowered by the Act.

Section 44 of the 1996 Act provides for the court to have power to make orders in relation to the following matters, unless otherwise agreed by the parties:

"(a) the taking of the evidence of witnesses;
(b) the preservation of evidence;
(c) making orders relating to property which is the subject of the proceedings or as to which any question arises in the proceedings—

(i) for the inspection, photographing, preservation, custody or detention of the property, or
(ii) ordering that samples be taken from, or any observation be made of or experiment conducted upon, the property;

and for that purpose authorising any person to enter any premises in the possession or control of a party to the arbitration;

(d) the sale of any goods the subject of the proceedings;
(e) the granting of an interim injunction or the appointment of a receiver."

The courts' powers are further limited to acting in cases of urgency, or otherwise with the permission of the tribunal or agreement of the parties. The court will not act if the tribunal or any arbitral institution has the power and is able to act. These provisions emphasise the extremely narrow circumstances in which the court will now take "supportive" action. All the above measures are within the list of powers that may, in the absence of contrary agreement, be exercised by the arbitrator (see section 38) save for the powers under (e) to grant an interim injunction or to appoint a receiver. The parties must therefore consider at the outset whether they wish the court to retain these potentially valuable powers. In addition, section 43 permits a party to make use of the powers of the court to secure the attendance of a witness, but only with the permission of the tribunal or the agreement of the parties. Again, this is a matter to be considered in the arbitration agreement or in any incorporated rules.

Powers in case of default

The 1979 Act contained cumbersome powers[38] by which a party could apply to the court to vest the arbitrator with enhanced powers to deal with a default by the other party. This is now replaced by section 41 of the 1996 Act which provides that the parties are free to agree default powers. In the absence of agreement, the tribunal has power, where a party fails to comply with an order or direction without sufficient cause, to make a "peremptory order", giving a time for compliance. If the party then fails to comply then the tribunal may do any of the following:

(a) direct that the party in default shall not be entitled to rely upon any allegation or material which was the subject matter of the order;

(b) draw such adverse inferences from the act of non-compliance as the circumstances justify;

(c) proceed to an award on the basis of such materials as have been properly provided to it;

(d) make such order as it thinks fit as to the payment of costs of the arbitration incurred in consequence of the non-compliance.

Alternatively, unless the parties have excluded the power, either the tribunal or the other party with the permission of the tribunal, may apply to the court for an order requiring compliance with the peremptory order. In addition to the peremptory order procedure, section 41 re-enacts the power[39] to dismiss a claim where the claimant has been guilty of inordinate and inexcusable delay; and also confirms the power of the tribunal to continue with the proceedings in the absence of a party after being duly notified.[40]

ARBITRATION RULES

No uniform system of procedure applies in arbitration comparable to the Civil Procedure Rules of the courts. It is therefore natural

---

[38] s. 5.
[39] Originally s. 13A of the Arbitration Act 1950.
[40] *Bremer Vulkan v. South India Shipping* [1981] A.C. 909, *per* Lord Diplock.

that various trade and professional bodies should consider creating their own rules of procedure. The courts have adopted a supportive approach to rules in the past. Under the 1996 Act they are essential to make effective use of the available powers.

The need for procedural rules is also shown by the wide diversity of arbitrations. At one extreme, there are commodity arbitrations concerning consignments of goods, where the arbitrator may be required to form a rapid opinion as to the quality of perishable goods (sometimes referred to as a "look-sniff" dispute). At the other extreme there are commercial disputes, for example in the field of re-insurance, involving contractual argument, many documents and lengthy evidence, which are conducted in a manner similar to disputes in the High Court. While many construction disputes tend to resemble more the latter, there is often a substantial element of the former, and it is therefore the more important that construction arbitrators should be in a position to exercise effective control with flexibility. Under the pre-1996 legislation several sets of rules existed for construction disputes, particularly the ICE Arbitration Procedure (1983) and the JCT Arbitration Rules (1988). With the impending arrival of the Arbitration Act 1996, steps were taken to draw up a set of rules for adoption throughout the United Kingdom construction industry. Under the title Construction Industry Model Arbitration Rules, a document was issued in draft in 1997 and finally published in February 1998 with the endorsement of all the major bodies within the construction industry. The Rules are reviewed below. Other industries have adopted or adapted their own rules and other sets of rules may occasionally be encountered. Thus, the Chartered Institute of Arbitrators has drawn up a set of rules which are partly based on the CIMA Rules. The ICE, prior to the introduction of CIMAR, produced an amendment to the ICE Arbitration Procedure which is still available as an alternative to CIMAR.

Construction Model Arbitration Rules

The Rules are arranged in roughly the same logical order as the Act, although preliminary matters such as joinder are necessarily dealt with at the outset. After Rules 1 to 3 concerning the setting up of the arbitration, Rules 4, 5 and 6 deal generally with powers and procedure. Rules 7, 8 and 9 set out three alternative forms of procedure. Rule 10 deals with provisional relief and Rule 11 with default powers. Rules 12 and 13 deal with the award and costs. The following particular points may be noted:

> RULE 1: OBJECTIVE AND APPLICATION—repeats the general objectives of the Act and provides that the parties may not,

without agreement of the arbitrator, amend the rules after his appointment.

RULE 2: BEGINNING AND APPOINTMENT—for the first time, makes provision for a mechanism whereby the appropriate appointor must give consideration to whether or not the same arbitrator is to be appointed in two related arbitrations. This is achieved by a contractual requirement (forming part of the arbitration agreement) that the appointor should "give due consideration". The appointor is not, of course, a party to the agreement. Nevertheless, an effective sanction exists in that a failure to give consideration may result in an appointment which is not in accordance with the parties' agreement.

RULE 3: JOINDER—covers all different forms of joinder— multiple claims brought by the claimant—cross claims by the respondent— additional disputes raised by either party—joinder of related arbitral proceedings in which the same arbitrator is appointed—consolidation of related proceedings (with consent). The objective is to secure the efficient resolution of all related disputes so far as practically possible. The arbitrator is, accordingly, given various discretions as to allowing (or not allowing) additional disputes to be added.

RULE 4: PARTICULAR POWERS—includes provision as to the grounds upon which the arbitrator may order security for costs, based on practice in the courts. The Act itself is silent as to the grounds. Rule 4.7 encourages (but does not bind) the arbitrator to give reasons. The parties may well wish to amend this rule.

RULE 5: PROCEDURE AND EVIDENCE—while providing that the arbitrator is not bound by the strict rules of evidence, requires a formal record to be made of the evidence tendered in cases of (a) an application for Security for Costs; (b) an application to strike out; (c) an application for Provisional Relief; and (d) any other instance where it is appropriate. Section 34 (2) (which contains wide procedural powers), is incorporated.

RULE 6: FORM OF PROCEDURE AND DIRECTIONS—requires the arbitrator, with information provided by the parties, to consider the appropriate procedure as soon as he is appointed. This may involve adopting the procedure set out in Rules 7, 8 or 9, or any part of those procedures, or any other procedure he considers appropriate. The arbitrator is thus given the widest possible discretion.

RULE 7: SHORT HEARING—is appropriate where there is to be a hearing of not more than one day, the award to be made within one month of the conclusion of the hearing.

RULE 8: DOCUMENTS ONLY—is appropriate where there is to be no hearing. The arbitrator is nevertheless empowered to direct a hearing of not more than one day, but must otherwise make his award within one month.

RULE 9: FULL PROCEDURE—sets out a formal procedure involving exchange of pleadings, statements and expert reports. The arbitrator is to fix the length of the hearing and may require any matter to be put into writing.

RULE 10: PROVISIONAL RELIEF—puts into effect the machinery of section 39, whereby the arbitrator is empowered to order on a provisional basis any relief which he would have power to grant in a final award. This rule contains equivalent powers to those in Civil Procedure Rules, Parts 24 and 25 which, in the light of *Halki Shipping v. Sopex*[41] now represent the only means of obtaining a summary decision which is (i) clearly enforceable, and (ii) not subject to appeal save in the course of subsequent arbitration proceedings (see further below).

RULE 11: DEFAULT POWERS AND SANCTIONS—this puts into effect the provisions of section 41—power to dismiss a claim for inordinate delay and power to proceed in the absence of a party in default; also the power to give a peremptory order which may be enforced by the court under section 42. Rule 11.3 empowers the arbitrator to order direct sanctions against non-compliance, including the drawing of adverse inferences.

RULE 12: AWARDS AND REMEDIES—gives effect to sections 47, 48, 49 and 57 empowering the arbitrator to make awards on different issues, to grant a wide range of remedies, to award simple or compound interest and to correct an award.

RULE 13: COSTS—gives effect to section 63 and 65 (power to limit costs). The rule also sets out the matters which the

---

[41] [1998] 1 W.L.R. 726.

arbitrator should have regard to in awarding costs, on the general principle that they should be borne by the losing party. By rule 13.4 the arbitrator, in imposing a limit on recoverable costs, is to have regard primarily to the amounts in dispute. Provision is also made for taking account of offers of settlement.

RULING 14: MISCELLANEOUS—deals, *inter alia*, with service and reckoning of time periods.

A review committee has been established which will keep under review the Rules and published notes, with a view to considering amendments or additional notes for guidance in the light of cases and experience with the Rules.

JURISDICTION AND COMPETENCE

Arbitration, unlike litigation, is subject to a number of constitutional difficulties which stem from the essential nature of the process. Foremost among these is the extent of the arbitrator's jurisdiction and the power he may have to keep the arbitration going when challenges are made. What is the position if one party asserts that the arbitrator does not have jurisdiction over some dispute which the other party wishes to bring forward? At common law, while the arbitrator can and should consider such a challenge in order to decide whether to go on with the arbitration or not,[42] only the court can finally decide on jurisdiction. The same principle applies in a case where the existence of the arbitration agreement is put in issue, which may be because one party contends that the underlying contract was never concluded. In both respects, the 1996 Act clarifies the position and gives substantial support to the ability of the arbitrator to continue with the arbitration despite such challenges. The question whether and to what extent an arbitrator can make decisions bearing on his own jurisdiction is often given the German label *kompetenz kompetenz* or, in French *competence de la competence*. The first advance in the 1996 Act is the adoption of a provision from the UNCITRAL model law requiring a party to raise the question of jurisdiction timeously:

"**31**—(1) An objection that the arbitral tribunal lacks substantive jurisdiction at the outset of the proceedings must be raised by

---

[42] *Christopher Brown v. Genossenschaft Oesterreichischer* [1954] 1 Q.B. 8.

a party not later than the time he takes the first step in the proceedings to contest the merits of any matter in relation to which he challenges the tribunal's jurisdiction.

A party is not precluded from raising such an objection by the fact that he has appointed or participated in the appointed of an arbitrator.

(2) Any objection during the course of the arbitral proceedings that the arbitral tribunal is exceeding its substantive jurisdiction must be made as soon as possible after the matter alleged to be beyond its jurisdiction is raised.

(3) The arbitral tribunal may admit an objection later than the time specified in subsection (1) or (2) if it considers the delay justified."

Unless otherwise agreed by the parties, the arbitrator is given power under section 30 of the 1996 Act to rule on its own jurisdiction including the question whether there is a valid arbitration agreement or what matters have been submitted to arbitration. This ruling, if given in the form of an award, may be subject of an application for leave to appeal (see below). Alternatively, the court may be asked to rule on the question of jurisdiction under section 32. Where a party does not raise a challenge to jurisdiction timeously, he may lose the right thereafter to raise objection.[43] The 1996 Act thus establishes a means whereby challenges to jurisdiction should be resolved at an early stage, usually by the arbitrator.

Separability

This is another concept well known under civil law systems but relatively novel under English law. The issue arises most frequently in construction disputes where the parties continue to negotiate a contract despite commencement of the work and end up in disagreement as to whether a contract has been concluded. Simple analysis suggests that if an arbitrator appointed under such an arrangement were to decide that no agreement had been made, his jurisdiction would thereby disappear along with the arbitration clause contained in the non-existent contract. Such a conclusion, however, is neither necessary nor even logical. It was held by the Privy Council in *Heyman v. Darwins*[44] that an arbitration clause survived termination of the contract through repudiation; and in *Ashville v. Elmer*[45] it was held that an arbitration clause might empower an arbitrator to rectify the contract. Both these cases are examples of the arbitration clause being seen as separate from the

---

[43] Arbitration Act 1996, s. 73.
[44] [1942] A.C. 356.
[45] [1989] Q.B. 488; 37 B.L.R. 55.

underlying contract. Section 7 of the 1996 Act has now placed the doctrine of separability beyond doubt as follows:

> "Unless otherwise agreed by the parties, an arbitration agreement which forms or was intended to form part of another agreement shall not be regarded as invalid, non-existent or ineffective because that other agreement is invalid, or did not come into existence or has become ineffective, and it shall for that purpose be treated as a distinct agreement."

Thus either the court or the arbitrator[46] may, if the matter is placed in issue, decide whether the parties have entered into a separate arbitration clause, in the event that the underlying contract was never concluded.

AWARDS

No particular form is required for an award, but it should ordinarily be in writing and should be carefully drafted and checked by the arbitrator, since it cannot be altered later. The essential requirements of an award are that it should decide the matters submitted and no others. It must be certain in its effect, and it must be consistent with any other findings or awards of the arbitrator in the same matter (see Subsequent Claims, Chapter 2). Once the arbitrator has expressed a decision in an award, he ceases to have further jurisdiction over that matter and is said to be *functus officio*. Thus, the arbitrator must be careful not to decide matters which either party may wish to argue further. He should not express decisions on matters where he may wish to alter his view and he should never express an opinion on a matter which has not been brought forward or argued.

The 1996 Act clarifies the range of remedies which the arbitrator, subject to the agreement of the parties, may award. These may include the following:

> "**48**—(3) The tribunal may make a declaration as to any matter to be determined in the proceedings.   .
>
> (4) The tribunal may order the payment of a sum of money, in any currency.
>
> (5) The tribunal has the same powers as the court—
>
> (a) to order a party to do or refrain from doing anything;

---

[46] Pursuant to the 1996 Act, s. 30.

(b) to order specific performance of a contract (other than a contract relating to land);

(c) to order the rectification, setting aside or cancellation of a deed or other document."

Under the Arbitration Act 1950 arbitrators were empowered to make an "interim" award, which meant no more than an award which dealt with some but not all of the issues submitted, but was otherwise final as to the matters decided. The term gave rise to confusion and has now been dropped. Section 47 of the 1996 Act provides that, unless the parties otherwise agree, the arbitrator may make more than one award at different times on different aspects of the matters to be determined. All that is required is that the award should specify the issue or claim or part of a claim which is the subject of the award. No particular label is given to these awards and it may be that the term "interim" will continue to be used. The French, in the ICC rules, use the term *"Sentence Partielle"*, which is confusingly translated into English as "Partial award". Again, it means no more than an award on some on the issues which is final. The term "Final award" is usually reserved to the last award in time, *i.e.* that which deals with all outstanding matters. The French version *"Sentence Definitive"* is potentially confusing, but its meaning is clear.

The question lying behind the somewhat confusing terminology referred to above, is whether an arbitral tribunal can grant a partial remedy, for example, in circumstances similar to those in which the English court can grant summary judgment on the basis that the claim is worth "not less than" the figure ordered. The jurisdiction of arbitrators to grant such a remedy has never been clear, although such a power was contained in the ICE Arbitration Procedure (1983).[47] This particular question has been resolved by the 1996 Arbitration Act where the parties may empower the arbitrator, to grant by "provisional" award or order, any relief which can be granted in an award which is final.[48] The Act makes clear that a provisional award is one which may be varied by the tribunal's subsequent (final) adjudication. To clarify these new powers, the Act states that a provisional award may include "a provisional order for the payment of money or the disposition of property as between the parties". The CIMA Rules (see above) make provision for the exercise of these important powers.

While the question of the arbitrator's jurisdiction to grant "provisional" relief is now clarified, the position of the courts

---

[47] r. 14.
[48] Arbitration Act 1996, s. 39(2).

remains uncertain. For many years prior to the 1996 Act there had been a practice of applying to the court for summary judgment in respect of claims to which it was contended there was no defence and therefore no dispute. The courts, either on the basis that there was no dispute or in the exercise of the court's discretion to refuse to stay the proceedings, adopted the practice of granting summary judgment in respect of indisputable claims and ordering a stay for arbitration in respect of the balance.[49] After the passing of the 1996 Act and the removal of the court's discretion to refuse a stay (see above), the question remained whether the court would regard a claim to which there was no defence as creating a "dispute" which the parties remained bound to take to arbitration. In other words, is arbitration now to be regarded as the exclusive remedy for any claim within the arbitration agreement? In *Halki Shipping v. Sopex Oils*[50] the Court of appeal answered the question affirmatively, by refusing an application for summary judgment. A strong dissenting judgment was given by Hirst L.J., but leave to appeal the House of Lords was not pursued. These difficult questions therefore still await final resolution.

Equity clauses

Another major advance of the 1996 Act is to recognise, finally, the validity of an "equity" clause empowering the arbitrator to decide the issues not in accordance with strict legal rules but following equitable principles, "good conscience" or other like expressions. There has been long debate in judicial as well as academic circles as to the acceptability of such clauses and particularly as to whether an award made under such a contract would be enforceable in England.[51] The 1996 Act provides for such clauses in the following terms:

"**46**—(1) The arbitral tribunal shall decide the dispute—

    (a) in accordance with the law chosen by the parties as applicable to the substance of the dispute, or

    (b) if the parties so agree, in accordance with such other considerations as are agreed by them or determined by the tribunal."

During the drafting of the Act consideration was given to descriptions such as *ex aequo et bono,* or the French *amiable compositeur,*

---

[49] See particularly *Ellis Mechanical Services v. Wates Construction* (1976) 2 B.L.R. 57, and *The Kostas Melas* [1981] 1 Lloyds Rep. 18.
[50] [1998] 1 W.L.R. 726.
[51] See *DST v. RAKOIL* [1987] 2 Lloyds Rep. 246; *Home & Overseas v. Mentor* [1989] 1 Lloyds Rep. 473.

but plain English prevailed in the form above. One of the problems generated by equity clauses is the practical impossibility of judicial review of any legal decision embodied in the award. The result of giving statutory recognition to these clauses is that an appeal on law from such a decision will not be available. Such an award would, however, still be open to question on the ground of serious irregularity (see above).

Award of costs

It has always been the accepted practice that in dealing with costs, the arbitrator should adhere broadly to the principles adopted in the High Court, *i.e.* the successful party should receive his costs unless there are proper reasons for departing from this order. Where there is a claim and counterclaim, each must be considered in relation to costs, but it is frequently found convenient to reflect all the matters in one global order, such as that one party is awarded a proportion of his costs.

The above broad principles are now codified to an extent and the undoubted discretion of the arbitrator clarified in the 1996 Act thus:

> "**61**—(2) Unless the parties otherwise agree, the tribunal shall award costs on the general principle that costs should follow the event except where it appears to the tribunal that in the circumstances this is not appropriate in relation to the whole or part of the costs."

By section 60 of the 1996 Act agreement that the parties shall pay the whole or part of their costs of the arbitration in any event is valid only if made after the dispute. This, therefore, precludes such an agreement being placed in a standard form of contract or in an arbitration agreement made in advance of the dispute. However, the parties may enter into such an agreement once the dispute has arisen and the new Act encourages the parties and the arbitrators to adopt measures likely to reduce or control the expenditure of costs. In addition to the general requirements as to avoiding unnecessary expense (see sections 1(a) and 33(1)(b)) section 65 allows the arbitrator to place a limit upon the amount of costs which can be recovered as follows:

> "**65**—(1) Unless otherwise agreed by the parties, the tribunal may direct that the recoverable costs of the arbitration, or any part of the arbitral proceedings, shall be limited to a specified amount.
>
> (2) Any direction may be made or varied at any stage, but this must be done sufficiently in advance of the incurring of costs

> to which it relates, or the taking of any steps in the proceedings which may be affected by it, for the limit to be taken into account."

The process formerly referred to as "taxation" of costs is now renamed the determination of recoverable costs. Once the arbitrator has determined in principle who should recover costs and in what proportion, the arbitrator himself may be asked to determine what is recoverable. If he does not do so, either party may apply to the court for a determination. In either event, but subject to any agreement between the parties, the recoverable costs are to be determined in the same way as in court (see Chapter 2):

> "**63**—(5) Unless the tribunal or the court determines otherwise
>
> > (a) the recoverable costs of the arbitration shall be determined on the basis that there shall be allowed a reasonable amount in respect of all costs reasonably incurred, and
> >
> > (b) any doubt as to whether costs were reasonably incurred or were reasonable in amount shall be resolved in favour of the paying party."

Recovery of interest

In awarding interest, the arbitrator should also follow similar principles to those applied in the High Court. He should normally award interest by allowing a realistic rate on the sum awarded from such date as the money ought ordinarily to have been paid, *i.e.* for the period the successful party has been deprived of the sum awarded.

The powers of the arbitrator are now set out in section 49 of the Arbitration Act 1996. Subject to agreement of the parties, the power of the arbitrator are as follows:

> "**49**—(3) The tribunal may award simple or compound interest from such dates, at such rates and with such rests as it considers meets the justice of the case—
>
> > (a) on the whole or part of any amount awarded by the tribunal, in respect of any period up to the date of the award;
> >
> > (b) on the whole or part of any amount claimed in the arbitration and outstanding at the commencement of the arbitral proceedings but paid before the award was made, in respect of any period up to the date of payment.
>
> (4) The tribunal may award simple or compound interest from the date of the award (or any later date) until payment, at such rates and with such rests as it considers meets the justice

of the case, on the outstanding amount of any award (including any award of interest under subsection (3) and any award as to costs)."

The power to award interest on a sum claimed but paid before award is to the same effect as the rule applying in the High Court. In both cases, a defendant gains no advantage by deferring payment. However, both in court and in arbitration, there is no power to award interest only, *i.e.* on a sum in dispute which is paid before proceedings are commenced.

The above section includes two important new provisions. First, the arbitrator may be empowered to award compound interest, a power not available to judges except where interest is claimed as damages or is payable under the terms of the contract (see Chapter 2). Secondly, the arbitrator is empowered to award interest to be payable in the future after the date of the award. Awards previously carried interest automatically, but this is now within the arbitrator's discretion.

JUDICIAL REVIEW OF AWARDS

Section 24 (power to remove arbitrator) and section 68 (challenge for serious irregularity) of the 1996 Act are dealt with above; also challenges on the ground of lack of jurisdiction. This section deals with appeal on a point of law. This issue has a long history which is bound up with the essential nature of English arbitration.

Before 1979, points of law arising in a reference could be referred to the High Court under a procedure known as "case stated".[52] The decision of the court could be obtained during the reference, or a case could be stated at the end, the award depending upon the opinion of the court. This is the process which allowed English arbitration, unlike most civil law countries, to develop the use of "trade" arbitrators, who were not legally qualified. However, in the 1960s the procedure began to produce a substantial number of references to the High Court, such that the decision of the arbitrator could no longer be regarded as normally final. To remedy the situation and to secure the future of London as a venue for international disputes, the Commercial Court Committee promoted reforming legislation which became the Arbitration Act 1979. This Act provided for appeal on the point of

---

[52] Arbitration Act 1950, s. 21.

law conditional upon consent of the parties or the court granting leave. Shortly after the Act came into force, the House of Lords decided in *BTP Tioxide v. Pioneer Shipping (The Nema)*[53] that leave should be given only in very limited circumstances. The grounds upon which leave would or would not be granted became highly complex in the light of further cases. The 1996 Act has taken the bold step of codifying (and to some extent modifying) the existing rules which are now as follows:

"**69**—(1) Unless otherwise agreed by the parties, a party to arbitral proceedings may (upon notice to the other parties and to the tribunal) appeal to the court on a question of law arising out of an award made in the proceedings . . .

(2) An appeal shall not be brought under this section except—

(a) with the agreement of all the other parties to the proceedings, or

(b) with the leave of the court.

The right to appeal is also subject to the restrictions in section 70(2) and (3).

(3) Leave to appeal shall be given only if the court is satisfied—

(a) that the determination of the question will substantially affect the rights of one or more of the parties,

(b) that the question is one which the tribunal was asked to determine,

(c) that, on the basis of the findings of fact in the award—

(i) the decision of the tribunal on the question is obviously wrong, or

(ii) the question is one of general public importance and the decision of the tribunal is at least open to serious doubt, and

(d) that, despite the agreement of the parties to resolve the matter by arbitration, it is just and proper in all the circumstances for the court to determine the question."

Section 70(2) and (3) require respectively that the applicant must first exhaust any available means of the correcting the award and that the application must be brought within 28 days of the award. Subsection (3)(c) of section 69 succinctly expresses the effect of *The Nema* and other decisions under the 1979 Act and should result in simplification of the process of applying the leave, which will now be dealt with in most cases on an administrative basis by the judge, without a hearing.[54]

Section 69(1) recognises that the parties may make an agreement to exclude the court's jurisdiction to consider an appeal on

---

[53] [1982] A.C. 724.
[54] s. 69(5).

law, and an agreement to dispense with reasons for the award is to be treated as having the same effect. Under the Arbitration Act 1979, such an "exclusion agreement" could be made in advance in respect of an international arbitration, and the ICC Rules have been held to contain such an exclusion.[55] As regards domestic arbitration, the 1979 Act required that an exclusion agreement must be entered into after the dispute in question had arisen. It was intended that the 1996 Act would have the same effect[56] but the relevant provision was considered potentially discriminatory (see above) and was not brought into effect. In the result, an agreement to exclude the court's jurisdiction in regard to appeals will be effective whenever made. Similarly, it appears that the courts will recognise an agreement made in advance for giving consent to the bringing of an appeal. Such an agreement is found in section 41.6 of the JCT Conditions.

An alternative means of obtaining the decision of the court on a point of law is available under section 45 of the 1996 Act. This enables an application to be made during the course of arbitral proceedings to determine a question of law arising. Such application is limited by the following requirements:

"**45**—(2) An application under this section shall not be considered unless—

    (a) it is made with the agreement of all the other parties to the proceedings, or

    (b) it is made with the permission of the tribunal and the court is satisfied—

        (i) that the determination of the question is likely to produce substantial savings in costs, and

        (ii) that the application was made without delay."

This procedure is not greatly used but remains potentially useful. The arbitration proceedings may be continued, if appropriate, while the court is considering the matter.

Reasons

Under the old law, an award could be set aside for "error on the face". Consequently, arbitrators went to great length to avoid giving reasons with the award. This ground of setting aside was abolished in the 1979 Act, and there is now no bar to giving full reasons with an award. Ordinarily the parties will wish to know the

---

[55] *Marine Contractors v. Shell Nigeria* [1984] 2 Lloyd's Rep. 77.
[56] See s. 87(1).

arbitrator's reasons and, if there is such a request, the arbitrator may be bound as a matter of contract to give reasons. In any event, it is normally considered desirable so that the losing party can know why his case has failed. The arbitrator must, however, retain a wide discretion over the extent of detail which he includes in his award. For example, when giving an award on some issues, where others remain to be argued, the arbitrator will wish to avoid giving any reasons which might prejudice one of the parties in arguing subsequent issues. Reasons including findings of fact will be relevant to an appeal on a point of law (see section 69(3)(c) above). The court has power under section 70(4) to order the arbitrator to state reasons or further reasons for the award. Where the parties agree to dispense with reasons, this will be treated as an agreement to exclude the right of appeal.[57]

International Arbitration

There is no uniform definition of international arbitration; the term is used here to mean arbitrations between parties based in different states. Viewed from the United Kingdom, international arbitration has a number of different aspects. There are wholly foreign cases (for example an arbitration between Greek and German parties, conducted in Rome) in which a person from the United Kingdom might be appointed as a neutral arbitrator. Many foreign construction disputes are conducted in the English language which is frequently also the language of the contract. Alternatively, the same wholly foreign dispute might be heard in London, perhaps with one or more English arbitrators. In either case, the parties might choose to instruct English lawyers to conduct the arbitration or English experts to give evidence. There may also be arbitrations in either category in which a United Kingdom company is one of the parties.

In any international arbitration, the parties face the same procedural problems as in domestic cases, but the barriers of distance and language, as well as cultural and other differences, give rise to a rich variety of other problems. The arbitration clause in the contract will usually deal with the means of appointing arbitrators and may state the venue for the hearing. Often the contract will specify the applicable arbitration rules such as those

---

[57] s. 69(1).

of the ICC or perhaps the LCIA. These and other rules are promulgated by bodies who undertake supervision of arbitrations which are then referred to as "institutional". The ICE and RIBA limit their functions to the appointment of arbitrators and an arbitration under their rules is not institutional. International arbitration institutions exist in many parts of the world including Stockholm, Hong Kong, Singapore, Kuala Lumpur and in the Middle East. A set of rules is issued by UNCITRAL, but these require to be administered through one of the existing institutions.

There are in theory four different laws which may affect the conduct of an international arbitration[58]:

(i) the law governing the underlying contract, applicable to the merits of the case (substantive law);

(ii) the law governing the agreement to arbitrate (which will be the same as (i) if contained in the contract);

(iii) the law governing the arbitration proceedings (the procedural law);

(iv) the law governing the submission to arbitration (which may in theory be different from (i) to (iii)).

The applicable or substantive law (i) is a matter of the express or implied choice of the parties. In most construction contracts there will be an express choice and this is usually found to be the law of the state of the employers, at least where the contract concerns major construction work abroad (see Chapter 1).

Some academics argue that international disputes should be subject to international concepts of commercial law collectively referred to as *lex mercatoria*. This is a body of principles not deriving from the laws of any particular state. The English courts have shown no enthusiasm for adopting this approach. However, in the light of section 46(1)(b) of the Arbitration Act 1996 there is now no reason why an arbitration which is subject to the Act should not adopt *lex mercatoria*. The Channel Tunnel contract adopted a similar approach, stipulating that the contract should be interpreted in accordance with *"the principles common to both English law and French law and in the absence of such common principles, by such general principles of international trade law as have been applied by national and international tribunals"*.

The procedural law is also subject to the express choice of the parties. In theory this may differ from the law of the place of the

---

[58] See *Black Clawson v. Papier Werk A.G.* [1981] 2 Lloyd's Rep. 446.

arbitration, but there are obvious practical difficulties about an arbitration held, for instance, in Brussels but subject to English procedural law which should include the powers exercisable by the English courts. Under English law a choice of venue implies a choice of the procedural law of that venue.[59] However, international arbitrations take place at different locations at different times, and a more convenient concept is that of the "seat". This is now recognised by the Arbitration Act 1996 and defined as the "juridical seat of the arbitration" which may be designated by the parties or by any institution having the power to do so, or by the arbitrators if so authorised, or otherwise determined (section 3). The seat will thus designate the procedural law and the national court having supervisory jurisdiction. This cannot, however, preclude the possibility that the court of some other state might also assert jurisdiction over the arbitration proceedings, which may then create a serious and currently insoluble conflict.[60] There has also been much academic discussion on the possibility of floating or "transnational" arbitration, not being subject to the procedural law of any one state. English law has so far rejected this concept on the basis that procedure must be referable to the laws of one state and to the courts of that state.

Where an international arbitration is subject to the procedural law of one particular state, it may be found that that state has a separate part of its domestic law designed to apply to international arbitration. Switzerland and France, which are major centres for international arbitration have such provisions within their domestic law. England, conversely, has never had any body of distinct rules applying to international arbitration. The Arbitration Acts 1975 and 1979 attempted to create such distinctions in relation to the power of the court to refuse to stay proceedings and in relation to exclusion clauses. Both of those distinctions have now disappeared and there is substantially no difference under English law between the procedures applicable in domestic and in international arbitration. As noted above, Scotland has adopted the UNCITRAL model law and therefore has a dual law system for international and domestic arbitration. It remains to be seen whether the English Act of 1996 will be followed by an Act governing international arbitration in England.

International arbitration has a distinct advantage over litigation in that it avoids the perceived bias of selecting, as the forum, the court of one state, which is necessarily more closely connected with

---

[59] *Miller v. Whitworth Street Estates* [1970] A.C. 583.
[60] *Rupali v. Bunni* [1995] Con.L.Yb. 155.

one of the parties. An international arbitration tribunal of three (or more) members can properly reflect the national and cultural balance of the parties as well as selecting genuinely neutral members. The record of enforcement of arbitration awards is at least as good as that of court judgments being enforced in different countries.

ICC rules of arbitration

The ICC rules are referred to in the arbitration agreement contained in the international FIDIC form of contract, and in other standard forms as well. The International Chamber of Commerce, located in Paris, in addition to other commercial interests, administers a substantial body of international arbitration, much of it in the field of construction. ICC arbitration is different in a number of essential respects from English domestic arbitration. The ICC rules, which were revised in 1998, require the service of a "request for arbitration" and "answer" which may be accompanied by a counterclaim (Articles 4 and 5) before the tribunal is fully constituted, these steps being taken under the administration of the ICC Court. These initial documents are intended to contain a much fuller statement of each party's case than a conventional English pleading. They include the nomination by each party of an arbitrator. After these steps have been taken, a chairman is appointed (or in an appropriate case, a sole arbitrator). The full tribunal of three arbitrators is then required, with the parties, to draw up terms of reference (Article 18). These define the jurisdiction of the tribunal, the applicable procedural rules and other matters that may be necessary or desirable. Thus, unlike English Domestic Arbitration (where the arbitrator is appointed at the outset with jurisdiction deriving from the request for arbitration), an ICC tribunal defines its jurisdiction through the terms of reference after the parties have stated their cases.

The procedure is left for the parties and the arbitral tribunal to decide. Depending on the wishes and expectation of the parties, there may be formal oral hearings, or the case may be conducted largely on documents. The question whether full or partial disclosure is to be ordered will again depend on the procedural rules to be agreed or settled by the arbitrators (Article 15). The place or seat of the arbitration is to be fixed by the court unless the parties agree (Article 14). When the arbitrators make their award, it is required to be submitted to the ICC Court for scrutiny (Article 27). The ICC controls the administrative costs and arbitrators' fees, by requiring deposits from the parties at the outset. The arbitrators are given power to award costs (Article 31).

ENFORCEMENT OF AWARDS

An arbitration award does not of itself compel the losing party to comply with its terms. The aid of the court must be invoked, and this may be done in two ways. First, under section 66 of the Arbitration Act 1966, the award may, by leave of the High Court, be enforced as a judgment. Secondly, the party seeking enforcement may bring an action on the award as a contract if the award is for a sum of money, the claimant may seek to enter summary judgment for the amount awarded. In either case the losing party may object to enforcement, for example, on the ground that the arbitrator had no jurisdiction.[61]

Where enforcement of a foreign award is sought, either in England or abroad, the right of enforcement depends upon the domestic law of the country in which enforcement is sought. It depends in particular on whether that country has acceded to the relevant international conventions on enforcement which limit the grounds on which an award may be challenged. The Geneva Convention of 1927 was ratified in the United Kingdom in 1930.[62] It provides for enforcement of awards in convention countries provided that both parties are subjects of, and the award is made in, convention countries. These do not include much of the Middle East and Africa, where a great deal of international construction work is carried out. A later convention, drawn up by the United Nations, is known as the New York Convention of 1958. It was ratified in the United Kingdom in 1975.[63] It provides for reciprocal enforcement where an award is obtained in a convention country. Enforcement is possible only in a convention country, but the nationality of the parties is immaterial. Thus it can be used to enforce an award against a non-convention national who has assets in a convention country. The list of states which have acceded to the New York Convention is more extensive than those which have adopted the Geneva Convention, and now includes the great majority of countries which engage in international trade.

The New York Convention is arguably the single most important instrument in the whole sphere of international arbitration. Without the convention international arbitration would hardly exist, having regard to the difficulties of enforcement of awards and the

---

[61] Arbitration Act 1996, s. 66(3).
[62] Arbitration (Foreign Awards) Act 1930, now the Arbitration Act 1950, Pt II, which continues in force.
[63] Arbitration Act 1975.

variety of national concepts of arbitration. The convention provides an exhaustive definition of the grounds upon which enforcement may be refused. The grounds are[64]:

(a) a party to the arbitration was under some incapacity;

(b) the arbitration agreement was not valid under the applicable law;

(c) the party against whom enforcement is sought had no proper notice of the proceedings;

(d) the award is outside the terms of the submission;

(e) the tribunal or the procedure was not in accordance with the agreement or with the law of the country where the arbitration took place;

(f) the award has not become binding or has been set aside in the country in which or under the law of which it was made.

The last of these grounds (f) has given rise to some international conflict. The courts in Indià have purported to question and refuse enforcement to an international award made under a contract whose substantive law was the law of India,[65] the arbitration having been conducted abroad and subject to other procedural law. It is clear that ground (f) refers to the country whose procedural law applies, and not that whose substantive law applies, the object of the convention being to restrict the influence of the courts of the country in which enforcement is sought. The effect of enacting the New York Convention into English law is illustrated by the case of *Kuwait Ministry of Public Works v. Snow*,[66] in which enforcement was sought in England of an award made in Kuwait in 1973. The United Kingdom acceded to the convention in 1975 and Kuwait in 1978. The House of Lords held that the subsequent accession of Kuwait nevertheless had the effect of making the existing award "a convention award" so that direct enforcement in the United Kingdom was then available.

Enforcement of a foreign award without the benefit of either the Geneva or the New York Convention depends on the domestic law of the country in which enforcement is sought. In theory, an international award could be enforced in England under section 66

---

[64] Arbitration Act 1975, s. 5, re-enacted as 1996 Act, s. 103.

[65] *National Thermal Power v. Singer, Supreme Court of India* (1992) ICCA Yb. XVIII.

[66] [1984] A.C. 426.

of the Arbitration Act 1996. The conventions, however, are more restrictive of the grounds of objection. Where the conventions do not apply it remains the case that the courts in some foreign states would allow the award to be reopened on the merits, thereby negating the whole process of arbitration. This has been the situation in a number of countries in the Middle East, Africa and Asia most of which, happily, have now adopted the New York Convention.

CHAPTER 4

PARTIES AND STATUS

THE principles of substantive law apply to an individual of full age
and legal capacity. While most persons involved in the construction
industry will have attained the age of majority (now 18) they will
usually be involved as employees or representatives of some larger
body whose legal capacity and liability is limited. In this chapter
the role and status of the different parties who make up the
construction industry is examined. Then the legal capacities and
liabilities of those bodies most commonly encountered is discussed.

PARTIES IN THE CONSTRUCTION INDUSTRY

The client

The most essential person is the client, who commissions the
work. He may be referred to as the building owner or promoter,
but the term "employer" is used in the JCT and ICE forms of
contract. The employer may have practically any status. He may be
a private individual, partnership, limited liability company, part of
local or central government or any other incorporated or unincorporated body. Contractors need to be concerned with the status of
the client since, in the absence of special provisions, the contractor
has no security in the work, once it becomes attached to land
owned by another person (see Chapter 14). Invariably a construction contract will contain provisions for payments on account
so that the contractor's exposure is limited. But there will be no
security in respect of a claim, beyond the financial worth of the
employer. In recent years it has become the fashion for the
"employer" to be a specially nominated company, often set up to
run the project on behalf of other backers. In PFI projects, the

employer may be specifically referred to as the SPV (Special Purpose Vehicle), the intention being that it should be owned in whatever proportions the party (including the prospective contractors) may agree. The viability of the whole scheme then depends on the creation of a chain of appropriate security. Another recent trend in general development work is for the nominated "employer" to assign its interest in the project during the course of the work, sometimes more than once. Contractors must be alive to the consequences of these devices.

Contractors and sub-contractors

Most building work in the United Kingdom and abroad is carried out under the system generally referred to as traditional general contracting. Under this system, the person who carries out the works is the main contractor, also referred to as the builder, building contractor, civil engineering contractor, etc. The employer and the main contractor are the two parties to the main or head contract, which may also be called the construction contract, or the building or engineering contract according to the nature of the works. Professional services, including design, are provided by other persons who may be named in the main contract but they are not parties to it. Their relationship is by separate contract with the employer.

The contractor, in all but the smallest jobs, sub-contracts (or sub-lets) parts of the work to one or more sub-contractors. Main contracts commonly provide for certain sub-contractors to be chosen by the employer to carry out certain work usually identified as "prime cost" (PC) work. They are usually called "nominated" sub-contractors and their status gives rise to particular issues under the standard forms of contract.[1] Sub-contractors who are not nominated are sometimes called domestic sub-contractors. Both the contractor and the sub-contractor will usually be a limited liability company although small concerns may be partnerships or even sole traders. A practical problem often met is that the contracting "company" is a group consisting of a "holding" company and several "subsidiary" companies. The holding company owns the shares in the subsidiaries, and often has most of the assets of the group. This arrangement has taxation advantages, but means also that a subsidiary can be allowed to be wound up to the detriment of creditors, without financial harm to the group. In such

---

[1] See JCT, cl. 35, 36; ICE, cl. 59.

circumstances it will be appropriate for the employer (in the case of a contractor) to consider requesting a parent company guarantee, and the same is frequently done by contractors in respect of their intended sub-contractors.

The professional team

In traditional general contracting, the task of designing the works and supervising their construction is usually carried out by the same person or body. Under a building contract he is the architect, and under a civil engineering contract, the engineer. The title "architect" is, in England, reserved by statute for those professionally entitled to it.[2] The same is not true for engineers, although in some countries, such as Italy and Germany, the title is protected. Statutory registration for engineers has been considered in the United Kingdom but not accepted by government. Instead a uniform system of qualification has been created for all professional engineers, by award of the title Chartered Engineer, abbreviated as "C.Eng." Also, under European Community legislation, engineers throughout the community can register and use the prefix title "Euro. Ing." Usually a specific person or firm is designated as the architect or the engineer under the main contract. The person so designated will be given certain powers and duties by the contract which he must exercise as the construction work proceeds.

The architect or engineer is not a party to the main contract nor to any sub-contract, but is engaged under his own contract with the employer. In building contracts where the employer engages an architect, a civil or structural engineer may be required to carry out part of the design work. He may be engaged directly by the employer or by the architect. Similarly, an architect may be brought in to assist in the design of civil engineering works. Engineers and architects have traditionally practised as partnerships. More recently, however, many firms of engineers and architects have set up as limited liability companies, which is now permitted by the professional bodies. The result is that some former professional firms are now "owned" by the their financial backers; their shares can be bought and sold and they are subject to take-overs. The latest development, aimed at preserving professional status, is the creation of a "limited liability partnership". This awaits legislation.

A quantity surveyor (QS) is often found on larger contracts. His principal function is to take off quantities from the drawings and

---

[2] Architects Registration Act 1938.

other technical descriptions of the intended work, and to prepare from them bills of quantities; and to carry out measurements and valuations. In the JCT form a quantity surveyor is named and given certain duties. He does not appear in the ICE form. His duties there are placed on the engineer, but are usually carried out by a QS. The quantity surveyor may be engaged by the employer or by the architect or engineer under a separate contract. Again, quantity surveyors have traditionally practiced as individuals or partnerships; but more recently many have set up as limited companies. This is particularly so among quantity surveyors who carry out post-contract work, who are sometimes referred to as "claims consultants".

In his capacity under the main contract, the architect or engineer is required to carry out functions as the employer's agent, when he must represent the interest of his employer. In addition, the architect or engineer is usually required to carry out particular duties, such as certification on the basis of his professional opinion, sometimes loosely referred to as acting "independently". In such cases while he remains the employer's agent, he is under a duty to hold the scales fairly between the two parties (see Chapter 9).

Project manager

In recent years new forms of procurement of construction works have emerged, involving new types of contract under which the roles of the parties differ from traditional general contracting. These are discussed in Chapter 8. The new contract forms have given rise to a new professional known as the project manager. Although his position may be defined in a particular case, he does not fulfil a fixed role in the way that the designer or supervisor does. Project management can be a separate professional role, dedicated to the achievement of cost, time and performance requirements, using programming and monitoring techniques. This type of service will be performed under a separate contract of engagement with the employer. A project manager is also appointed by the contractor under a management contract; and the contractor's agent under a conventional construction contract is sometimes called the project manager. No single definition of the role and status of the project manager can therefore be given.

In addition to these major participants, there is a group of persons who appear in building and engineering contracts with certain functions and powers. These include the engineer's representative, the clerk of works, the agent and the foreman. All these persons are individuals who represent one or other of the major parties; thus, the engineer's representative and the agent represent on site, respectively, the engineer and the contractor.

Joint ventures

With new forms of procurement and the ever increasing size of construction projects, it is becoming common for two or more parties to combine as a joint venture to fulfil the role of one of the parties in a project. The simplest form of joint venture involves two companies who contract on the basis that each of them takes on full responsibility (joint and several liability) so that one could drop out and the other complete the project. The two joint venturers themselves enter into a contract regulating their internal rights and liabilities. In the simplest case this may involve equal pooling of resources and sharing of costs and profits. But there may be a much more elaborate division of responsibility and sharing of profit or loss. For example, two contractors may divide up the work between them, but they will then need to make provision for a joint management structure and for making decisions which affect both parties. Such joint ventures operate as a partnership, limited to a specific project. A joint venture can also be formed between parties acting as joint employer. It is not necessary that joint venture partners make the same type of contribution to a project. A joint venture can be set up to perform a design and build contract between a designing company and a constructing company. A simple alternative to joint venturers entering into contracts in their individual names is to set up a jointly owned company, whose shares and assets are held in agreed proportions. In this event, where the holding company acts as contractor, the employer will invariably require guarantees from the participants in the joint venture.

Effect of status

The differing legal capacities and liabilities of those bodies most often encountered in the construction industry are discussed below. The most common is the limited company, while some professional bodies still operate as partnerships. There is a significant difference in the ability to enforce debts. For example, where a building owner has a claim against his contractor (a limited company) for bad workmanship, and against his architect (a partnership) for bad supervision, if the contractor is without assets, pursuit of the claim will lead to winding up the company with no benefit to the building owner, even though the shareholders and directors may have substantial assets. Conversely, the architect's firm will have no such protection. Even if the firm as such is insolvent, the partners will be liable to the limit of their personal possessions. But the enforcement of a substantial judgment may be dependent on insurance.

Professional firms must maintain professional indemnity insurance. This is ostensibly for the protection of the partners, but in practice it represents a further "asset" available to the client, should a loss occur either during or after completion of the project. Conversely, insurances provided by a contractor are for the most part maintained in force only during the course of the work, so that claims for latent defects will usually be dependent on the contractor's own assets (see Chapter 10).

LIMITED COMPANIES

The word "company" can embrace any body of persons combined for a common object, whether incorporated or not. But its commercial use is narrower and refers to an incorporated company, as opposed to a partnership. While a partnership is the product of an agreement between partners, an incorporated company is entirely the product of statute, which provides for the essential ingredient of limited liability. In the case of most commercial companies, limitation of liability attaches to the issued shareholding. Incorporated companies may, as an alternative, be limited by guarantee, which is often a convenient device for non-profit making companies, or they may be unlimited.

The essential feature of a company limited by shares is that it exists as a separate legal entity, distinct from its shareholders (members). The assets and debts belong to the company, which has perpetual existence, until it is dissolved. Changes of the directors or the members (shareholders) do not change the company. When a company contracts only the company can sue or be sued on the contract. If a wrong is done by or to a company, the proper party in any action is the company itself. A shareholder is generally not entitled to conduct an action even if he holds a majority of the shares. Reference to companies being taken over or bought and sold means only that the purchaser has acquired a majority of the shares in the company. Companies are taken over because of their assets and business. But they also take with them all debts and liabilities.

Ownership of companies

The assets of a new company are contributed by the members, who subscribe to purchase the shares or "equity" of the company.

The main advantage of a limited company, as opposed to a partnership lies in the ability to acquire a financial interest in the success of a commercial venture while not participating in the risks: the liability of shareholders is ordinarily limited to the value of the shares held.

The operation of companies is closely regulated by statute. Most of the law is consolidated into the Companies Act 1985. Under this Act, there are two types of company limited by shares: public and private companies. Private companies are usually much the smaller and are often family businesses, although they comprise by far the greater number of registered companies. In a private company the number of members is limited, and the shares cannot be freely transferred. But a private company enjoys certain privileges which make its operation simpler. A public company must be identified by the letters "plc" after its name. Its membership is unlimited; shares are quoted on the stock exchange and are freely transferable. Both private and public companies must file annual accounts which are open to public inspection. Only unlimited companies are not required to file accounts. The Companies Act 1989 introduced extensive new provisions relating to company accounts.

Any legal person, including another company, can buy shares. Subject to certain restrictions, a company may purchase its own shares. The capital of a company has a nominal or authorised limit, which in a small private company is often £100. Shareholders are paid an annual dividend out of the company's profits. Additional capital can be raised by selling unissued shares, or by a fresh issue of shares. Companies borrow money in their own right but lenders invariably require security. In the case of small companies, this is likely to be in the form of a personal guarantee given by the major shareholders, which effectively defeats the objective of limitation. In the case of companies with substantial assets, a loan may be secured by a debenture which is a charge over the company's assets. A person putting money into a company set up to pursue some new business (venture capital) has to choose between the higher security but lower reward offered by a debenture loan as against the risk and rewards of buying equity.

The company's business

Every company must have written rules for its operation, set out in the articles and memorandum of association. The articles regulate the internal management of the company, including the appointment and powers of directors. The memorandum sets out, *inter alia*, the objects for which the company was formed. A company is entitled to do only those things set out in the memorandum, and anything reasonably incidental to them. An act outside the company's objects is *ultra vires* and void. The effect of the *ultra vires* rule is substantially amended by the European Communities Act 1972, s. 9. Where a person deals with a company in good faith, a transaction is deemed to be within the capacity of the company. The modern practice is, nevertheless, to draft the objects clauses of commercial companies very widely, so that a building company, for example, may if it wishes, carry on business in property development or plant hire or financing.

Management of companies

A company is run jointly by its board of directors and by the members in general meeting. Prima facie, only the board has power to act for the company. But subject to the articles, it may delegate powers to a managing director and to other directors, who often hold paid employment in the company. Directors who are not employed by the company are sometimes called "non-executive" directors. They contribute their expertise to the running of the company in return for fees. Subject to the *ultra vires* rule a company can contract in the same way as an individual and will be equally bound by written or oral contracts provided they are entered into by an agent with authority (see Chapter 7). A company may be liable for torts, although they must necessarily be committed through its servants or agents. Certain statutory offences expressly provide for directors to be held personally liable including the possibility of imprisonment, for example under the Environmental Protection Act 1990 (see Chapter 15).

The primary duty of a director is to the company rather than to the shareholders. He must act in the best interests of the company, and must disclose his personal interest in any contract made. A director must act with reasonable skill and care, although he may delegate his duties to employees of the company. Although the liability of members is limited, a director may become personally liable, for claims against the company if he commits fraud of breach of duty. A company acts through its secretary who must put into effect decisions of the board. The secretary is also responsible for keeping proper records.

The members exercise their powers by voting in general meetings. The company must hold an annual general meeting to consider the accounts, the payment of a dividend, election of new directors and other matters. Any other meeting is called an extraordinary general meeting, and may be held to consider, for example, changing the name of the company, issuing new shares or winding up. Meetings must be conducted strictly in accordance with statutory procedure. Company management has in recent years become more closely controlled by statute. Particular matters covered include regulation of the conduct of directors who have personal interests. "Insider" share dealing is now unlawful. Directors are also required to have regard to the interests of employees in carrying out their functions.

Winding up

When a company is wound up its business is concluded by a liquidator who takes over the powers of the board. He collects in the debts which are owed to the company and, so far as he is able, pays off the creditors. He may have to decide whether an alleged liability should be settled, such as a pending action for damages against the company. When the debts are paid, any money left is distributed among the members. Finally the company is dissolved and ceases to exist.

Winding up may be compulsory or voluntary. Compulsory winding up is by order of the court, upon the petition, usually, of a creditor. In most cases the ground for winding up is that the company is unable to pay its debts. A company may be wound up voluntarily for any reason by the passing of a resolution in general meeting. This may be done, for example, to amalgamate with another company. If the company is insolvent the creditors control the winding up. For the consequences of insolvency see Chapter 9.

If a company is insolvent, secured debts such as debentures will be paid first upon winding up, and they may consume all or most of the assets. Ordinary debts include an unsatisfied judgment against the company. Thus, if a creditor is owed an undisputed ordinary debt, a judgment for the debt is of no advantage if the company goes into liquidation before it can be executed. Often, debenture holders will pre-empt a winding up by appointing a receiver to protect their security (see Chapter 10). The subsequent liquidation is then more of a formality.

Foreign companies

Many foreign companies operate within the United Kingdom, and many United Kingdom companies operate abroad. Some foreign countries apply restrictions to those entitled to carry on business. In England and Wales (Scotland and Northern Ireland have separate legal systems) any foreign company may carry on business and may sue or be sued in the English courts. The constitution of a foreign company is subject of the law of the place of incorporation.

Some provisions of English company law apply to any foreign company carrying on business within the jurisdiction, but particular provisions apply where a company sets up a place of business in England and Wales. The company must then register and file certain publicly available documents similar to those required of English companies. This includes a list of directors and their interests, accounts and company reports and details of the means of serving notices or legal process on the company. A registered foreign company may then be subject to winding up proceedings in England.

Particular provisions apply to companies within the European Community. The idea of a "European company" is still under development but Regulations provide for a new form of entity called a European Economic Interest Grouping.[3] This is intended to be an amalgamation of companies, or other bodies, located in different E.C. Member States, based on similar provisions under French law.

PARTNERSHIPS

A partnership is an unincorporated body of persons combined for a common object. While the incorporation or dissolution of a company is an unequivocal act, it can be difficult to determine whether or not a partnership exists. There is often a written partnership agreement or articles of association, which may be in the form of a deed. Professional firms such as architects or consulting engineers will invariably have their constitution set out in such a document. However, a partnership agreement may be oral or even inferred from the acts of the parties. The essential feature which distinguishes a partnership is the carrying on of a business in common with a view to profit. There must be a sharing of net profits, although the shares need not be equal and it is unnecessary for all the partners to take part in running the firm.

A partnership, unlike a limited company, is not a separate legal entity. It is owned by the partners in common, and the partners are liable for the firm's debts. The capital of the firm is contributed by

---

[3] E.C. Regulation 2137/85.

the partners in any proportions they agree, so that one partner may contribute only capital and another only his expertise. They may agree to share profits in any proportions and prima facie losses must be shared in the same proportions. The question whether a partnership exists may have important consequences, for instance, in relation to loans. If A lends £100 to B to help finance B's business, then depending on the circumstances, there may be a partnership between A and B so that A might, in addition to losing his £100, become liable for B's business debts.

In the absence of a contrary agreement, a partnership ceases on the death, bankruptcy or retirement of a partner and must be dissolved. A partnership agreement therefore usually provides for the firm to be carried on by the surviving or remaining partners. Much of partnership law is codified in the Partnership Act 1890.

Management of a firm

Partners, as between themselves, must act with the utmost good faith *uberrima fides*. A partner may not make a private profit from the firm's business. Decisions must be made by a majority of the partners, but changes in the constitution of the firm, such as taking in a new partner, must be made unanimously. Every partner is prima facie the agent of the firm and can make binding contracts on its behalf. If a partner commits a tort in the course of the firm's business, the firm will be liable.

When a firm is liable in contract or for a debt the partners are jointly liable. All or any of the partners may be sued in their own names or alternatively they may all be sued by bringing proceedings against the firm. A judgment may be enforced against any of the partners who have been sued, but a partner who was not joined in the original action cannot subsequently be sued. In tort, however, partners are jointly and severally liable so that they may be sued together, or sued separately until a judgment is satisfied.

Dissolution

A partnership, unlike a limited company, may be dissolved without the assurance of the courts. If formed for an indefinite period, a partnership is dissolved by one partner merely giving notice to the other of his intention to dissolve it. If the partnership is for a fixed and unexpired term it can only be dissolved by a decree of the court on the grounds, for example, that one of the partners is guilty of prejudicial conduct, or that the business can only be carried on at a loss. While the dissolution of a company takes place after winding up and distribution of assets, the dissolution of a partnership is the first act. This is followed by the winding up of the business, for which purpose the partner's authority continues, but may be limited by the appointment of a receiver.

OTHER CORPORATE BODIES

Local authorities

Local authorities are corporate bodies whose constitution and powers derive directly or indirectly from statutes. Constitutional and general matters are found principally in the Local Government Act 1972. The powers and duties of local authorities are laid down in many statutes. Examples of particular importance are the Public Health Acts 1936 and 1961, the Education Act 1944, the Highways Act 1980 and the Town and Country Planning Act 1990. Every part of the country is within the jurisdiction of one or more authority. The distribution of functions between different local authorities and between the authorities and central government varies according to district and according to the service in question. Different local authorities may combine to provide services by setting up joint committees or a permanent joint board.

The Local Government Act 1972 brought about a massive reorganisation of local government. In addition to the re-drawing of boundaries, the Act created a two-tier structure of local government throughout England and Wales. In any area, the primary local authority was either a county or a metropolitan borough council. Below this were district councils, many having as their base the former county boroughs or boroughs. Further changes introduced in 1996 have abolished some second-tier councils and re-established unitary authorities largely based on the old counties.

London has traditionally occupied a special position. Its local authority system was laid down in the London Government Act 1963. It consists of 32 London Borough Councils, including the City. The Greater London Council, which formerly presided over the London Boroughs has now been abolished, together with the metropolitan county councils, thereby reducing the number of tiers of local government.

The powers of a local authority to enter into contracts are similar to those of an incorporated company. A local authority will be bound by a contract whether written or oral, provided it is made by an agent acting with authority. However, since the powers of all local authorities derive directly or indirectly from statute, their capacity to contract is limited by these powers in the same way that an incorporated company is limited by its objects. Any contract which a local authority purports to make for a purpose beyond such powers is *ultra vires* and void.

In entering into a contract a local authority must also comply with its own standing orders, unless they have been suspended for the purpose. In the case of *R. v. Hereford Corporation, ex p. Harrower*,[4] the Council sought to negotiate a contract with the Electricity Board and failed to invite tenders in accordance with the standing orders, because the Board were to prepare the design. The Court of Appeal rejected this as a ground for non-compliance, holding that there was a statutory duty to observe the standing orders. The question then arose whether the applicants had the right to apply to the court. Lord Parker C.J. held:

"The mere fact that these applicants were electrical contractors does not, in my judgement, of itself give them a sufficient right. But if, as I understand they or some of them are rate payers as well, then, as it seems to me, there would be a sufficient right to enable them to apply for mandamus."

However, a contract once entered into is valid despite any breach of standing orders. Any member of a local authority having an interest in a contract made or proposed must disclose the fact in the same way as a director of a company. Such a member may not, however, take part in discussion or voting connected with the contract.

As an alternative to using commercial contractors for construction work, many local authorities have set up "direct labour" organisations whereby they employ their own workforce to carry out construction work. In some cases this has led to major projects being undertaken by councils acting, in effect, as their own main contractor. Direct labour organisations also carry out work for bodies other than their parent authorities. The practice is now controlled by the Local Government Planning and Land Act 1980, which restricts the power of local authorities to enter into agreements to carry out work for other bodies and regulates the way in which direct labour organisations carry out work for their parent authorities. Competitive tenders must be obtained, and authorities

---

[4] [1970] 1 W.L.R. 1424.

are required to publish accounts and to show a return on capital employed. Recent developments in local government financing have led to "privatisation" of many traditional local government services, which are now provided through competitive tendering. In these areas, local authorities have become much more active as employers of services. Conversely, in the construction field, changes in local government financing have severely reduced local authority housing projects.

The Crown

The word "Crown" has several different meanings. It is used here to denote the sum of governmental powers exercised through the civil service, that is, central government as opposed to local government. It is not synonymous with the monarch. But historically, governments have found it convenient to invest themselves and their executive departments with the privileges and immunities attaching personally to the monarch, and so the term "Crown" is apt. Formerly the Crown enjoyed general immunity in tort and could only be sued in contract by a special process. This was radically changed by the Crown Proceedings Act 1947, which allows the appropriate government department, or the Attorney-General, to be sued by ordinary process of law.

In contract the Crown is bound by any agreement made on its behalf by an agent having authority. But if a contract provides for funds to be voted by Parliament, an affirmative vote is a condition precedent to liability. With some exceptions, principally relating to the armed forces, the Crown is liable in tort as if it were a private person of full age and capacity, and it can be made liable for the acts of its servants or agents. By virtue of its residuary immunities the Crown cannot be restrained by injunction, nor can it be deprived of property. The Crown also has a far-reaching privilege to restrain disclosure of documents in legal proceedings, whether or not it is a party to the proceedings.

Building and engineering contracts in which the employer is a government department are often subject to one of a series of standard forms known as GC/Works. In keeping with government privatisation policy, the latest edition of the forms has been produced commercially and is offered for general use in a private version known as PC/Works (see Chapter 11). Some government departments favour the use of private sector standard forms, notably the Department of Transport (now operating through the Highways Agency), which has for many years favoured the use of the ICE Form of Contract but with a number of amendments.

All governments throughout the world operate through various departments or ministries presided over by ministers or Secretaries of State. The question sometimes arises whether different ministries comprise separate legal entities or whether they are equivalent, for example, to different departments of the same company. In some countries, notably France and others whose constitution derives from that of France, government departments do indeed comprise separate legal entities. In the United Kingdom, however, the Crown and the Government are regarded in law as indivisible. While the Crown Proceedings Act stipulates for particular ministers or Secretaries of State to be sued they are, in law, one with the government. The authority of a Secretary of State extends, in law, over other ministries so that one can represent others.

There are other important contrasts between the constitution position of the Crown and that of the government of other countries. In the United Kingdom, despite the recent upsurge of public law rights, many of which are exercisable against the Government (see Chapter 1), the Crown, in general, exercises no rights in regard to commercial contracts beyond those exercisable by any ordinary legal person. Where the Government wishes to exercise special powers, it does so through Parliament by passing legislation granting such powers. There are many examples which have operated, for instance during war-time or other emergencies. In general, however, the Government has no special commercial powers. This is in sharp contrast to the position in most other countries. In France there is a well-developed body of rights which the Government is entitled to exercise in the national interest. For example, under the doctrine of *imprévision*, where unforeseen circumstances arise which might result in a contract not being performed, the Government is entitled to require performance to continue on different terms. In *Compagnie Generale d'Eclarage de Bordeaux*[5] the contract for gas supply to Bordeaux became economically impossible to perform owing to huge increases in coal prices as a result of French coal fields being occupied by the German army during the First World War. The result would have been to cut off lighting to the city. The company was ordered to continue to perform the contract, but at a substantially increased price to take account of the price of coal.

Whenever it is sought to bring an action against a foreign government, whether the action is brought here or abroad, the position may be very different from an action against the British

---

[5] Conseil d'Etat, March 30, 1916.

Government. As a general rule a foreign sovereign state is immune from action brought in this country, whether in civil or criminal law. Further, if there is no local equivalent to the Crown Proceedings Act, it may not be possible to bring proceedings in the country in question. Parties contracting with foreign states should therefore give serious consideration to the question of guarantees or performance bonds.

Public utilities and privatisation

Until 1945 "public" services throughout the United Kingdom were provided by private enterprise although many of the major industries such as coal, steel and the railways had been taken over by the Government during the two world wars. The Labour Government of 1945 introduced, for the first time, the concept of "nationalisation" under which these major industries were formally brought under permanent state control in the form of public corporations with a monopoly, under the responsibility of a minister answerable to Parliament. Subsequent governments de-nationalised certain industries, such as steel and the airlines. During the 1980s, however, an entirely new policy was embarked upon by which virtually the whole of the former public utilities have been systematically "privatised" by breaking them up into commercial organisations for sale as public companies. In some cases the Government has temporarily retained a substantial shareholding. But in the main the process has involved the Government entirely divesting itself of ownership and setting up, in place of the former government control, a series of "watchdog" bodies intended to represent the public interest. These bodies are known as the Office of Water Services or OFWAT, OFTEL (telecommunications), OFGAS (gas), etc., and are given regulating powers under the statutes which prescribe the duties of each supply company. In the case of the water industry these are the Water Industry Act and the Water Resources Act, 1991.

The result of the privatisation programme is the setting up of a large number of substantial companies, which operate the utilities and other basic industries on a commercial basis, and whose constitution does not differ essentially from that of any other public limited company. Shares are quoted and traded in the same way as other commercial companies and, as has been seen in recent years, the companies are susceptible to takeovers and mergers. Where an apparent monopoly has existed, government policy has been to insist on the creation of competition which now applies (despite the use of common mains) in telecommunications, electricity and gas supply. Many of these newly emerging commercial enterprises are in process of developing appropriate commercial forms of contract and dispute resolution procedures for their operations. Many of these are based on the principles of construction law.

CHAPTER 5

THE LAW OF OBLIGATIONS

In English law, obligations arise in a variety of ways which sometimes overlap. For purposes of analysis they are divided into discrete categories and they are so dealt with in this book. However, it is important to be aware that these categories can be artificial or, more strictly, are imposed as part of a logical order. In this chapter, obligations are considered more broadly and some of the areas of overlap are explored.

In this chapter obligations are divided primarily between those arising as a matter of agreement and those imposed by virtue of status. Within the former, it is necessary to distinguish between circumstances in which the law will imply a "contract" which conventionally must have ascertainable terms, and those in which, in the absence of contract, the law imposes an obligation, for example to account for benefits received. In the latter category are obligations arising under the law of tort and through statutory duties imposed on defined persons or bodies? Also included is the law of trusts, a wholly separate area of law which continues to be of considerable utility in the world of commerce and business.

OBLIGATIONS THROUGH AGREEMENT

English law in general rarely finds difficulty in constructing a contract from any circumstances which evidence a mutual exchange of obligations. It may be said that English law is more concerned with identifying circumstances in which an apparent exchange of obligations does not create an enforceable contract (as to which see Chapter 6).

The way in which the law identifies and categorises agreements was described in *New Zealand Shipping Co. v. Satterthwaite & Co.*[1]

---

[1] [1975] A.C. 154.

This case concerned whether a negligent stevedore could rely on the terms of the bill of lading issued by agents of the carrier for whom the stevedore carried out unloading. The question was, therefore, who were the relevant parties to the transaction? Lord Wilberforce's judgment included this:

> "It is only the precise analysis of this complex of relations into the classical offer and acceptance, with identifiable consideration, that seems to present difficulty, but this same difficulty exists in many situations of daily life, *e.g.* sales at auctions; supermarket purchases; boarding an omnibus; purchasing a train ticket; tenders for the supplies or goods; offers of rewards; acceptance by post; warranties of authority by agents; manufacturers' guarantees; gratuitous bailments; bankers' commercial credits. These are all examples which show that English law, having committed itself to a rather technical and schematic doctrine of contract, in application takes a practical approach, often at the cost of forcing the facts to fit uneasily into the marked slots of offer, acceptance and consideration."

In the construction field, the case of *Shanklin Pier v. Detel Products*[2] is regarded as a landmark decision establishing that oral representations about a product made to a prospective user could found a claim in breach of warranty. The user, who was the pier owner, specified the product under a contract made with the main contractor, who placed an order for the product. This has given rise to the now common practice of employers entering into express direct warranties with suppliers or sub-contractors. But in a number of cases both preceding and following the *Shanklin* case, the courts have held that statements may give rise to an enforceable contract without the need for any other direct relationship between the parties in question. In *Andrews v. Hopkinson*[3] a second hand car dealer made the memorable statement "It's a good little bus. I'd stake my life on it. You will have no trouble with it". This was followed by the plaintiff entering into a hire purchase contract with a finance company to whom the dealer sold the car. The vehicle had a serious steering defect which led to an accident. The statement was held to be an enforceable warranty which entitled the plaintiff to recover damages both as to the difference in value of the car and damages for personal injury. The *Shanklin* case can therefore be seen as an application of existing principles.

Unilateral contracts

Despite the importance of mutuality in the law of contract there are many situations in which the law has to find a consistent and

---

[2] [1951] 2 K.B. 854.
[3] [1957] 1 Q.B. 229; approved in *Yeoman Credit v. Odgers* [1962] 1 W.L.R. 215; and see also *Brown v. Sheen* [1950] 1 All E.R. 1102.

rational answer to transactions which lack mutuality. They are sometimes referred to as unilateral contracts, although they must be bilateral in the sense that there must be two parties and obligations which each of them takes on. The classic unilateral contract is one in which an offer is made to the world through advertisement expressed to be capable of acceptance by any individual who fulfils the stated requirements. A common example is offers of reward, which may lead to dispute if, for example, several people fulfil the stated conditions. The leading case, now over a century old, is *Carlill v. Carbolic Smoke Ball Co.*[4] in which the defendant, who sold a patent cure, offered to pay £100 to any person who used the cure and then caught influenza. This was held to be an offer capable of acceptance particularly as the defendant also advertised that £1000 had been deposited with their bankers. More difficult questions arise where there are multiple claims, or where the person fulfilling the condition was unaware of the offer.

An example of a unilateral contract in the field of construction is the instruction of an estate agent to negotiate the sale of a house on commission. The estate agent does not undertake to do anything and will not be under any liability for failing to achieve a sale. Unless a separately enforceable option is granted, the client can revoke the agent's instructions at any time. Yet if a sale is negotiated the client becomes liable for the fee: it is a matter of construction whether the fee is to be payable upon sale or upon introduction of a willing purchaser. And a person appointed sole agent may be entitled to maintain a claim for damages if the client then sells through another agent.

A common device used in the construction field is the "call off" contract by which the client or employer enters into what appears to be a formal contract containing rates and prices for specified and described items of work or goods and materials. There are usually detailed provisions regulating the ordering of work and there may be a retainer or other fee payable in any event. The obligation to do the work or supply the goods, and the corresponding obligation to pay, arises only as and when orders are placed.[5] An example of this type of contract is one for routine repairs to public services, such as water pipes and apparatus beneath the highway. Water companies, as an alternative to employing their own maintenance gangs, currently let term contracts on this basis within defined areas.

---

[4] [1893] 1 Q.B. 256.
[5] See *Brogden v. Metropolitan Railway* (1877) 2 App. Cas. 666.

Letters of intent

Frequently, parties create a situation in which they intend no contract to come into existence, sometimes by expressly stating their intention to enter into a contract in the future. Such an intention, provided that it was honestly held, creates no obligation. Nevertheless, it is common for such a letter of intent to be accompanied by words authorising certain work to be carried out and this may give rise to what is sometimes called an "if" contract, *i.e.* a contract under which A requests B to carry out work on the basis that if he does so he will receive appropriate payment. This is another example of a particular type of unilateral contract. In *British Steel v. Cleveland Bridge*[6] the defendant wrote a letter to the plaintiff stating that it was their intention to enter into a sub-contract, on the basis of which the plaintiff arranged for the manufacture of steel work. The parties continued to negotiate and the steel work was progressively delivered, later than the dates in the contract under negotiation, which was never signed. It was held that there was no "if" contract. Both parties expected there to be a formal contract which would have governed their rights. But in the absence of a contract being signed, there was simply an obligation in law on the defendant to pay a reasonable sum for work done at its request.

Good faith, best endeavours and fair dealing

The question examined here is whether and to what extent the law of contract recognises enforceable obligations as to the way in which the parties are to behave. Where the parties act in a manner which conveys an intention to be bound, can there be an enforceable obligation to negotiate in good faith? In *Walford v. Miles*[7] the defendant was negotiating with a number of people, including the plaintiff, for the sale of a company. The defendant orally agreed with the plaintiff that he would terminate negotiations with any other prospective purchaser in return for the plaintiff furnishing a "comfort" letter from the bank stating that all necessary resources were available for the purchase. The defendant, however, unilaterally decided not to proceed with the plaintiff and sold to another party. The plaintiff brought proceedings alleging, *inter alia*, breach of an enforceable obligation to negotiate in good faith. The House of Lords rejected the claim, Lord Ackner holding:

---

[6] (1981) 24 B.L.R. 94.
[7] [1992] 2 A.C. 128.

"The reason why an agreement to negotiate, like an agreement to agree, is unenforceable, is simply because it lacks the necessary certainty. The same does not apply to an agreement to use best endeavours. This uncertainty is demonstrated in the instant case by the provision which it is said has to be implied in the agreement for the determination of the negotiations. How can a court be expected to decide whether, objectively, a proper reason existed for the termination of the negotiations? The answer suggested depends upon whether the negotiations were determined in 'good faith'. However, the concept of a duty to carry on negotiations in good faith is inherently repugnant to the adversarial position of the parties when involved in negotiations. Each party to the negotiations is entitled to pursue his (or her) own interest, so long as he avoids making misrepresentations. To advance that interest he must be entitled, if he thinks it appropriate, to threaten to withdraw from negotiations or to withdraw in fact, in the hope that the opposite party may seek to re-open negotiations by offering him improved terms."

The obligation to use best endeavours referred to by Lord Ackner may be thought equally vague. But such an obligation may be enforceable, given the existence of other *indicea* of contract, on the basis that the parties are not in opposed (adversarial) positions but rather taking on obligations of mutual support. The difficulty of proving breach of an obligation to use best endeavours, and the difficulty of establishing damage flowing from such a breach do not prevent such an obligation being given legal effect.

Another aspect of the question arises in the context of competitive tendering. Does this create any obligation on the party soliciting the tenders to act fairly or properly? In *Blackpool & Fylde Aeroclub v. Blackpool B.C.*[8] the defendant, which operated the local airport, invited tenders for a concession to operate pleasure flights. The form of tender stated that the Council "do not bind themselves to accept all or any part of any tender. No tender which is received after the last date and time specified shall be admitted for consideration". The plaintiff submitted a proper tender but owing to a mistake by the Council it was not considered and the defendant accepted a less favourable tender. The plaintiff claimed damages for breach of contract. Bingham L.J. commented on the fact that a tendering procedure was heavily weighted in favour of the invitor. He went on to say:

"But where as here the tenders are soliciting from selected parties all of them known to the invitor, and where a local authority's invitation prescribes a clear orderly and familiar procedure—draft contract conditions available for inspection and plainly not open to negotia-

---

[8] [1990] 1 W.L.R. 1195.

tion, a prescribed common form of tender, the supply of envelopes designed to observe the absolute anonymity of tenderers and clearly to identify the tender in question, and an absolute deadline—the invitee is in my judgement protected at least to this extent: if he submits a conforming tender before the deadline he is entitled, not as a matter of mere expectation but of contractual right, to be sure that his tender will after the deadline be opened and considered in conjunction with all other conforming tenders or at least that his tender will be considered if others are."

It has to be emphasised, however, that there is no general principle of good faith under English law outside particular contracts such as those of insurance (see Chapter 7). In *Interfoto v. Stiletto*[9] (see Chapter 6 for facts) Bingham L.J. said:

"In many civil law systems, and perhaps most legal systems outside the common law world, the law of obligations recognises and enforces an over-riding principle that in making and carrying out contracts, the parties should act in good faith . . . English law has, characteristically committed itself to no such over-riding principle but has developed piecemeal solutions to demonstrated problems of unfairness. Many examples could be given, thus equity has intervened to strike down unconscionable bargains. Parliament has stepped in to regulate the imposition of exemption clauses and the form of certain hire purchase agreements. The common law has also made its contribution, by holding that certain classes of contract require utmost good faith, by treating as irrecoverable what purport to be agreed estimates of damage but are in truth disguised penalty for breach and in many other ways."

Third party rights in contract

The spectrum of "obligations" includes the possibility of one party to a contract being under a duty to perform or to pay damages for non-performance to a third party who is not an original party to that contract. The existence of such a duty in tort is mentioned below. A duty of this sort also arises as a result of an express or implied collateral contract with the third party (see above). What is considered here is the creation of contractual "third party rights", by which the contract itself could be enforceable by a stranger.

English common law has consistently rejected the notion of third party rights in contract. Substantially the only rights acknowledged were those arising under the law of trust (see below). In *Beswick v.*

---

[9] [1989] Q.B. 433.

Beswick[10] a nephew bought his uncle's coal business and promised to pay £6-10s *per* week for life and thereafter £5 *per* week to his widow. The nephew refused to pay the widow who was held, under the doctrine of privity, to be unable to recover the money in her own name. However, the House of Lords held that, as the aunt was also the personal representative, she could enforce the contract in the name of her deceased husband. The case represents the high point of the doctrine of privity, and seems an affront to common sense.

However, two recent developments have suggested major inroads into this area in which English law appears to stand aloof from the rest of the world. First, the Law Commission, after some years of consultation have published a report in which it is proposed that the present exceptions to the law of privity be replaced by a general right of a third party to enforce a contract in which the parties expressly or impliedly intend that he should receive a benefit.[11] An amended version of the Bill accompanying the report has now been introduced into the House of Lords as the Contracts (Rights of Third Parties) Bill (see further Chapter 6).

The second development is in three construction cases, concerning the right of action where damages had become separated from the legal right of recovery. In *Linden Gardens v. Lanesta Sludge*,[12] it was held that a purported assignment of the benefit of a large construction contract was ineffective in the absence of consent from the contractor. The result was that the party entitled to enforce the contract (the original employer) was not the party who had suffered the loss. The House of Lords rejected a submission that the loss therefore disappeared into a "black hole" and held that the parties were to be treated as having entered into the contract on the basis that the employer would be entitled to enforce rights against the contractor on behalf of those who would suffer from deficient performance. This finding was recognised as an exception to the general rule of privity. Lord Griffiths, however, proposed a wider test as follows:

"In everyday life contracts for work and labour are constantly being placed by those who have no proprietary interest in the subject matter of the contract. To take a common example, the matrimonial home is owned by the wife and the couple's remaining assets are owned by the husband and he is the sole earner. The house requires a new roof and the husband places a contract to carry out the work.

---

[10] [1968] A.C. 58.
[11] Law Commission Report No. 242, 1996.
[12] Heard with *St. Martin's v. McAlpine* [1994] 1 A.C. 85.

The husband is not acting as agent for his wife, he makes the contract as principal because only he can pay for it. The builder fails to replace the roof properly and the husband has to call in and pay another builder to complete the work. Is it to be said that the husband has suffered no damage because he does not own the property? Such a result would in my view be absurd and the answer is that the husband has suffered loss because he did not receive the bargain for which he had contracted with the first builder and the measure of damages is the cost of securing the performance of the bargain by completing the roof repairs by the second builder."

In *Darlington v. Wiltshire*[13] the building owner faced a similar difficulty arising, not from commercial transfer of rights but from the well-known device of using a separate financier (Morgan Grenfell) to enter into the building contract which was then to be assigned to the local authority as the true owner. There were defects in the work, but the contractor contended that the loss was suffered by Morgan Grenfell prior to the assignment and the local authority had therefore acquired no rights. The Court of Appeal held the plaintiff entitled to recover, basing themselves on the exception to the rule of privity which had been applied in the *Lanesta* case. Steyn L.J. would have decided the case on a broader basis. He said:

"It is, of course, manifest that the Council, as third party, accepted the benefit of the building contract. But for the rule of privity of contract the Council could simply have sued on the contract made for its benefit. The case for recognising a contract for the benefit of a third party is simple and straightforward. The autonomy of the will of the parties should be respected. The law of contract should give effect to the reasonable expectations of contracting parties. Principle certainty requires that a burden should not be imposed on a third party without his consent. But there is no doctrinal, logical or policy reason why the law should deny effectiveness to a contract for the benefit of a third party where that is the expressed intention of the parties. Moreover, often the parties and particularly third parties, organise their affairs on the faith of the contract. They rely on the contract. It is therefore unjust to deny effectiveness to such a contract.

We do well to remember that the civil law legal systems of other members of the European Union recognise such contracts. That our legal system lacks such flexibility is a disadvantage in the single market. Indeed it is a historical curiosity that the legal system of a mercantile country such as England, which in other areas of the law of contract (such as, for example, the objective theory of the interpretation of contracts) takes great account of the interest of third parties, has not been able to rid itself of this unjust rule deriving from a technical conception of a contract as a purely bilateral *vinculum juris*."

---

[13] [1995] 1 W.L.R. 68.

The third case of *Alfred McAlpine v. Panatown*[14] involved the making of a building contract with a nominee company within the same group as the company which owned the site, for reasons of VAT liability. Again, the right of action appeared to be vested in a party other than the party which had suffered damage. The Court of Appeal held that the contracting party (*Panatown*) was entitled to recover substantial damages on the basis that this was "intended or contemplated" by the parties.

Status of offers

The principles of offer and acceptance are normally applied with some rigidity so that the parties move at the moment of acceptance from there being no contract to full and binding contractual relations. There are, however, a number of situations in which actions short of acceptance of an offer may give rise to enforceable rights. Some of these are discussed below in relation to *quantum meruit* and equity. In addition, under the law of some countries the offer itself is regarded as giving rise to legal obligations if relied upon. In *Northern Construction v. Gloge Heating*[15] and a number of similar cases the Canadian courts have held that an offer made by a sub-contractor to a main contractor could not be withdrawn (save on grounds which would vitiate a contract) after the main contractor had relied on the offer in making his own tender. Similar principles exist in some civil law countries. Under English law, however, an offer creates no obligation unless accompanied by an enforceable option or an agreement not to revoke the offer.

RESTITUTION AND QUANTUM MERUIT

The absence in English law of any general principle of fair dealing or good faith is commented on above. Yet in two respects at least, English law does afford remedies based on such principles. One is in the area of law loosely referred to as Equity and more specifically the law of trusts, which is dealt with below. The other, dealt with here, is an area of law which has come to be referred to compendiously as Restitution. This area of law has grown out of the common law principle of quasi-contract in which the law afforded the remedy of repayment of money received or payment

---

[14] (1998) 88 B.L.R. 67.
[15] (1986) 27 D.L.R. 265; (1984) 1 Const. L.J. 144.

of a reasonable sum, in the absence of a contract. The principle has now developed into a general principle of affording restitution where other remedies are unavailable.[16]

The leading cases in the construction field start with *William Lacey v. Davis*[17] in which a contractor, having tendered for building work and in anticipation of obtaining the contract, prepared further calculations and particulars for use by the employer in obtaining a ward damage claim. No contract was awarded and the contractor sued for compensation for the services rendered. Barrie J. held:

> "In my judgement, the proper inference from the facts proved in this case is not that the work was done in the hope that this building might possibly be reconstructed and that the plaintiff company might obtain the contract, but that it was done under a mutual belief and understanding that this building was being reconstructed and that the plaintiff company was obtaining the contract. . . . The court should imply a condition or imply a promise that the defendant should pay a reasonable sum to the plaintiff for the whole of these services which were rendered by them.

This principle must be limited, however, to circumstances in which a benefit is conferred subject to a mutual implied expectation of payment. In the old case of *Sumpter v. Hedges*,[18] a building contractor who became insolvent abandoned a contract leaving partly completed work, some of which had not been paid for. His action to recover payment for the work failed on the basis that the defendant had no choice but to accept what was attached to his land. In the more recent case of *Regalion Properties v. London Dockland Development Corporation*[19] the plaintiff had incurred substantial costs in anticipation of a development contract which did not materialise. These costs would have been recovered out of the income from the development. Rattee J. held:

> "Each party to such negotiations must be taken to know (as in my judgement Regalion did in the present case) that depending on the conclusion of a binding contract any cost incurred by him in preparation for the intended contract be incurred at his own risk, in the sense that he will have no recompense for those costs if no contract results. I accept . . . . that by deliberate use of the words 'subject to contract' with the admitted intention that they should have

---

[16] See generally Goff and Jones, *The Law of Restitution* (4th ed.); and *Legal Obligations in Construction* (CCLM, King's College), papers 11 and 12.
[17] [1957] 1 W.L.R. 932.
[18] [1898] 1 Q.B. 673.
[19] [1995] 1 W.L.R. 212.

their usual effect, LDDC and Regalion each accepted that in the event of no contract being entered into, any resultant loss should lie where it fell."

A potentially important question in relation to work performed in the absence of a formal contract, where the contractor is entitled simply to payment for the work done, is whether the person commissioning the work (the would-be employer) is entitled to credit for defective performance of the work. In the absence of a contract, there could not be a cross-claim, but there could, in theory, be a set-off (see Chapter 2). As regards physical defects in the work, there is no difficulty in principle about reflecting this in the value of what is to be paid for. Difficult questions of fact might arise where the defect relates to suitability rather than quality. Such issues would depend on what the contractor had been requested and had agreed to undertake. More difficult is the question of delay. In the absence of any contract there cannot be terms controlling the performance of the work in question. There could not, therefore, be a fixed completion date nor can there be any implied obligation to perform within a reasonable time.[20] The possibility that the value of the work should be reduced by reason of tardy performance cannot, however, be ruled out.[21]

Quantum Meruit under contracts

Where no price is stated for work carried out within an existing contract, the employer will be obliged to pay a reasonable sum, which may be regarded as a species of *quantum meruit*. More difficult, and of considerable importance in relation to construction contracts, is whether and in what circumstances, the contractor may be entitled to claim *quantum meruit* when there is an existing agreed pricing mechanism. Such a remedy is occasionally claimed on the footing that "the basis of the contract has changed" or that circumstances have arisen which were not contemplated by the parties. Some support for this approach appears in the old case of *Bush v. Whitehaven Trustees*[22] where variations led to a summer contract being turned into a winter contract. However, the authority of the case has been soundly repudiated by the House of Lords.[23] The case was again relied on by the official referee in *McAlpine Humberoak v. McDermott (No. 1)*, but again repudiated

---

[20] *Sanders & Foster v. Monk* (1980) [1995] Con. L.Yb. 189.
[21] *Crown House Engineering v. AMEC* (1990) 48 B.L.R. 32.
[22] *Hudson's Building Contracts* (4th ed.), Vol. 2, p. 120.
[23] *Davis Contractors v. Fareham* UDC [1956] A.C. 696 at 732.

by the Court of Appeal in the same case.[24] There is, it is submitted, still a basis in law for such a contention. In *Thorn v. London Corporation*[25] the House of Lords recognised, *obiter*, that circumstances might arise in which the contractor was effectively being required to perform a different contract: *"non haec in foedera veni"*. And in *Parkinson v. Commissioners of Works*,[26] the Court of Appeal allowed a contractor to recover *quantum meruit* in part where the original contract, which had been subjected to extensive variations, was renegotiated with a provision for a fixed profit. This was held to limit the amount of work which could be ordered so that work in excess of such limit gave rise to a *quantum meruit*.

An important and unresolved question is whether a contractor, whose contract is brought to an end by the repudiation of the employer which is accepted by the contractor, may claim *quantum meruit* in respect of the whole of the contract works, as an alternative to claiming the value of work done together with loss of profit. There is high authority in support of such right, specifically the Privy Council decision in *Lodder v. Slowey*.[27] However, the principle has been doubted in *Keating on Building Contracts* as being inconsistent with both earlier[28] and later[29] authority. The issue is of the highest commercial importance, since it would in theory permit a contractor to escape entirely from an unprofitable contract. The issue can be resolved only by a decision of the House of Lords. It is to be noted that United States and Commonwealth decisions have upheld the right to *quantum meruit* without question[30]; while on the other hand, the principle appears to rest on the theory that a contract which has been terminated following repudiation disappears, relieving the parties from any onerous obligations thereunder. This principle is demonstrably contrary to many authorities which have held that the contract terms survive repudiation.[31]

---

[24] (1992) 58 B.L.R. 1.
[25] (1876) 1 App. Cas. 120.
[26] [1949] 2 K.B. 632.
[27] [1904] A.C. 442 and see *Chandler Bros v. Boswell* [1936] 3 All E.R. 179.
[28] *Ranger v. G.W. Railways* (1854) 5 H.L.C. 72.
[29] *Johnson v. Agnew* [1980] A.C. 368, *per* Lord Wilberforce at p. 396.
[30] See *Morrison-Knudsen v. British Columbia Hydro Authority* (1978) 85 D.L.R. 3d 186; 7 Const. L.J. 227.
[31] *e.g. Heyman v. Darwins* [1942] A.C. 356; *Suisse Atlantique v. N.V. Rotterdamsche Kolen* [1967] 1 A.C. 361.

Obligations Through Status

The most common circumstance in which the law imposes obligations on a person by reason of status, rather than agreement, is under the law of tort. The essence of tortious obligations is that a person is required to avoid causing harm to another. The widest area of duty is in relation to the law of negligence where persons in particular circumstances are regarded in law as owing a duty of care in relation to particular types of loss. Thus, a person will be regarded as owing a duty of care not to cause physical harm to anyone whom he ought reasonably to foresee as being affected by his actions. This test will readily cover drivers of motor vehicles and people in many other circumstances. A person will owe a duty to prevent economic harm to a much more limited class, typically those with whom there exists some special relationship of proximity. This was the situation in the celebrated case of *Hedley Byrne v. Heller*[32] in which it was held that a bank might be liable in tort (but for a disclaimer) in circumstances close to, but falling short of contract.[33] In all such cases, the obligation is to take reasonable care to avoid causing harm. Other areas of tort impose absolute obligations to make good loss, for example where it arises from intentional rather than accidental acts. This is the subject of the law of trespass to goods and nuisance in relation to land (see further Chapter 14).

An important question which has been the subject of much apparently conflicting authority is whether a plaintiff may take advantage of a potential right in tort when there exists a contract between the same parties. The question is important because it may be possible to bring a claim in tort after the limitation period in the contract has expired. Before the rapid expansion of the tort of negligence in the 1970s, the law appeared to be that rights in contract and tort were mutually exclusive. However, there have been a number of cases in which parallel remedies have been allowed.

An example of this is the case of *Batty v. Metropolitan Realisations Limited*,[34] where a developer was held liable in breach of contract for having sold to the plaintiff a house which was not fit for habitation because it had been built at the top of a potentially unstable slope. When the question arose whether the plaintiffs

---

[32] [1964] A.C. 465.
[33] See Chap. 14, Negligent Misstatement.
[34] [1978] Q.B. 554.

were entitled also to have judgment entered in tort Megaw L.J. held:

> "In my judgement the plaintiffs were entitled here to have judgement entered in their favour on the basis of tortious liability as well as on the basis of breach of contract, assuming that the plaintiffs had established a breach by the first defendant of the common law duty of care owed to the plaintiffs. I have no doubt that it was the duty of the first defendants, in the circumstances of this case, apart altogether from the contractual warranty, to examine with reasonable care the land, which in this case would include adjoining land, in order to see whether the site was one on which a house fit for habitation could safely be built. It was a duty owed to prospective buyers of the house."

This case has subsequently been doubted by the House of Lords, because the plaintiffs in *Batty* had suffered no physical damage; but the decision as regards parallel duties in contract and in tort remains applicable. The existence of parallel duties has been considered in other situations, widely different from that of building developer and purchaser. In the leading case of *Lister v. Romford* Ice & *Cold Storage*[35] the House of Lords had to consider the following facts. A lorry driver employed by the defendant company took along his father to act as mate. While negligently driving the lorry, the son injured his own father who succeeded in recovering damages against the company for the son's negligence. The company claimed indemnity against the son, and because of the then restrictive rules on contribution the court had to consider whether the potential liability of the son arose in contract or in tort. The majority opted for contract and held the son liable to the company and not entitled to be indemnified by them or their insurer (so that in the result the family recovered nothing). In a dissenting judgment, Lord Radcliffe said:

> "Since in any event the duty in question is one which exists by imputation or implication of law and not by virtue of any express negotiation between the parties, I should be inclined to say that there is no real distinction between the two possible sources of obligation. But it is certainly, I think, as much contractual as tortious."

This case was considered in *Tai Hing v. Liu Chong Hing Bank*,[36] where the Privy Council had to consider the position of a bank customer where a bank clerk had fraudulently drawn and cashed

---

[35] [1957] A.C. 555.
[36] [1986] A.C. 80.

cheques against the customer's accounts, and the customer had
failed to detect or notify the bank about the losses. It was argued
that, apart from the terms of contract between the customer and
the bank, the customer owed a duty in tort to prevent such losses
to the bank (on the footing they were otherwise liable to repay the
money). The Privy Council decided the case in contract, holding
that the bank's terms were not sufficient to impose liability on the
customer for the loss. In regard to the tort claim, it was said:

> "Their Lordships do not, therefore, embark on an investigation as to
> whether in the relationship of banker and customer it is possible to
> identify tort as well as contract as a source of the obligations owed by
> the one to the other. Their Lordships do not, however, accept that
> the parties' mutual obligations in tort can be any greater than those
> to be found expressly or by necessary implication in their contract . . .
> the bank cannot rely on the law of tort to provide them with greater
> protection than that which they have contracted."

This observation is in line with a number of other cases,
including *Junior Books v. Veitchi*,[37] a case much criticised for its
decision, but interesting in having raised the question of the
possible impact on tort of terms in another contract. In that case,
Lord Roskill said:

> "During the argument it was asked what the position would be in a
> case where there was a relevant exclusion clause in the main contract.
> My Lords, that question does not arise for decision in the instant
> appeal, but in principle I would venture the view that such a clause
> according to the manner in which it was worded might in some
> circumstances limit the duty of care just as in the Hedley Byrne case
> the plaintiffs were ultimately defeated by the defendants' disclaimer
> of responsibility."

Two other construction cases have touched on another aspect of
the contract-tort relationship. In *Greater Nottingham Co-op v.
Cementation*[38] a piling sub-contractor had caused damage to an
adjoining property, and this had resulted in losses and claims as
between the main contractor and the employer. The question arose
whether the employer could claim these losses against the sub-
contractor in tort, where the subcontractor had been required to
enter into the usual direct warranty covering design but not the
execution of the work. The Court of Appeal held that the fact that
these parties had deliberately made a contract (the warranty) which

---

[37] [1983] A.C. 520.
[38] [1989] Q.B. 71.

excluded the work in relation to which the tortious duty was alleged, was sufficient to exclude the existence of a duty in tort to fill the gap. Conversely, in the case of *Warwick University v. McAlpine*,[39] a similar point was resolved the other way. The unusual facts of this case were that McAlpine were carrying out remedial work arising from an earlier contract. During the course of the work it was decided to bring in Cementation Chemicals Limited (C.C.L.) to carry out specialist work. The university had the opportunity of employing C.C.L. direct, but instead requested McAlpine to employ them, without taking a warranty. The tortious duty alleged was "failing to warn the plaintiffs or the defendants of the damage that would result from the use of such materials and services." Garland J. held as follows:

> "If this is the duty, then the university are not seeking to establish an essentially contractual one, fitness for purpose, but a normal tortious one-to take reasonable care to avoid reasonably foreseeable damage to a sufficiently proximate plaintiff. The fact that the university might have created a contractual duty by an express warranty is not something which in my view should negative the existence of any duty. Where there is a direct warranty which omits to provide for a particular category of damage, the omission may lead to the conclusion that any duty in tort or the consequences of a breach should be correspondingly restricted. In my view there was a duty in the terms I have set out."

The decision in this case was reversed by the Court of Appeal on the facts but the legal discussion in the judgment remains relevant.

A further development occurred in the House of Lords' decision on claims brought against underwriters by Lloyds "names", some of whom had direct contracts and others not, so that their claims were in tort. It was held that claims could proceed on either basis but that the contractual arrangements in other cases might prove inconsistent with an assumption of responsibility in tort. Lord Goff drew an analogy with attempts by a building owner to bring a claim in tort against a sub-contractor:

> "But if the sub-contracted work or materials do not in the result conform to the required standard it will not ordinarily be open to the building owner to sue the subcontractor or supplier direct under the Hedley Byrne principle, claiming damages from him on the basis that he has been negligent in relation to the performance of his functions. For there is generally no assumption of responsibility by the sub-contractor or supplier direct to the building owner, the parties having

---

[39] (1988) 42 B.L.R. 1.

so structured their relationship that it is inconsistent with any such assumption of responsibility."[40]

In addition, the House of Lords held in *White v. Jones*[41] that a solicitor who negligently failed to draw up a will owed a duty, not only in contract (and tort) to the intended testator, but also in tort to the intended beneficiary. The testator, before his death, had given instructions to the solicitor who failed to prepare a new will. Lord Goff held as follows:

"Your Lordship's House should in cases such as these extend to the intended beneficiary a remedy under the *Hedley Byrne* principle by holding that the assumption of responsibility by the solicitor towards his client should be held in law to extend the intended beneficiary who, (as the solicitor can reasonably foresee) may, as a result of the solicitor's negligence, be deprived of his intended legacy in circumstances in which neither the testator nor his estate will have a remedy against the solicitor".

Statutory duty

This is an area of law closely related to tort in which an injured plaintiff seeks to rely on breach by the defendant of a duty under statute, rather than breach of duty imposed by common law. In fact, there are many statutory duties which represent codified common law duties, such as the duty of occupiers to take reasonable care (see Chapter 14). The discussion here is concerned with duties which arise only by reasons of some duty or power created by statute.

The modern trend is to state expressly in a statute whether it is intended to create any, and if so what, right of action in favour of individuals who may suffer loss. This applies in the case of the Building Act 1984 where section 38 provides expressly for breach of a duty imposed by building regulations to be actionable. However, this section has not yet been brought into force so that no direct right of action presently exists. The debate at present is whether there is a duty at common law in relation to the exercise of statutory powers. Health and safety legislation creates statutory duties and also provides criminal sanctions for breach. There is generally also a civil right of action in favour of individual workmen injured through breach of the relevant statutory duty (see Chapter 16).

---

[40] *Henderson v. Merrett Syndicates* [1995] 2 A.C. 145.
[41] [1995] 2 A.C. 207.

Building control legislation places upon local authorities powers to enforce compliance with building regulations. In *Murphy v. Brentwood D.C.*[42] the House of Lords left open the question of whether these statutory powers gave rise to a common law duty to owners or occupiers (see also Chapter 14). However, in a New Zealand case[43] it was held that the local authority was liable for the negligent exercise of such statutory powers of inspection. The decision was upheld by the Privy Council[44] who emphasised, however, that the common law could adapt itself to the different circumstances of the countries in which it had taken root. In New Zealand there was a significant expectation of reliance on the local authority. The case does not, therefore, resolve the question of whether local authorities in England will be held to owe the same duty.

The House of Lords further held, by a majority, that the powers available under the Highways Act 1980 (see Chapter 16) did not give rise to a duty on the highway authority. The authority could not, therefore, be held liable for a major accident arising from a danger of which it had knowledge.[45]

The law of trusts

A trust is a binding arrangement under which property is held by trustees for the benefit of specified beneficiaries. The law of trust is the creation of equity which has developed a substantial body of law, initially through decisions of the courts, much of which is now codified, notably in the Trustee Act 1925. The objective of the law is to impress a high degree of security on the trust property and corresponding duties on the trustees, such that the trust fund is insulated from the rights of the trustees and, for example, survives intact despite their bankruptcy. In order for a trust to be created there must be identified or appropriated trust property together with a declaration or other act creating or setting up the trust.

As in other areas of the law where the court will, for instance, imply the existence of a contract, the courts may declare the existence of an "implied trust" where one person receives money or property which is to be held for the benefit of another. Thus where a person is possessed of property of another, as an alternative to the court declaring the recipient under a duty to make

---

[42] [1991] 1 A.C. 398.
[43] *Invercargill City Council v. Hamlin* (1994) 3 N.Z.L.R. 513; 11 Const. L.J. 249.
[44] [1996] A.C. 624.
[45] *Stovin v. Wise* [1996] A.C. 923.

restitution of the property, it may declare the property subject to an implied trust and therefore insulated from adverse claims against the trustee in his personal capacity. One of the novel aspects of the law of trusts is that it permits wholesale departure from the common law rule of privity in that A may transfer property to B on trust for C who may himself declare a sub-trust in favour of D and so on. These arrangements are enforceable at the suit of the beneficiary.

The mere holding of property on trust is sometimes called a "bare" trust. Expressly created trusts will be subject to conditions laying down the powers of the trustees and the rights of the beneficiaries, for example as to the conditions under which they are eligible to receive the trust funds. In the construction industry express trusts have become a familiar device for seeking to secure the interest of a contractor or sub-contractor in the retention fund. This is referred to in clause 30.5 of the JCT form of contract which states that "the employer's interest in the retention is fiduciary as trustee for the contractor and for any nominated sub-contractor". The words impose an obligation on the employer to appropriate and set aside a sum equivalent to the retention money in a separate trust fund.[46] Once created such trust will survive employer's insolvency. However, in *MacJordan Construction v. Brookmount*[47] it was held that there was no trust until a sum of money was set aside in a separate account and until that was done the contractor was merely an unsecured creditor. While the contractor could obtain a mandatory injunction ordering the employer to set up the trust,[48] no rights in the fund were required until this had been done. In the case of a retention trust fund, the contractor's interest remains subject to the terms of the contract. The trust funds are therefore not payable until the retention becomes payable under the contract, and the employer retains all rights of set-off available under the contract. This topic formed part of the proposals of the Latham Report[49] but was not carried forward into the Housing Grants, etc., Act 1996.

The law of trusts can have a wider application in relation to construction work. In *Hussey v. Palmer*[50] a mother-in-law who lived with the family paid for building work to the house. Although there was no enforceable loan, it was held that the value of the work

---

[46] *Wates v. Franthom* (1991) 53 B.L.R. 23.
[47] (1991) 56 B.L.R. 1.
[48] *Rayack Construction v. Lampeter Meat Co.* (1979) 12 B.L.R. 30.
[49] *Constructing the Team* (1994).
[50] [1972] 1 W.L.R. 1286.

done was held on trust for the mother-in-law. In giving judgment, Lord Denning said:

> "By whatever name it is described, it is a trust imposed by law whenever justice and good conscience require it. It is a liberal process founded upon large principles of equity, to be applied in cases where the legal owner cannot conscientiously keep the property for himself alone, but ought to allow another to have the property or the benefit of it or a share in it. The trust arises at the outset when the property is acquired, or later on, as the circumstances may require. It is an equitable remedy by which the court can enable an aggrieved party to obtain restitution."

Another application of the principles of equity being applied in a contractual situation is the Australian case of *Walthons v. Maher*[51] in which the plaintiff builder was the owner of an old building. The defendant proposed to take a lease of a new building to be erected on the site and contracts were drawn up which the plaintiff thought had become binding. He demolished the old building and began constructing the new one. The defendant then pulled out of the arrangement. The High Court of Australia held that the defendant had so acted as to encourage in the plaintiff an assumption that a contract would come into existence and that it would be unconscionable to permit the defendant to depart from that assumption. Compensation was awarded for the detriment suffered by the plaintiff.

---

[51] (1988) 164 C.L.R. 387.

CHAPTER 6

CONTRACT: GENERAL PRINCIPLES

ENGLISH law of contract is contained principally in case law. It is only during the present century that statutes have begun to play any significant part. Historically the law of contract has been built up by the judges as a coherent whole so that there exists a body of principles which apply generally to all contracts including building and engineering contracts. In this chapter the general principles are discussed under the headings, (1) the formation of a contract, (2) contracts which though validly formed may not be binding, and (3) the discharge of contracts. In later chapters there are considered some particular types of contract including building and engineering contracts. These particular contracts, in addition to the general principles set out in this chapter, have their own characteristics, and some are governed by individual statutes.

The law of contract is based on the mutual exchange of obligations, in that each side must contribute something to the agreement to make it binding. The only exception to this principle is a contract made by deed. Such contracts were formerly referred to as "under seal", but seals are now abolished[1] and replaced by the simpler requirement for signature in the presence of a witness. It may be that references to contracts under seal will continue for some time in view of their long history. A contract by deed binds its maker without need of any exchange of obligations. Contracts other than those made by deed are called simple contracts, whether made orally or in writing. Although the practice is not universal, the term "agreement" can be used to denote a mutual understanding between the parties, and the term "contract" for an agreement which is binding in law. In such terms there can be an agreement without a contract, but every contract must embody an agreement, except a contract by deed.

---

[1] Law of Property (Miscellaneous Provisions) Act 1989.

Parties to a contract are in general free to make any terms they choose, but certain limits may be placed upon them by the common law and by statute. For example, terms may be implied into a contract which will mitigate the severity of an agreement; or one party may have relief against the other for a misrepresentation outside the terms of the contract itself. Apart from such limits, the function of the courts is to enforce contracts according to the terms agreed. However unjust the terms are, or however unjust they may become, the courts have no power to rewrite the terms of an agreement. Thus, if a contractor has contracted to carry out works at such prices that he is bound to make a loss, he must still carry out the works or pay damages for breach of contract.

FORMATION OF CONTRACT

If a simple contract is to be legally binding, there must be an offer from one party which is accepted by the other, and each party must contribute something to the bargain. The contribution is called consideration. If a contract exists the courts will determine what its terms are, for instance, when part of the agreement is in writing and part oral, or when there are implied terms. These points are considered in order.

Offer and acceptance

An offer must consist of a definite promise to be bound on specified or ascertainable terms, and it may be made to a particular person or class of persons, or even the public at large (see Chapter 5). The exhibition of goods for sale is not an offer but an invitation to make an offer. A shopkeeper may therefore accept or reject an offer from a customer to buy. He is not bound to sell the goods at the price shown. The same applies to an invitation to tender for the construction of building works. The invitation to tender, whether to the public or to an individual builder, is no more than an offer to negotiate. The contractor's tender constitutes an offer which the client may accept or not. Once accepted it forms a binding contract. This is so, despite any provisions as to subsequent execution of formal documents.[2] A proviso that the client is not bound to accept the lowest or any tender is generally unnecessary.

---

[2] See ICE, cl. 9.

The offer and the acceptance may be in writing or oral, or may even be inferred from the parties' conduct. A person who goes into a hairdresser's shop and sits down in the chair will be bound to pay for the ensuing haircut even though nothing is said. If a particular method for communicating acceptance is prescribed, it must normally be adopted. But an equally expeditious method may be sufficient. For example, if an offer requires acceptance by return of post, a fax is likely to be held sufficient. When the post is used the rule is that the acceptance is effective and the contract made at the moment of posting. Silence cannot normally constitute acceptance. But there is an exception when goods are taken on a sale or return basis. There will be an implied acceptance if they are not returned within a reasonable time.

An acceptance must be unqualified. A conditional acceptance or a counter offer may destroy the original offer so that it cannot be accepted later. The traditional form of acceptance "subject to contract" is not binding at all.

Essential terms

Frequently a contract will be concluded after a period of negotiation involving offers which are accepted in part, so that the applicable terms are gradually agreed. Particularly in the case of construction contracts, it is often found that the parties begin to perform the contract on the assumption there is, or will shortly be, a concluded agreement. Subsequently, the parties may contend that an important term has not been agreed and that there is, in consequence, no contract. The attitude of the courts is that a contract will be upheld if the parties have agreed upon the essential terms, such as the price, scope of works, commencement date and duration, etc. Minor omissions will not prevent a contract coming into existence. In a number of cases the TCC judges have upheld apparently binding contracts even though terms which might be considered important were not agreed. In *Drake & Scull v. Higgs & Hill*,[3] lengthy negotiations between the parties resulted in agreement of all matters except Drake & Scull's daywork rates. It was held that a contract came into existence despite this, on the basis that a term could be implied that the sub-contractor would be paid a reasonable sum. In *Mitsui Babcock Energy v. John Brown Engineering*,[4] negotiations led to the signing of a contract despite the failure to agree on a term covering performance tests and

---

[3] (1995) 11 Const. L.J. 214.
[4] (1996) 51 Con. L.R. 129.

liquidated damages. The clause was struck out and noted as "to be discussed and agreed". It was held that a contract nevertheless came into existence on the basis that the parties had "made a coherent and workable contract". It remains the case, however, that failure to agree on one or more essential terms may prevent an apparent agreement having legal effect.

Retrospective acceptance

Acceptance of a contract may have retrospective effect if this is the intention of the parties. In a case where a contractor was instructed to proceed and started work while the contract for the works was still under negotiation, it was held that the parties had intended such works to be governed by the contract as eventually made: *Trollope & Colls v. Atomic Power Construction.*[5] The judgment of Megaw J. included the following:

> "Frequently, in large transactions a written contract is expressed to have retrospective effect, sometimes lengthy retrospective effect; and this in cases where the negotiations on some of the terms have continued up to almost, if not quite, the date of the signature of the contract. The parties have meanwhile been conducting their transactions with one another, it may be for many months, on the assumption that a contract would ultimately be agreed on lines known to both the parties, though with the final form of various constituent terms of the proposed contract still under discussion. The parties have assumed that when the contract is made—when all the terms have been agreed in their final form—the contract will apply retrospectively to the preceding transactions . . . I can see no reason why, if the parties so intend and agree, such a stipulation should be denied legal effect."

In some types of building contract (and many other commercial transactions) the principles of implied or retrospective acceptance need often to be applied to identify the legal basis of a contract which neither party has ever doubted was binding. A problem which frequently occurs is where the parties enter into correspondence as to the precise terms on which they are to contract. This happens often between main contractor and sub-contractor. The contractor places an "order" on his standard terms and the sub-contractor "accepts" on his standard terms, which are inconsistent with the order. Correspondence follows in which some terms are agreed and others not. At some point the sub-contractor starts the work. The principles to be applied to such problems are that the

---

[5] [1963] 1 W.L.R. 333.

last letter is deemed to be accepted if the recipient then starts or continues the work (or permits the other party to do so). But if the parties show by their continuing negotiation that they do not regard themselves as bound, there may be no contract. Equally, if the parties are not agreed as to some important term there will be no contract. In the absence of a binding contract, a party who has carried out work at the request of the other will be entitled to payment of a reasonable sum (see Chapter 5).

Revocation

Revocation of an offer is effective only when it reaches the offeree. A promise to keep an offer open for a certain period does not prevent the offer from being revoked prematurely, unless the promise is itself a binding contract, such as an option to purchase shares. An agreement for the periodic supply of goods to order has the legal effect of a standing offer from the supplier which creates a binding contract each time goods are ordered. The offer may accordingly be revoked by the supplier except in respect of orders already placed. If there is no fixed period an unaccepted offer may lapse after a reasonable time. In some countries there exists a doctrine under which an offer may not be revoked once the offeree has relied on it, for example, through a main contractor tendering on the basis of sub-contract tenders. No such doctrine exists in English law, under which contracts are either binding or not and parties to would-be contracts seeking redress must bring their claims within other legal principles (see Chapter 5).

Standard terms of business

Most construction work is undertaken through tendering based on one party's standard conditions of contract. Questions which arise regarding formation of the contract are often concerned with whether other documents (such as programmes or qualifying letters) have been incorporated, in addition to the standard terms. In other transactions, typified by sub-contract orders, wider questions arise when standard terms of business are used by both parties. These may be incorporated by reference or printed on the back of order forms or "acceptance" forms. Such terms may compete with each other, for example, each purporting to exclude the other; and there may be particularly onerous provisions hidden within such standard clauses.

The attitude of the courts to such problems has a long history, covering many types of transaction involving tickets, receipts and the like which contain or refer to standard conditions. The courts have evolved principles requiring that particularly onerous clauses should be brought fairly to the attention of the party adversely affected. This area of law is now fundamentally affected by the Unfair Contract Terms Act (see below), which refers to the question whether the "customer" knew or ought reasonably to have known about the terms being relied on. Where the Act does not apply the common law principles will still be effective, permitting the court to refuse to enforce onerous conditions. In a case decided before the Unfair Contract Terms Act,[6] Lord Denning remarked that:

> "Some clauses which I have seen would need to be printed in red ink on the face of the document with a red hand pointing to it before the notice could be held to be sufficient."

The notice principles were applied in the case of *Interfoto v. Stiletto*,[7] in which the defendant hired 47 transparencies from a lending library, the transaction being subject to printed conditions which required return within 14 days or a fee of £5 a day plus VAT for each one retained. The defendant, who had not read the conditions, returned the transparencies four weeks later and was given a bill for £3783.50. Judgment was given for the plaintiff in the county court, but the Court of Appeal held that the plaintiff had failed to show that the relevant clause had been brought fairly to the defendant's attention, and therefore substituted a reasonable charge of £3.50 *per* transparency *per* week. Dillon L.J. in giving judgment said:

> "In the ticket cases the courts held that the common law required that reasonable steps be taken to draw the other party's attention to the printed conditions or they would not be part of the contract. It is, in my judgment, a logical development of the common law into modern conditions that it should be held . . . that if one condition in a set of printed conditions is particularly onerous or unusual, the party seeking to enforce it must show that that particular condition was fairly brought to the attention of the other party. In the present case, nothing whatever was done by the plaintiffs to draw the defendant's attention particularly to condition 2; it was merely one of four columns' width of conditions printed across the foot of the delivery note. Consequently condition 2 never, in my judgment, became part of the contract between the parties."

---

[6] *Spurling v. Bradshaw* [1956] 1 W.L.R. 461.
[7] [1989] Q.B. 433.

Battle of forms

Where each of the parties is trying to impose its terms on the other, the question of notice is unlikely to be relevant. This exchange of standard conditions is sometimes referred to as "the battle of the forms" and the principles which are applied here are simply those of offer and acceptance which, however, may be complicated by the conditions themselves.

The general principle was stated in *Butler Machine Tool Co. v. Ex-cell-o Corp.*,[8] where the plaintiff gave a quotation providing that orders were accepted only on terms of the quotation, which included a price variation clause. The defendant gave an order subject to their own terms and conditions, having no price variation clause, but having a tear-off acknowledgment for signature and return which accepted the order "on the terms and conditions thereon", The plaintiff signed and returned the acknowledgment but with a covering letter stating that delivery was to be "in accordance with our revised quotation", The Court of Appeal construed the acknowledgment as an acceptance which did not bring back the plaintiff's price variation clause. Lord Denning M.R. explained the law as follows:

> "It will be found that in most cases where there is a "battle of the forms" there is a contract as soon as the last of the forms is sent and received without objection being taken to it . . . the difficulty is to decide which form, or which part of which form, is a term or condition of the contract. In some cases the battle is won by the man who fires the last shot. He is the man who puts forward the latest terms and conditions: and if they are not objected to by the other party, he may be taken to have agreed to them . . . in some cases the battle is won by the man who gets the blow in first. If he offers to sell at a named price on the terms and conditions stated on the back: and the buyer orders the goods purporting to accept the offer—on an order form with his own different terms and conditions on the back— then if the difference is so material that it would affect the price, the buyer ought not to be allowed to take advantage of the difference unless he draws it specifically to the attention of the seller. There are yet other cases where the battle depends on the shots fired on both sides. There is a concluded contract but the forms vary. The terms and conditions of both parties are to be construed together."

A further matter of potential difficulty is provisions which are inconsistent with or repugnant to the remainder of the document. The leading case is *Glynn v. Margetson*,[9] which concerned the

---

[8] [1979] 1 W.L.R. 401.
[9] [1893] A.C. 351.

shipment of oranges from Malaga to Liverpool. The bill of lading provided for liberty to proceed to a variety of other ports for any purpose. The port of shipment was left blank and was filled up in writing. The ship deviated to another Spanish port with the result that, when the oranges were delivered to Liverpool, they were damaged. The case is authority on the question of a written clause prevailing over printed clauses. But the House of Lords also dealt with the question of clauses inconsistent with the main purpose of the contract. Lord Halsbury said:

> "It seems to me that in construing this document, which is a contract of carriage between the parties, one must in the first instance look at the whole of the instrument and not at one part of it only. Looking at the whole of the instrument, and seeing what one must regard . . . as its main purpose, one must reject words, indeed whole provisions, if they are inconsistent with what one assumes to be the main purpose of the contract. The main purpose of the contract was to take on board at one port and to deliver to another port a perishable cargo."

The repugnancy principle is not to be applied lightly, but does provide authority for enforcing the main purpose of a contract where other provisions are inconsistent.

Consideration

Each party to a contract (other than one made by deed) must provide consideration if the contract is to be binding. The most common forms of consideration are payment of money, provision of goods, or performance of work. But it may also consist in any benefit accruing to one party or detriment to the other. For example, A may promise to release B from a debt if he will dig A's garden. The courts are not concerned with whether the bargain was a good one. If B's debt was £1,000 and the garden small, that is still good consideration.

There are, however, certain acts and promises which cannot constitute good consideration. Anything which has already been done is no consideration. If B voluntarily dug A's garden yesterday, today's promise of reward is not binding because B gives no fresh consideration to the bargain. Further, if a party promises to do nothing more than he is already bound to do he provides no consideration. If B is A's gardener, A's promise of additional reward is not binding.

The courts have, however, on many occasions shown a highly liberal attitude towards what may constitute consideration. In *Williams v. Roffey & Nicholls (Contractors).*[10] the defendant, the

---

[10] [1991] 1 Q.B. 1.

main contractor, concerned that the plaintiff, a carpentry sub-contractor, might not be able to complete on time, orally promised additional payments if the work was completed on time. The Court of Appeal upheld the decision of the trial judge that this promise was enforceable, the consideration being the benefit, or the avoidance of detriment to the defendant. In any such case, the defendant may argue that the plaintiff has done nothing more that he was already bound to do. But if the sub-contractor agrees, for example, to accelerate (which he may have no obligation to do) this could amount to good consideration, and the court would not be concerned with the adequacy of the bargain.

Intention to be legally bound

Sometimes, despite the undoubted existence of offer, acceptance and consideration, one party may allege that the contract is not binding because there was no intention to create legal relations. This is not uncommon in family arrangements and it is presumed that domestic agreements are not intended to create legal relations. It is therefore up to the party seeking to enforce such a contract to rebut the presumption.[11] In commercial agreements there is naturally a strong presumption that there was an intention to create legal relations. Nevertheless, the intention may be rebutted. The parties may go further and make it an express condition that the contract is not to be binding in law. This is invariably a condition under which football pool companies accept entries, and the effect is to prevent an enforceable contract coming into existence. A similar result is achieved by a clause purporting to exclude the parties' rights to bring actions in the courts upon the contract. Such a clause will be treated as of no effect by the courts, but may make the contract void and unenforceable. It is to be noted, however, that the new Arbitration Act 1996 permits the parties to agree that their dispute is to be decided in accordance with "such other considerations as are agreed by them or determined by the tribunal",[12] as opposed to the principles of law. This refers to so-called "equity" clauses or other provision permitting arbtrators to apply principles of fairness of good conscience. The result of such an arbitration will be enforceable[13] but in court the parties are bound by the law as it stands, or by nothing.

---

[11] See *Hussey v. Palmer* noted in Chap. 5.
[12] s. 46(1).
[13] See Chap. 3 generally.

Form of contract

Simple contracts may, in general, be in any form and are enforceable despite a complete absence of documentation. But a few special types of contract are unenforceable unless evidenced in writing. These are, principally, contracts for the sale or disposition of land or an interest in land, and some others such as a contract of guarantee. Such a contract need not be made in writing, but some written evidence is necessary which must be signed by or on behalf of the defendant and which states the material terms. However, a contract which does not comply with these requirements may sometimes be enforceable in equity if there has been a part performance of the contract by the person seeking to enforce it, such as a buyer who has entered into possession of a house.

Terms of a contract

The final step in the formation of a contract is the identification of the terms and their effect. If the contract is wholly in writing, the problem is one of construction. But often there are additional terms. Statements made by the parties during their negotiations may have contractual effect. There may also be terms implied in the contract. An express term purporting to exclude or limit liability may raise special problems of interpretation. These points are discussed below.

A statement made during the negotiation of a contract may amount to a representation (see below) or it may become a term and have full contractual effect. There is no decisive test, but a statement is more likely to become a binding term if it is made immediately before agreement is reached, or if the maker of the statement had special knowledge, or if the contract itself was not reduced to writing.

Implied terms

In addition to the express terms, there may be other terms implied into a contract which, although not specified by the parties either in writing or orally, are nevertheless as binding as express terms. The leading case on the general implication of terms into contracts is *Liverpool C.C. v. Irwin*,[14] in which tenants in a multi-storey council block sought to establish against the local authority a

---

[14] [1977] A.C. 239.

duty to repair and maintain common parts, including lifts and staircases, which were frequently unusable because of vandalism and defects. The House of Lords, in holding the Council under a duty to take reasonable care, considered the contractual basis of the arrangement, which was based on a tenancy agreement silent as to the matters in issue. Lord Wilberforce dealt with the question of implication of terms as follows:

> "To say that the construction of a complete contract out of these elements involves a process of 'implication' may be correct; it would be so if implication means the supplying of what is not expressed. But there are varieties of implications which the courts think fit to make and they do not necessarily involve the same process. Where there is, on the face of it, a complete bilateral contract, the courts are sometimes willing to add terms to it, as implied terms: this is very common in mercantile contracts where there is an established usage: in that case the courts are spelling out what both parties know and would, if asked, unhesitatingly agree to be part of the bargain. In other cases, where there is an apparently complete bargain, the courts are willing to add a term on the ground that without it the contract will not work . . . There is a third variety of implication, that which I think Lord Denning M.R. favours, or at least did favour in this case, and that is the implication of reasonable terms. But though I agree with many of his instances, which in fact fall under one or other of the preceding heads, I cannot go so far as to endorse his principle; indeed, it seems to me, with respect, to extend a long and undesirable way beyond sound authority. The present case, in my opinion, represents a fourth category or I would rather say a fourth shade on a continuous spectrum. The court here is simply concerned to establish what the contract is, the parties not having themselves fully stated the terms. In this sense the court is searching for what must be implied."

An example of an attempt to imply a term in the third category, namely one that would be reasonable, occurred in the case of *Trollope & Colls v. North West Metropolitan Regional Hospital Board*,[15] where the parties had made a contract for construction work to be carried out in phases, but had omitted to make any provisions for the consequences of the first phase overrunning. It would doubtless have been reasonable to introduce a term which regulated the timing of subsequent phases, but the contract made no such provision and the judgment of the House of Lords illustrates the limitations on the power of the Court to do what is reasonable. Lord Pearson expressed himself thus:

> "The court does not make a contract for the parties. The court will not even improve the contract which the parties have made for

---

[15] [1973] 1 W.L.R. 601.

themselves, however desirable the improvement might be. The court's function is to interpret and apply the contract which the parties have made for themselves. If the express terms are perfectly clear and free from ambiguity, there is no choice to be made between different possible meanings: the clear terms must be applied even if the court thinks some other terms would have been more suitable. An unexpressed term can be implied if and only if the court finds that the parties must have intended that term to form part of their contract: it is not enough for the court to find that such a term would have been adopted by the parties as reasonable men if it had been suggested to them; it must have been a term that went without saying, a term necessary to give business efficacy to the contract, a term which, though tacit, formed part of the contract which the parties made for themselves."

An example of the application of established usage or trade custom occurred in the case of *William Lacey v. Davis*[16] (see also Chapter 5) where, in relation to the contractor's claims for costs of tendering work, the judge said:

"Mr. Daniel rightly conceded that if a builder is invited to tender for certain work, either in competition or otherwise, there is no implication that he would be paid for the work—sometimes the very considerable amount of work—involved in arriving at his price: he undertakes this work as a gamble, and its cost is part of the overhead expenses of his business which he hopes will be met out of the profits of contracts as are made as a result of tenders which prove to be successful. This generally accepted usage may also—and I think does also—apply to amendments of the original tender necessitated by bona fide alterations in the specification and plans."

There are certain types of contract into which terms are implied by statute, such as under the Sale of Goods Act (see Chapter 7). The principles of this Act have been extended to contracts for the supply of services, by the Supply of Goods and Services Act 1982 which applies, whether or not goods are also transferred, so that it will govern ordinary construction contracts. Sections 13, 14 and 15 of the Act provide that such contracts are subject to implied terms that the supplier will carry out a service with reasonable care and skill and that, in the absence of agreement, the service will be carried out within a reasonable time and for a reasonable charge. These terms may be negatived or varied by express agreement subject, however, to the effect of the Unfair Contract Terms Act (see below).

Implied terms in building contracts

Terms which have been implied in decided cases may act as precedents for similar contracts and therefore give rise to what may

---

[16] [1957] 1 W.L.R. 932.

be called a common law contractual duty, beyond the express terms of the contract. There are some important terms which are usually to be implied into building and engineering contracts. Such terms require that the building owner shall give possession of the site within a reasonable time, and give instructions and information at reasonable times.[17] Similarly the contractor must carry out his work with proper skill and care or, as sometimes expressed, in a workmanlike manner. Goods and materials must normally be of good quality and reasonably fit for their purpose.[18] However, there will be no implied term where the matter in question is dealt with by express terms. Some of the matters mentioned may be covered by express provisions in the building contract, so that there may be no case for further implication.

There are some notable terms which are normally not to be implied into building and engineering contracts. The employer gives no implied warranty of the nature or suitability of the site or subsoil, or as to the practicability of the design. Thus, where a contractor agreed to build a new bridge over the Thames using caissons, it was found they could not be used and the work proved much more expensive. There was held to be no implied warranty that the bridge could be built according to the engineer's design: *Thom v. London Corporation*.[19] In giving judgment the Lord Chancellor, Lord Cairns, observed:

"Can it be supposed for a moment, that the Defendants intended to imply any such warranty? My Lords, if the contractor in this case had gone to the Bridge Committee, then engaged in superintending the work, and had said: You want Blackfriars Bridge to be rebuilt; you have got specifications prepared by Mr. Cubitt; you ask me to tender for the contract; will you engage and warrant to me that the bridge can be built by caissons in this way which Mr. Cubitt thinks feasible, but which I have never seen before put in practice. What would the committee have answered? Can any person for a moment entertain any reasonable doubt as to the answer he would have received? He would have been told: You know Mr. Cubitt as well as we do; we, like you, rely on him—we must rely on him; we do not warrant Mr. Cubitt or his plans; you are as able to judge as we are whether his plans can be carried into effect or not; if you like to rely on them, well and good; if you do not, you can either have them tested by an engineer of your own, or you need not undertake the work; others will do it.

My Lords, it is really contrary to every kind of probability to suppose that any warranty could have been intended or implied between the parties; and if there is no express warranty, your

[17] *Merton L.B. v. Leach* (1985) 32 B.L.R. 51.
[18] *Young & Marten v. McManus Childs* [1969] 1 A.C. 454.
[19] (1876) 1 App. Cas. 120.

Lordships cannot imply a warranty, unless from the circumstances of the work some warranty must have been necessary, which clearly is not the case here, or, unless the probability is so strong that the parties intended a warranty, that you cannot resist the application of the doctrine of implied warranty."

The other side of the coin is that where a contractor builds in accordance with detailed instructions, he will give no implied warranty as to the fitness of the finished product. Thus, where a builder constructed, as specified, a solid brick wall without rendering which allowed rain to enter the house, it was held that the builder was not liable for the defect: *Lynch v. Thorne*.[20] In giving judgment, Lord Evershed held:

"If a skilled person promises to do a job, that is, to produce a particular thing, whether a house or a motor car or a piece of machinery, and he makes no provision, as a matter of bargain, as to the precise structure or article which he will create, then it may well be that the buyer of the structure or article relies on the judgment and skill of the other party to produce that which he says he will produce. That, however, is only another way of formulating the existence in such circumstances of an implied warranty. On the other hand, if two parties elect to make a bargain which specifies in precise detail what one of them will do, then, in the absence of some other express provision, it would appear to me to follow that the bargain is that which they have made; and so long as the party doing the work does that which he has contracted to do, that is the extent of his obligation."

Where the builder, in similar circumstances, is under an obligation to comply with the Building Regulations, it may be contended that there is an express obligation to ensure that the works are adequate. However, this is a matter of construction of each contract (see Chapter 16).

Exclusion clauses

It is common in standard form contracts for there to be a term excluding or limiting the liability of one party in the event of breach. This often applies in the case of supply of goods or services to exclude or limit liability for defective materials or work. Such provisions may be unenforceable by statute. However the courts have always been ready to find grounds on which exclusion clauses could be avoided at common law.

An exclusion clause must be carefully drafted since it will be construed against the party seeking to rely upon it (see Chapter 8).

---

[20] [1956] 1 W.L.R. 303.

General words are unlikely to exclude specific liability. If the term is contained in a document forming part of the contract then the party assenting to such a document is bound by its terms, whether or not he reads them and whether or not he signs the document. But a written exclusion clause may be over-ridden by an oral statement. Thus, where a customer took a dress to be dry cleaned, she was asked to sign a document excluding all liability; the assistant, however, said that the exclusion covered only damage to buttons. It was held that the cleaners could not rely on the clause to exclude liability: *Curtis v. Chemical Cleaning Co.*[21] Denning L.J. in the Court of Appeal held:

> "By failing to draw attention to the width of the exemption clause, the assistant created the false impression that the exemption related to the beads and sequins only, and that it did not extend to the material of which the dress was made. It was done perfectly innocently, but nevertheless, a false impression was created. . . . It was a sufficient misrepresentation to disentitle the cleaners from relying on the exemption, except in regard to the beads and sequins. In the present case the misrepresentation was as to the extent of the exemption. In other cases it may be as to its existence. For instance, if nothing was said by the assistant, this document might reasonably be understood to be, like a boot repairer's receipt, only a voucher for the customer to produce when collecting the goods, and not to contain conditions exempting the cleaners from their common law liability for negligence."

Standard terms are often designed to avoid this result by providing that no statement is to affect the conditions unless confirmed in writing. But such a term is likely to be effective only where the parties communicate primarily in writing; it would be unlikely to affect the decision in the *Curtis* case above.

Where an exclusion clause is exhibited in an hotel or a garage stating "The management accepts no responsibility . . ." the term is not binding unless the party must be taken to have known of, and agreed to it, before entering into the contract. Even when such a term is effective, it can normally protect only the parties to the contract, so that the negligence of servants, agents or subcontractors may not be excluded. Nevertheless, the terms of the contract may be relevant to duties in tort undertaken both by the immediate parties to the contract and by others (see Chapter 14).

While liability may be limited or excluded, it can generally be done only by contract or in situations akin to that of contract. The exhibition on a motor car of a sign saying "no liability for negligent

---

[21] [1951] 1 K.B. 805.

driving" would not prejudice the rights of the public at large. But where liability in tort would arise, for example, from free advice given negligently, such liability may be excluded or limited by an appropriate written or oral statement.[22] While clauses excluding all liability are construed most strictly by the courts, clauses which seek only to cap liability at a specified or determinable sum, (for example, the contract price) will be construed more liberally and are more likely to be given effect.

Statutory alteration of contract terms

The above represents the common law position on exclusion and limitation clauses. The fundamental principle of freedom and enforceability of contract terms has been the subject of statutory intervention. Provisions introduced in 1973[23] require that clauses excluding or restricting the implied obligations of the seller of goods as to their conformity with description or sample, quality or fitness, should be void in a consumer sale; and in a non-consumer sale, *i.e.* one between commercial parties, such terms are enforceable only in so far as it is fair and reasonable (see Chapter 7).

By the Unfair Contract Terms Act 1977 these provisions were extended in their application to other contracts involving the provision of goods (section 7), which include building contracts. The Act contains other far-reaching provisions governing exclusion clauses and notices. Liability for death or personal injury may not be excluded where resulting from negligence, which includes an obligation to exercise care or skill in contract or tort (section 2(1)). Liability for other loss resulting from negligence may be excluded only so far as it is fair and reasonable (section 2(2)). The effect of a contractual term excluding or restricting liability for breach of contract depends on the relative position of the parties. Where the innocent party deals as a consumer or on the other party's written standard terms of business, such a term is enforceable only so far as it is fair and reasonable (section 3). A private employer under a building contract may deal as a consumer, and therefore be entitled to the protection of section 3. It is not clear to what extent other employers under building contracts would be regarded as dealing on the contractor's standard terms where, for example, these are the JCT or ICE conditions.

Where an exclusion clause is required to be fair and reasonable, it is to be given effect even when the contract has been terminated

---

[22] *Hedley Byrne v. Heller* [1964] A.C. 465.
[23] Re-enacted in the Sale of Goods Act 1979.

by acceptance of repudiation (section 9). The application of the Act is restricted in regard to certain types of contract, including contracts of insurance, and contracts with a foreign element. The Act is not to apply where English law applies merely as a choice of proper law in a foreign contract. However the Act cannot be excluded by stipulating a foreign proper law in an otherwise English contract.

An example of the application of the test of fairness occurred in the case of *George Mitchell v. Finney Lock Seeds*,[24] where similar provisions in earlier legislation were considered in relation to a contract for sale of late cabbage seed. The seed delivered was of the wrong variety, and had to be ploughed in, after 60 acres had been planted and had germinated. The buyer claimed loss of profit. The contract limited the right of the buyer to replacement or refund of the price paid. It was shown in evidence that the sellers could have insured against the risk of crop failure at little additional cost, and that it was the practice of seed merchants to attempt to negotiate and settle claims by farmers. The buyers had no opportunity to negotiate the terms offered, which were common to all seed suppliers. In those circumstances, the Court of Appeal and the House of Lords held the clause limiting liability to be unfair and consequently unenforceable. Lord Bridge also observed:

> "The only other question of construction debated in the course of the argument was the meaning to be attached to the words 'to the extent that' in sub-section (4) and, in particular, whether they permit the court to hold that it would be fair and reasonable to allow partial reliance on a limitation clause and, for example, to decide in the instant case that the respondents should recover, say, half their consequential damage. I incline to the view that, in their context, the words are equivalent to 'in so far as' or 'in the circumstances in which' and do not permit the kind of judgment of Solomon illustrated by the example."

An application of the Unfair Contract Terms Act in the context of construction occurred in *Smith v. E. S. Bush*,[25] a case concerning the purported exclusion of liability by surveyors, acting for a building society, whose report was shown to and relied on by the house purchaser. Both the Court of Appeal and the House of Lords held that it would not be fair and reasonable to allow reliance on the disclaimer included in the report. Lord Griffiths in the House of Lords observed that while it was impossible to draw

---

[24] [1983] 2 A.C. 803.
[25] [1990] A.C. 831.

up an exhaustive list of factors to be taken into account, the following matters should always be considered:

(1) Where the parties are of equal bargaining power the requirement of reasonableness will be more easily discharged than in a case where the purchaser has no effective power to object.

(2) In the case of advice, it is relevant to consider whether it is reasonably practicable to obtain advice from an alternative source.

(3) Where the task undertaken is difficult or dangerous, with a high risk of failure, that is a pointer towards exclusion of liability being reasonable.

(4) The practical consequences of excluding liability should be considered, including the question whether either party can insure, and at what difficulty and cost.

CONTRACTS WHICH ARE NOT BINDING

A number of situations exist where, although a contract has been made, one or even both parties cannot enforce the agreement. Those most commonly encountered are when one or both parties make a mistake of fact, or when the contract is induced by misrepresentation. Other contracts which may not bind the parties include those which involve illegality, and contracts where one party is under some legal incapacity. These are discussed below.

Mistake

There are two distinct categories of mistake. First, where both parties make the same common mistake, the existence of an agreement is undisputed, but one party may say that the mistake has deprived the contract of its efficacy. Where the mistake relates to the existence of the subject matter the contract is void, as in the case of a sale of goods which have perished at the time of sale, or which never existed. But less fundamental mistakes may not be sufficient. Thus, the sale of a painting by Constable was held not to be void when the picture turned out to be by a lesser artist: *Leaf v. International Galleries*.[26] Denning L.J. in the Court of Appeal observed:

---

[26] [1950] 2 K.B. 86.

> "There was a mistake about the quality of the subject-matter because both parties believed the picture to be a Constable; and that mistake was in one essential or fundamental. But such a mistake does not avoid the contract: there was no mistake at all about the subject-matter of the sale. It was a specific picture 'Salisbury Cathedral.' The parties were agreed in the same terms on the same subject-matter and that is sufficient to make a contract."

The second category of mistake arises when the parties have different intentions. Whether they are both mistaken or whether the mistake is unilateral (one party merely acquiescing in the other's mistake) the law considers this as a question of offer and acceptance. The contract is void only if the mistake prevents one party from appreciating the fundamental character of what he is offering or accepting. A mistake which affects only motives, as when one party thinks he had a much better bargain than was the case, cannot affect the validity of the contract.

If the parties had different intentions the court will, if possible, ascertain the true meaning of the contract. If this cannot be done the contract is void. Thus, where parties contracted for the sale of cotton identified as being delivered by a ship named Peerless from Bombay, it transpired that there were two such ships out of Bombay; the buyer intended one and the seller the other. It was held there was no binding contract.[27] The difference from the "Constable" case above was that the parties there were agreed on one picture as the subject-matter of the contract. Where only one party was mistaken the question is whether there was an acceptance of what was offered. If there was such acceptance the contract is not void, but could still be voidable if induced by misrepresentation (see below).

In addition to common law remedies for mistake, other relief may be available in equity. Principally, if both parties, or even one party, intended something different from that which the documents record, rectification may be available (see Chapter 8).

Misrepresentation

A representation is a statement relating to a contract made by one party which does not become a term of the contract. If it is untrue, whether fraudulent or innocent, it is a misrepresentation and the general effect is to render the contract voidable. A voidable contract may be renounced by the injured party, but until renounced it is valid and binds both parties. Only when a misrepresentation has induced mistake could the contract be declared void.

---

[27] *Raffles v. Wichelhaus* (1864) 2 H. & C. 906.

In order to constitute a representation, the statement must be made before or at the time of contracting; and it must be a statement of fact, not opinion or mere "puff". An estate agent's description of the desirable qualities of a house is not to be taken as a statement of fact unless it gives specific information such as sizes. Silence may amount to a misrepresentation, as when a previous statement becomes false before the contract is concluded. Further, a misrepresentation does not make a contract voidable unless it induced the contract. Thus, the injured party must have relied on the statement and not on his own knowledge, and it must have been a material cause of his entering into the contract.

The rights and remedies which flow from a misrepresentation depend upon whether it was fraudulent or innocent. If the person who made the representation did not honestly believe it to be true then, whatever his motive, it is fraudulent. This gives the other party a right in tort to damages for deceit, and a further right to elect either to affirm the contract (when it will continue for both parties) or to rescind. Rescission involves cancellation of the contract and restoration of the parties to their situation before the contract was made.

A misrepresentation is innocent if the maker honestly, although carelessly, believed it to be true. The principal remedy for the other party is then under the Misrepresentation Act 1967 whereby damages can be recovered unless the maker proves that he had reasonable grounds for believing that the facts represented were true. The person to whom the innocent misrepresentation is made may also sue for rescission, as well as for damages. But the court (or an arbitrator) may, in its discretion, award damages in lieu of rescission.

The right to rescind a contract for misrepresentation, innocent or fraudulent, is available only if restoration of the parties to their former situation is possible, and if no innocent third party would suffer. A party who affirms the contract loses the right to rescind. An alternative remedy for misrepresentation may be available under the law of tort for negligent misstatement.[28] Further, a representation which has become a term of the contract will give rise to the usual remedies for breach of contract if it proves untrue (see below).

Illegal contracts

Contracts which contravene the law are in general void and no action may be brought upon them. The illegality may involve doing

---

[28] *Hedley Byrne v. Heller* [1964] A.C. 465.

an act prohibited by statute, such as building contrary to the Building Regulations; or it may consist in a project not as such prohibited, for example, an agreement to commit a crime or a tort or some immoral act. Such contracts cannot be enforced by either party.

The law draws a distinction between a contract which is illegal in its inception and one which is merely performed in an unlawful way. A contract illegal in its inception is totally void and no action can be brought by either party. Property transferred under the contract cannot generally be recovered. But when a contract is illegal only as performed, the effect depends upon the extent of the illegality. If this goes to the core of the contract then a guilty party will have no remedy, while an innocent party may have the usual contractual remedies. If, however, the illegality is not essential to the performance of the contract, both parties may have their normal remedies. Thus, the illegal overloading of a ship in the course of a voyage did not deprive the owners of their right to recover the freight charges, since the overloading was not an essential incident of the contract.[29] Devlin J. in giving judgment held:

> "There is . . . a distinction between the contract which has as its object the doing of the very act forbidden by the statute and the contract whose performance involves an illegality only incidentally. . . . There is no doubt that the plaintiffs cannot succeed if their claim for freight involves showing that they carried the goods in an overloaded ship. But in my judgment, the plaintiffs need show no more in order to recover their freight than that they delivered to the defendants the goods they received in the same good order and condition as that in which they received them."

Where building work is carried out in contravention of statutory provisions, such as the Building Regulations or Planning Acts, the above principles apply in determining whether the builder can recover the price of the work (apart from the question of statutory powers as to enforcement and penalties). Thus where on the face of the contract the work must contravene the law, the contractor cannot recover payment. In *Townsend v. Cinema News*[30] building work which was so designed as to comply with the law was carried out in contravention of a building bye law. It was held that the contractor could recover payment for the work since there was no fundamental illegality. But in such circumstances the building

---

[29] *St John Shipping v. Rank* [1957] 1 Q.B. 267.
[30] [1959] 1 W.L.R. 119, more fully reported at 20 B.L.R. 118.

owner will usually be entitled to set off a counterclaim for any work necessary to make the original work comply with statutory require- ments. Difficulties will arise, for instance, where the fault lies in the foundations, which can be cured only by demolition and rebuilding of parts which are properly built. Such problems depend upon whether the local authority seek to enforce compliance, whether the building owner has acquiesced in the breach and also on the express terms of the contract.

Incapacity of parties

Certain parties are restricted in their contractual capacity and liability. The most important of these are corporate bodies and infants. A corporate body (such as a limited company or a local authority) can make contracts only within its specific powers. A contract outside these powers is void (but see Chapter 4).

An infant occupies a privileged position under the law of contract. He is, in general, bound only by contracts which are substantially for his benefit. Thus, if an infant contracts to purchase goods, he is only liable to pay for them if they are necessary and suitable for his requirements, and he is liable to pay only a reasonable price. An acquisition of property such as land or shares is binding, but may be avoided by an infant before or within a reasonable time after reaching majority (*i.e.* 18 years). Despite the infant himself not being bound, he may still take advantage of the contract, by suing the supplier of defective (though non-necessary) goods. Contracts with infants should therefore be approached with caution.

Privity

The common law rule of privity is that a contract cannot be enforced by or against a person who is not a party to that contract. For example, a clause in a building contract enabling the employer to pay money direct to a sub-contractor may be used by the employer, but cannot be enforced by the sub-contractor, who is not a party to the main contract. There are, however, exceptions both general and specific. The law of agency is a general exception, for the principal may sue and be sued on contracts made by his agent (see Chapter 7). Specific contracts on which a stranger may sue include (by statute) a contract under seal respecting land or other property, and a third party insurance policy. Contracts which can be enforced against a stranger are practically limited to restrictive covenants over land, which on certain conditions are enforceable against a person who subsequently acquires the land (see Chapter 15).

An important aspect of the law of privity is whether, and in what circumstances, a third party can take advantage of an exclusion or limitation clause in a contract which binds the party seeking to make the claim. The leading case is *Scruttons v. Midland Silicones*,[31] in which the House of Lords held that a negligent stevedore was not entitled to rely on a limitation clause in the bill of lading. The House left open the possibility that the head contract containing the relevant clause could be expressed in such terms as to permit the third party, in effect, to set up a contract directly with the party suffering the loss. Such an argument was advanced, unsuccessfully, in relation to the taking over certificate in the Model Form A contract,[32] which states that for the purpose of the clause, the contractor contracts "on his own behalf and on behalf of and as trustee for his sub-contractors". In *Southern Water Authority v. Lewis & Duvivier (No. 2)*,[33] Judge Smout said:

> "I must be cautious before extending into a wider field those decisions insofar as they apply the principle of unilateral contract to the specialised practice of carriers and stevedores in mercantile law. To my mind, the principle of unilateral contract does not, taken by itself, fit easily on to the accepted facts in the instant case and it strikes me as uncomfortably artificial."

Judge Smout found that the conditions laid down in the *Midland Silicones* case were not satisfied, but nevertheless held that the taking over certificate could be relied on as a defence in tort (see Chapters 5 and 14).

Law Reform on Privity

For many years, judges and academics have expressed dissatisfaction with the English rule on privity, some suggesting that earlier authorities had been misunderstood, and others drawing attention to more liberal laws applicable in virtually every other country, including Scotland. Reform was called for so that the law on privity should give effect to the reasonable expectations of contracting parties. The issue was passed to the Law Commission who produced a consultation paper in 1991 and a report in 1996,[34]

---

[31] [1962] A.C. 446.
[32] cl. 30.
[33] (1984) 1 Const.L.J. 74.
[34] Law Commission Report No. 242, "Privity of Contract: Contracts for the Benefit of Third Parties".

in which recommendations were made and a draft Bill put forward. The report produced much comment and an alternative, simpler Bill was introduced into Parliament in December 1998. The structure of the Bill is as follows:

> Clause 1 sets out the circumstances in which a third party may have the right to enforce a term of the contract. This is where the contract confers or purports to confer a benefit on the third party. The third party must be identified by name, class or description. The right can be enforced only subject to other terms of the contract.
>
> Section 2 restricts the way in which the original contracting parties can alter the third party's entitlement, by cancelling or varying the contract without consent, but subject to any express terms under which the contract may be cancelled or varied.
>
> Section 3 deals with defences available to the party against which the third party seeks enforcement.
>
> Section 4 provides that the right of the original contracting party to enforce the contract is not affected.
>
> Section 5 deals with the situation in which the original contracting party has already recovered in respect of the third party's loss.
>
> Section 6 contains restrictions on the rights created by the Act.
>
> Section 7 provides that other rights of the third party are generally unaffected.

The effect of the Contracts (Rights of Third Parties) Bill, if passed into law, will be to allow parties, other than the immediate parties to a contract, to enforce particular rights where the contract expressly so provides. This aspect of the new law is relatively uncontroversial. Given that the third party will effectively stand in the shoes of the original contracting party and be subject to such defences as were available against him. The controversial element of the Bill is in the possibility of a contract being so construed as to create rights in favour of a third party, where this is not expressly set out. The possibility of disputes as to whether one party or another is entitled to enforce a particular right, perhaps in circumstances where the original contracting party has already enforced or sought to enforce that right, will inevitably give rise to complex disputes and unanticipated consequences. Typically, such a dispute will involve three parties (at least), being the original parties to the contract and the third party. There may be more than one potential third party claimant. In these circumstances, the status and effect of the arbitration clause in the original contract becomes of direct relevance, given that an arbitration clause, under English law, is regarded as personal and not readily transferable (see Chapter 3). The solution adopted under the original draft Bill was to provide that the third party would acquire no right under an arbitration agreement. This provision has, however, been omitted from the Parliamentary Bill, leaving the position unclear. It may be that the third party could acquire the right to enforce the arbitration clause in his own right, if the terms of the contract are apt to achieve such a result. Otherwise, the third party would be able to enforce his right only by action, but in this case the original contracting parties would remain bound by the arbitration clause. In relation to building contracts, where rights might typically be created in favour of third parties who are intended to, or who subsequently acquire an interest in the property, separate and parallel proceedings to enforce the same right are a serious probability.

Cases in which the courts have considered rights arising from assignments are dealt with in Chapter 5.

PROCUREMENT AND COMPETITION RULES

These rules exist primarily through European Community law and have the effect of imposing restrictions or sanctions on parties where they apply.

Procurement

The European Treaty requires free movement of goods and freedom to provide services for the purpose of achieving the Common Market. Specifically, Article 30 of the Treaty provides:

> "Quantitative restrictions on imports and all measures having equivalent effect shall, without prejudice to the following provisions, be prohibited between member states."

This article was the subject of direct enforcement in the case of the *Commission v. Ireland*[35] which involved the Dundalk Water

---

[35] Case No. 45/87, (1988) 44 B.L.R. 1.

Supply contract. The contract required pipes which complied with an Irish standard specification and required Irish certification. A tenderer who offered to use pipes of Spanish manufacture which did not have the Irish certification was rejected. The tenderer complained to the Commission who brought proceedings against the Irish Government on the basis that they were liable for the acts of the employer, the Dundalk Urban District Council. The European court rejected a series of arguments seeking to justify the actions of the UDC. The judgment of the court included the following:

> "The Irish government further maintains that protection of public health justifies the requirement of compliance with the Irish standard insofar as that standard guarantees there is no contract between the water and the asbestos fibres in the cement pipes, which would adversely affect the quality of the drinking water. That argument must be rejected. As the Commission has rightly pointed out, the coating of the pipes, both internally and externally was the subject of a separate requirement in the invitation to tender. The Irish government has not shown why compliance with that requirement would not be such as to ensure that there is no contact between the water and the asbestos fibres, which it considers to be essential for reasons of public health."

The court went on to declare that the government of Ireland had failed to fulfil its obligations under Article 30 of the Treaty.

Specific requirements as to procurement have been set out in a number of European Council Directives. Within the United Kingdom these have been implemented by a series of regulations[36] which apply to all works contracts exceeding five million ECU let by a public body or utility. These regulations make detailed provisions for tendering and award of contracts. Requirements as to advertising for tenders must be complied with including placing notices in the Official Journal of the European Community.

Breach of the procurement requirements may lead to enforcement measures in the relevant state court, including proceeding for interim measures. Under the United Kingdom regulations, once an offending contract has been concluded, the only enforcement measure is the award of damages.

Competition law

The objective is again to promote the free movement of goods and provision of services as required by the European Treaty.

---

[36] Public Works Contracts Regulations 1991; Public Supply Contracts Regulations 1991; Public Services Contracts Regulations 1993; Utilities Supply and Works Contracts Regulations 1992.

Agreements in restraint of trade are subject to the common law[37] and to statute law, principally the Fair Trading Act 1973, and the Competition Acts 1980, and 1998 under which the Competition Commission is given powers of investigation.

Articles 85 and 86 of the European Treaty prohibit agreements and practices which prevent or distort competition and further prohibit abuse of a dominant position having this effect. Breach of these principles may be enforced directly under English law by virtue of the Competition Act 1998; or by the European Commission which may itself investigate, issue directions and impose substantial fines. In recent years very heavy fines amounting to millions of pounds sterling have been imposed on cartel groups within the United Kingdom construction industry, notably those concerning road surfacing. Under English law, the task of investigating the breach of competition legislation falls on the Office of Fair Trading.

DISCHARGE OF CONTRACTS

Discharge is a general term for release of contractual obligations, when the parties become freed from obligation to do anything further under the contract. This must generally be brought about by some act of the parties. Contracts do not end automatically, unless perhaps by becoming statute barred (see below). Once a contract is discharged neither party can rely on its terms but can only enforce whatever rights may arise from the discharge. It is therefore important to know whether or not a contract is discharged (this may well be an issue between the parties). Determination of the contractor's employment under a provision in a building contract does not determine the contract itself and the parties remain bound by all its relevant terms.[38]

Discharge of a contract may be brought about in four ways:

(1) if the parties perform all their obligations the contract is said to be discharged by performance;

(2) if an event during the course of the contract renders performance impossible or sterile, it may be frustrated;

(3) serious breach by one party may lead to the contract being discharged;

---

[37] *Esso v. Harpers Garage* [1968] A.C. 269.
[38] See ICE, cl. 63; JCT, cll. 27 and 28.

(4) where one party commits a breach the other party may recover damages in satisfaction of the failure to perform.

These methods of discharge and also the wider topic of recovery of damages for breach of contract are discussed below. A fifth method of discharge is by express agreement; this is discussed later under variation of contracts.[39]

Performance

In general only exact and complete performance of contractual obligations can discharge the contract and a party who has only partially performed his obligations cannot recover payment. This rule is mitigated in a number of cases. Where a contract has been substantially performed, payment may be due with an allowance for deficiencies. Further, when a contract is divisible, either expressly or impliedly, payment will be due for parts which have been completed.[40]

A building or engineering contract will be discharged by performance on the part of the contractor when all the work has been completed, including obligations as to maintenance, and when the architect or engineer has issued all requisite certificates; and on the part of the employer when he has paid all sums due. If undisclosed (latent) defects are later discovered the contract has not been performed. The employer will retain the right to sue for breach during the period of limitation, subject to the effect of any final certificate[41] under the contract.

Frustration

As a general rule contractual obligations are regarded as absolute and a party is not absolved because performance becomes difficult or even impossible. A party who contracts to do the impossible is liable for failing to do it, unless he has excluded such liability.[42] If, however, without default of either party, the circumstances change so that performance of a contractual obligation becomes radically different from that undertaken, the contract may be frustrated and thereby automatically discharged.

Examples of situations which have constituted frustration are:

---

[39] See Chap. 8.
[40] As to performance of building contracts, see Chap. 9.
[41] See JCT Contract, cl. 30.9; also Chap. 12.
[42] See ICE, cl. 13.

(1) a building in which one party is to carry out work for the other is accidentally destroyed by fire;

(2) seats are sold to view a public event which does not take place;

(3) government action which prohibits performance of contract for a substantial period.

If a term of the contract provides for the contingency which has occurred, it is a question of construction whether it covers the particular circumstances, and thus keeps the contract in being. In *Metropolitan Water Board v. Dick Kerr*[43] the contractor had agreed in 1914 to construct a reservoir in six years, with a provision for extensions of time for various delays. In 1916, due to the war, the Ministry ordered the work to cease. It was held that the interruption was likely to be so long that the contract would be radically different, and the extension of time provision did not prevent frustration. Lord Dunedin, in the House of Lords, said:

> "The order pronounced under the Defence of the Realm Act not only debarred the respondents from proceeding with the contract, but also compulsorily dispersed and sold the plant. It is admitted that an interruption may be so long as to destroy the identity of the work or service, when resumed, with the work or service when interrupted. But quite apart from mere delay it seems to me that the action as to the plant prevents this contract ever being the same as it was."

Examples of building or engineering contracts being frustrated are extremely rare. Contracts are not frustrated by the work proving more difficult or costly than could have been anticipated, in any degree, unless the difficulty arises from some change of circumstance or supervening event. In the case of *Thorn v. London Corporation*,[44] the contract was not frustrated when the engineer's design for a new bridge over the Thames, with piers constructed inside caissons, proved impossible to construct. The contractor had taken the risk as to the method of construction and remained liable to carry out the work by whatever means were necessary, at no extra cost.

The legal effect of frustration is that the contract is discharged as to the future. Money paid before frustration is recoverable and money payable ceases to be payable. But the court may permit a party to retain or recover a sum to compensate him for any

---

[43] [1918] A.C. 119.
[44] (1876) 1 App. Cas. 120.

expense incurred, or for any benefit to the other party before the time of frustration.[45] A contractor whose contract becomes frustrated may therefore be unable to recover any payment if the employer gets no benefit from the work which has been completed. These rules are, however, subject to the provisions of the contract, and will usually be mitigated by insurances.

Breach of contract

Breach of contract occurs when a party fails to perform some primary obligation under the contract, for example when goods are not delivered on the date fixed, or when delivered work does not conform to contractual requirements. Under building and engineering contracts, defective work will not necessarily be a breach of contract when done if the contractor is not bound to execute particular work at a specified time. A breach would, however, occur if the contractor refused to obey a proper instruction to remove defective work or if work were not completed according to the contract by the date when it should have been completed.[46]

It is important, in dealing with claims under a building or engineering contract, to distinguish between a claim for breach of contract and a claim under the contract. A claim under the contract arises when an event occurs (which may or may not be a breach of contract) for which the contract provides a remedy. The remedy is usually payment of a sum of money to or by the contractor, or it may be some other benefit such as an extension of time. Often the same event will give rise to claims both under the contract and for breach. But the consequences of the two heads of claim are different. For example, consequential damages may be recovered for a breach; but under the contract only such remedies as are provided can be recovered. A claim under a contract is a way of enforcing its provisions. This section is concerned only with breach.

Breach of contract may have two principal consequences. First, every breach entitles the innocent party to sue for damages. Secondly, if the breach is sufficiently serious it gives the innocent party an option to treat the party in breach as having repudiated the whole contract. In this case the innocent party may bring the contract to an end by accepting the repudiation; or he may, at his option, treat the contract as subsisting, when it will then continue to bind both parties.

---

[45] Law Reform (Frustrated Contracts) Act 1943.

[46] *Kaye v. Hosier & Dickinson* [1972] 1 W.L.R. 146 and see also *Lintest v. Roberts* (1980) 13 B.L.R. 38.

Repudiation

A repudiation may consist in an express or an implied refusal by one party to perform the contract; or it may be a serious breach which goes to the root of the contract. The latter type of breach may be called a fundamental breach, or breach of a fundamental obligation. In each case the question whether a breach is to be taken as repudiation depends upon the importance of the breach in relation to the contract as a whole. Repudiation of a building contract may occur where a contractor carries out defective work and fails or refuses to comply with a proper instruction to rectify the work. Alternatively there may be a repudiation by failure to proceed with due expedition giving rise to substantial delay.

In both the examples given, the building contract will contain alternative express remedies, for example, to bring in another contractor to rectify defective work or to deduct liquidated damages. Contracts also variably contain powers to determine the employment of the contractor in cases of such default. In neither instance does the existence of these terms detract from the right of the employer to accept the contractor's repudiation as terminating the contract. The employer may give notice which may operate alternatively under the contract or as a common law acceptance of repudiation.

It is possible for the employer to commit a fundamental breach which may at the same time entitle the contractor to terminate his employment under the contract. The contractor may also in such a case serve notice which will operate on either basis in the alternative (see Chapter 9). Conduct by the employer which could amount to repudiation might be a failure to make several payments due, or a serious interference with the course of the works. Both in the case of employer and contractor, acceptance of repudiation brings the contract to an end, as to the future, while termination of employment does not.

Under the codified law applying to the sale of goods (see Chapter 7) certain obligations, such as correspondence with sample, are designated as "conditions", a breach of which is to be regarded as repudiation allowing the buyer to reject the goods and terminate the contract. Terms whose breach does not amount to repudiation are referred to as "warranties". This simple and convenient division of obligations cannot be applied to construction contracts and it would be necessary in each case to ascertain whether the particular breach could be regarded as fundamental.

It has been held that a fundamental breach could itself bring the contract to an end, on the basis that the other party had no choice but to accept the situation. This was in *Harbutts Plasticine v. Wayne Tank*,[47] where a defective pipe led to the destruction of factory premises by fire. The case has been criticised and should not be seen as affecting the primary rule that the innocent party is entitled to elect whether to terminate the contract or not. Where the innocent party does elect to terminate he must expressly or impliedly tell the other party. His choice is binding and will bring the contract to an end as from the election. If the innocent party chooses to treat the contract as subsisting, all the terms will continue to bind both parties.

Effect on exclusion clauses

When a contract is terminated by acceptance of repudiation, the question arises whether the party in breach may nevertheless rely on an exclusion or limitation clause. In a number of cases it was held that such clauses were destroyed with the contract. But it is now settled that this is not so. In *Photo Productions v. Securicor*[48] the defendant's employees, who were meant to guard the plaintiff's premises, entered and lit a fire which destroyed the factory. It was held by the House of Lords that Securicor were entitled to rely on the clear words of an exclusion clause which exempted them from responsibility for default of their employee unless due to want of care on their part. In giving judgment, Lord Diplock distinguished between the "primary obligation" of a contract, being the services to be provided, and the "secondary obligations" which arose upon breach of the primary obligations. Every failure to perform a primary obligation would be a breach of contract, and this would give rise to secondary obligations such as the payment of monetary compensation for loss sustained. The termination or rescission of a contract for repudiation brings to an end the primary obligations but leaves intact the secondary obligations. Lord Diplock went on to hold that exclusion clauses were to be applied according to their proper construction:

> "In commercial contracts negotiated between businessmen capable of looking after their own interests and of deciding how risks inherent in the performance of various kinds of contract can be most economically borne (generally by insurance), it is, in my view, wrong to place a strained construction upon words in an exclusion clause which are clear and fairly susceptible of one meaning only even after due

---

[47] [1970] 1 Q.B. 447.
[48] [1980] A.C. 827 and see *Suisse Atlantique v. N.V. Rotterdamsche Kolen* [1967] 1 A.C. 361.

allowance has been made for the presumption in favour of the implied primary and secondary obligations."

The continued existence of an exclusion clause is also confirmed by the Unfair Contract Terms Act 1977, which provides that such clauses are still to be given effect even when the contract has been terminated by acceptance of repudiation (see below).

REMEDIES FOR BREACH

If the innocent party properly elects to treat the contract as discharged, he is relieved from further liability. He may then claim damages, both as to loss flowing from the breach and loss flowing from the termination. The latter will usually include the additional cost of completing the contract. If the contract is not discharged, (whether or not he could have elected to terminate) the innocent party may claim damages. The right to recover damages is subject to a number of rules which restrict the monetary loss recoverable which are dealt with below. Alternatively, in very limited circumstances, the court may order specific performance, that is, it will compel the defendant to do what he has contracted to do (see below). The Arbitration Act 1996 confirms that an arbitrator may also award specific performance (section 48(5)).

Damages and remoteness

Not every loss which flows from a breach of contract is recoverable in damages. A claim can succeed only in respect of damage which is, in law, not too remote. Further, the innocent party must take all reasonable steps to mitigate his loss (see below). Damage is not too remote if at the time of the contract the parties ought reasonably to have contemplated that loss of that kind would be likely to occur. The common law has, further, developed two categories into which recoverable damage may fall, deriving from the leading case of *Hadley v. Baxendale*.[49] In this case, the plaintiffs, who were mill owners, contracted with the defendant to carry a broken crankshaft to the makers at Greenwich. There was delay in transit which resulted in the mill remaining idle so that the plaintiffs claimed loss of profit. Baron Alderson held that:

---

[49] (1854) 9 Ex. 341 and see *Suisse Atlantique v. N.V. Rotterdamsche* [1967] 1 A.C. 361.

"Where two parties have made a contract which one of them has broken, the damages which the other party ought to receive in respect of such breach of contract should be such as may fairly and reasonably be considered either arising naturally, *i.e.* according to the usual course of things, from such breach of contract itself or such as may reasonably be supposed to have been in the contemplation of both parties, at the time they made the contract, as the probable result of the breach of it."

This classic formulation led to the two grounds for recovery of damages becoming known as the first and second "limb" of *Hadley v. Baxendale*. Thus, the first limb covers such loss as will arise from the natural or usual course of events, each party being deemed to be aware of such matters. The second limb (separated by the word or in the quotation) covers such additional or special consequences as may arise from the actual events provided, however, that the parties were at the time of the contract aware of such consequences. This division of recoverable damages has been accepted throughout the common law world, including the United States of America, where *Hadley v. Baxendale* is still cited. The second limb is further illustrated by the more modern case of *Victoria Laundry v. Newman*,[50] where the plaintiff ordered a new boiler from the defendant for the purpose of taking on new work of an exceptionally profitable nature. The boiler was not delivered and the work was lost. It was held that the plaintiff could recover only the normal profit to be expected, since the defendant had no actual knowledge, at the time the contract was entered into, of the proposed new work.

Foresight

The defendant is liable only for the foreseeable consequences of his breach. This gave rise to particular difficulty in *South Australia v. York Montague*[51] in which the House of Lords considered three appeals relating to negligent valuation of properties where, not only had the defendant surveyors over-valued, but the plaintiff had suffered substantial additional losses as a result of the fall in the property market. In two of the cases where the over-valuation was marginal compared to the sum advanced by the plaintiff, the House of Lords held the surveyors not liable for the additional drop in market value. In the remaining case, the House held that the plaintiff had £10 million less security than they thought. If they had

---

[50] [1949] 2 K.B. 528.
[51] see *Banque Bruxelles v. Eagle Star Insurance* [1997] A.C. 191.

had this margin they would have suffered no loss. The whole loss (including the drop in market value) was therefore within the scope of the defendant's duty and was recoverable as damages (see further below).

The rules of remoteness have no application to a claim under a contract as opposed to a claim for breach. Thus if a dealer warrants that a motor car is in good condition and it breaks down, the purchaser may recover the cost of repairs and the cost of hiring an alternative vehicle, as damages for breach of contract. If, however, the dealer promises only to replace defective parts, the purchaser's entitlement under the contract is to the cost of repairs, and no question of consequential damage arises. In contracts of sale, an undertaking to replace defective parts is usually given in lieu of any warranty as to quality or fitness, so that the supplier is not in breach if the article is defective. In building contracts the position may be different. A so-called "defect liability" clause, which obliges the builder to put right defects, does not normally prevent the builder also being in breach so that consequential loss may be recoverable, even though the builder may carry out the necessary repair to the work (see Chapter 9).

Causation

The party claiming damages for breach of contract must also prove that the damage was caused by the breach. Such issues arise both under the law of contract and of tort, so that the authorities can apply in both areas. In *The Wilhelm*,[52] the master of a ship, at the onset of winter, delayed departure from port with effect that the ship became frozen in until the spring. The defendant was held liable for the whole of the delay since the possibility was apparent and could reasonably have been contemplated. Conversely, in *Associated Portland Cement v. Houlder*,[53] another shipping case, the defendant was in delay on a voyage during wartime to load the plaintiff's goods. The day after the due date, the ship was sunk en route by an enemy submarine. The plaintiff recovered damages in respect of the one day's delay; the event which caused the remainder of the damage was not reasonable to be contemplated. The question of causation in a construction case is illustrated by *Quinn v. Burch Bros*,[54] where the plaintiff, an independent plastering sub-contractor, was carrying out work for the defendant who

---

[52] (1866) 14 L.T. 638.
[53] (1917) 118 L.T. 94.
[54] [1966] 2 Q.B. 370.

was to supply necessary equipment. The defendant failed to supply a stepladder, as a result of which the plaintiff improvised a trestle, which collapsed causing him injury. Although the defendant was in breach of contract, Sellers L.J. in the Court of Appeal held:

> "This cannot be said to be an accident which was caused by the defendant's breach of contract. No doubt that circumstance was the occasion which brought about this conduct of the plaintiff, but it in no way caused it. It was in no way something flowing probably and naturally from the breach of contract."

Measure of damages

The measure of damages awarded is usually the actual monetary loss. The principle is that the innocent party should be restored to the position he would have been in had the other party performed his obligation. Thus where the plaintiff's factory was burnt down as a result of the defendant's breach and there was no reasonable alternative to rebuilding the factory, the plaintiff recovered the full cost of rebuilding: *Harbutts Plasticine v. Wayne Tank*.[55] Lord Denning held, in this case:

> "When this mill was destroyed the plasticine company had no choice. They were bound to replace it as soon as they could, not only to keep their business going, but also to mitigate the loss of profit. They replaced it in the only possible way, without adding any extras. I think they should be allowed the cost of replacement. True it is that they got new for old; but I do not think the wrong-doer can diminish the claim on that account. If they had added extra accommodation or made extra improvements, they would have to give credit. But that is not this case."

The normal measure of damage in respect of a breach affecting goods or property is the loss of value. However, where a contractor in breach of contract carries out defective work, the measure of damages is normally taken as the actual or estimated cost of reinstatement; and where the breach consists of not doing work the damages will normally be the extra cost of completing the work at the earliest reasonable time. Additional damages such as loss of rents and profits will depend upon the rules of remoteness set out above. Where the cost of remedial work is disproportionate or where other circumstances dictate that remedial cost is not the true measure of loss, the plaintiff may be compensated for the loss in value to the property. An interesting application of these principles

---

[55] [1970] 1 Q.B. 447.

arose in the case of *Ruxley Electronics v. Forsyth*[56] where the plaintiff contractor built a swimming pool for the defendant. The contract required a depth of 7ft 6in. in the diving area but on completion the pool was found to be only 6ft deep. There was no adverse effect on the value of the property nor on diving. The estimated cost of rebuilding to the specified depth was £21,560. Judge Anthony Diamond Q.C. awarded £2,500 for loss of amenity on the counterclaim, holding the cost of reinstatement to be an unreasonable claim. The Court of Appeal reversed the decision, awarding the owner the full cost of reinstatement. The House of Lords restored the original judgment on the basis the expenditure was out of all proportion to the benefit to be obtained. This was despite the fact that Mr Forsyth had stated his intention to rebuild the pool and offered undertakings as to use of the sum in question, if recovered. Lord Lloyd dealt with the point in the following terms:

> "In the present case the judge found as a fact that Mr Forsyth's stated intention of rebuilding the pool would not persist for long after the litigation had been concluded. In those circumstances it would be 'mere pretence' to say that the cost of rebuilding the pool is the loss which he has in fact suffered. This is the critical distinction between the present case and the example given by Staughton L.J., of a man who has had his watch stolen. In the latter case, the plaintiff is entitled to recover the value of the watch, because that is the true measure of his loss. He can do what he wants with the damages. But if, as the judge found, Mr Forsyth had no intention of rebuilding the pool, he has lost nothing except the difference in value, if any."

Inflation in building costs often results in argument as to the date at which repair costs should be assessed for the purpose of an award of damages. The position was clarified by the Court of Appeal in *Dodd Properties v. Canterbury C.C.*[57] The defendants were liable for damaging the plaintiff's garage premises in 1968. There was a delay in carrying out repairs until 1978, which resulted in a considerable increase in cost, for which the defendants disputed liability. The plaintiffs delayed doing the work because the cost would have resulted in financial stringency, and they were reluctant to lay out money before being sure of recovering it. It was held that the plaintiffs could recover the 1978 price of the work. The general position was restated as follows:

> "The general object underlying the rules for the assessment of damages is, so far as is possible by means of a monetary award, to

[56] [1996] A.C. 344.
[57] [1980] 1 W.L.R. 433.

place the plaintiff in the position which he would have occupied if he had not suffered the wrong complained of, be that wrong a tort or a breach of contract. In the case of a tort causing damage to real property, this object is achieved by the application of one or other of two quite different measures of damage, or, occasionally, a combination of the two. The first is to take the capital value of the property in an undamaged state and to compare it with its value in a damaged state. The second is to take the cost of repair or reinstatement. Which is appropriate will depend upon a number of factors, such as the plaintiff's future intentions as to the use of the property and the reasonableness of those intentions. If he reasonably intends to sell the property in its damaged state, clearly the diminution in capital value is the true measure of damage. If he reasonably intends to continue to occupy it and to repair the damage, clearly the cost of repairs is the true measure. And there may be in-between situations.

. . . a case in which the plaintiff has reinstated his property before the hearing, the costs prevailing at the date of that operation which were reasonably incurred by him are prima facie those which are relevant. Equally in a case in which a plaintiff has not effected reinstatement by the time of the hearing, there is a prima facie presumption that the costs then prevailing are those which should be adopted in ascertaining the cost of reinstatement. There may indeed be cases in which the court has to estimate costs at some future time as being the reasonable time at which to reinstate."

This case concerned a claim in tort, but the principles are equally applicable in contract.

Liability of surveyor

There is an apparent exception to the rule of reinstatement of the plaintiff's loss where a surveyor gives an erroneous report on the condition of the property. The purchaser relying on the report can recover, not the cost of repairing the undisclosed defects, but the difference in value between the property as reported and as it actually was: *Philips v. Ward*.[58] This case is not a true exception to the rule, however. The purchaser's position, had the surveyor performed his contract properly, would be that he knew the true value of the property, and should not pay more. The surveyor could be liable for the cost of rectifying defects only if he had given a warranty that the house was free of such defects.

The potential liability of a negligent valuer was considered in the *South Australia v. York Montague* (above). In that case, Lord Hoffmann dealt with an argument on behalf of the defendant that damages, following a negligent survey, should be limited to the

---

[58] [1956] 1 W.L.R. 471. See also *Watts v. Morrow* [1991] 1 W.L.R 1421.

excess over the highest valuation which would not have been negligent. The argument was dealt with as follows:

> "The valuer is not liable unless he is negligent. In deciding whether or not he has been negligent, the court must bear in mind that valuation is seldom and exact science and that within a band of figures valuers may differ without one of them being negligent. But once the valuer has been found to have been negligent, the loss for which he is responsible is that which has been caused by the valuation being wrong. For this purpose the court must form a view as to what a correct valuation would have been. This means the figure which it considers most likely that a reasonable valuer, using the information available at the relevant time, would have put forward as the amount which the property was most likely to fetch if sold upon the open market. While it is true that there would have been a range of figures which the reasonable valuer might have put forward, the figure most likely to have been put forward would have been the mean figure of that range. There is no basis for calculating damages upon the basis that it would have been a figure at one or other extreme of the range."

Mitigation of loss

A plaintiff seeking to recover damages is sometimes said to be under a "duty" to mitigate. This, however, belies the reality which is that the defendant has the burden of proving failure to mitigate. There are many cases in which the courts have rejected complaints of failure to mitigate, preferring indulgence to the innocent party rather than saving the money of the contract breaker or tortfeasor. An important aspect of this question arises where the innocent party has to decide upon the extent of remedial works to be carried out and the advisors of the defendant are contending that significantly less costly measures would be adequate. This was the case in *Great Ormond Street Hospital v. McLaughlin & Harvey*[59] where there was a major and public dispute between experts for the parties as to the extent of remedial piling work which was appropriate to the new cardiac wing. This debate took place before the work was carried out. The plaintiff acted on the advice of their expert. The advice and the expert were subjected to serious attack at the trial. Judge Newey rejected the challenge, holding that the natural consequences of the defendant's breach was that the plaintiff would take expert advice and would act upon it. He further held that the chain of causation might be broken if it were shown that the plaintiff's expert had acted negligently. But short of

---

[59] (1987) 19 Con. L.R. 25.

this, the fact that other experts held different opinions was not a matter which should result in the plaintiff recovering less than the sums actually expended.

Liquidated damages

In many commercial contracts, including construction contracts, provision is made for "liquidated" damages or sometimes "liquidated and ascertained" damages. Where the amount has been freely agreed between the parties, the sum in question will usually be enforced by the court, whether its effect is to obviate the need for proof or, conversely, to cap what would otherwise be a greater loss. In some circumstances, the liquidated damages may be declared unenforceable at the suit of the party being asked to pay (see Chapter 8) and it is theoretically possible they could be unenforceable as an exclusion of liability at the suit of the party suffering the loss.[60] Ordinarily, liquidated or prestated damages are a convenient device to avoid the need to prove loss. Other sums appearing in building contracts, for example rates and prices for elements of work do not constitute liquidated sums. The contractor may be entitled under the contract to be paid particular rates. But in a claim for damages, the actual loss must be established and the sums stated in the contract will constitute no more than evidence of the loss.

Joint liability

Where more than one person is potentially liable in respect of the plaintiff's damage, the plaintiff's right to recover the full loss from any one of the defendants is unaffected. However, as between the defendants who are each liable, the court has power to apportion liability. Formerly this power was available only between joint tortfeasors (see Chapter 14). However, it now applies whatever the legal basis of liability (including breach of contract) by virtue of the Civil Liability (Contribution) Act 1978. A defendant may bring another person who may be liable into an action as a third party; or separate proceedings may be brought against the other person later, to recover a contribution. The way in which the law of contribution applies in construction cases has been criticised in the DTI Report on Professional Liability.[61] The application of the rules is illustrated by the case of *Eckersley v. Binnie*.[62] The trial

---

[60] Unfair Contracts Act 1977.
[61] HMSO, 1989.
[62] (1988) 18 Con. L.R. 1; the facts are summarised in Chap. 9.

judge apportioned liability 55 per cent to Binnie, 30 per cent to the Water Authority and 15 per cent to the contractor. On appeal, the second and third defendants were found not liable at all, so that Binnie had to take 100 per cent of the responsibility. The same result would have obtained if some defendants had been unable to pay their share of the damages: the plaintiff would have been entitled to recover in full from any of them.

Where the fault lies partly with the plaintiff himself the damages recoverable may be reduced by the court if the claim is brought in tort.[63] For many years it was unclear whether contributory negligence applied also where the plaintiff sued only in contract. The law was finally clarified in *Vesta v. Butcher*,[64] where it was held that contributory negligence would apply where the defendant's liability in contract was the same as his liability in the tort of negligence. Where liability in contract was absolute, contributory negligence would not apply so that the plaintiff would be entitled to recover in full where negligence was proved against him. An intermediate category of obligations was also identified where it would still be open for the defendant to argue that there should be a deduction in respect of the plaintiff's negligence. In *Barclays Bank v. Fairclough*[65] the defendant was held responsible for causing asbestos pollution during building operations, but the trial judge held the plaintiff also to be negligent so that the damages recovered were reduced. The defendant was found to have breached both obligations of reasonable care (equivalent to liability in tort) and also absolute obligations. The Court of Appeal held that the plaintiff was entitled to recover in full for breach of the absolute obligations irrespective of the theoretical reduction in respect of lesser obligations.

Specific performance

Specific performance is an equitable remedy and is discretionary. It is not normally awarded if damages would be an adequate remedy, or if performance would require supervision by the court. Therefore, as a general rule, specific performance will not be ordered of a contract to build. The remedy may be available, for instance, when the plaintiff sells or leases land to the defendant with an obligation to build. Specific performance may be ordered if the following conditions are satisfied:

---

[63] Law Reform (Contributory Negligence) Act 1945.
[64] [1989] A.C. 852.
[65] [1995] Q.B. 214.

(1) the plaintiff has a substantial interest such that damages would not compensate him;

(2) the defendant is in possession of the land so that the plaintiff cannot do the work; and

(3) the work is adequately particularised.

Specific performance may be granted of a landlord's repairing covenant in a lease, when the tenant himself has no right to carry out the work, for instance, because it affects parts retained by the landlord.

LIMITATION PERIODS

A final condition which must be satisfied is that the claim must ordinarily be brought within the period of limitation. The effect of the limitation Acts is to bar the remedy and not to extinguish the right of action. Accordingly, limitation will be relevant only if raised by the defendant in his defence.

The present law of limitation is set out principally in the Limitation Act 1980, which provides that an action founded on simple contract must be brought within six years of the date on which the cause of action accrued (section 5) and a claim upon a contract by deed within 12 years (section 8). Most of the cases concern the date upon which the cause of action is to be taken as having accrued. In contract, the cause of action accrues on the date of the breach. The fact that damage occurs or is discovered at a later date does not give the plaintiff any further or other cause of action. In the case of a building contract, however, the date upon which the breach occurs may not be the same as the date upon which the acts complained of were carried out. Defective work may create one cause of action when done, but a further breach and a further cause of action may arise if the builder fails to comply with an instruction to rectify the defects. Yet a further breach may occur at the end of the maintenance period. Thus, it is necessary to read the contract and to consider the facts to ascertain the latest date upon which the plaintiff is entitled to commence proceedings.

There is an exception to the ordinary limitation periods under section 32 of the Limitation Act 1980 which provides for postponement of the limitation period where any fact relevant to the plaintiff's right of action has been deliberately concealed from him by the defendant. The period of limitation is not to begin until the plaintiff has discovered the concealment or could, with reasonable diligence, have discovered it. The section provides that deliberate commission of a breach in circumstances where it is unlikely to be discovered for some time amounts to deliberate concealment. The section codifies previous decisions of the court, which are illustrated by the case of *King v. Victor Parsons*.[66] In 1962 the plaintiff purchased from developers a plot of land on which the foundations and concrete oversite had already been laid with two courses of brickwork completed. The developers undertook to complete the house to the plaintiff's reasonable satisfaction. The plaintiff went into occupation in 1962. In 1968 large cracks developed and the plaintiff brought an action in breach of contract. The house had been built over a rubbish tip, and the developer had disregarded advice as to the type of foundation needed. It was held that the developer, or the builder as his agent, knew of these facts, and the failure to inform the plaintiff amounted to deliberate concealment, so that the plaintiff's cause of action arose in 1968 not 1962.

---

[66] [1973] 1 W.L.R. 29.

CHAPTER 7

SPECIAL CONTRACTS:

THIS chapter deals with special types of contract which are likely to be encountered in the construction industry, and which are governed by their own special rules in addition to the general principles set out in Chapter 6.

The topics covered are the sale of goods, which is perhaps the most universal form of legal transaction; the law of agency, which defines the position of architects and many other persons who act on behalf of another; and contracts of insurance, which are an incident to most building and engineering contracts. Finally, the section on sale of dwellings covers a combination of contractual and other relationships which may be encountered when a recently built house or flat is acquired.

In addition to the matters covered in this chapter, reference should be made to Chapter 4, which deals with the legal status of parties involved in the construction industry. These may also embody a special type of contract; for example, that between a company and its members or directors, and between the partners of a firm. Chapter 9 should be consulted for the special features of building and engineering contracts.

SALE OF GOODS

Contracts for the sale of goods are governed by the Sale of Goods Act 1979. This field covers a multitude of transactions ranging from retail purchases in shops to the sale of articles of great value or rarity, and may include contracts under which the goods are to be specially made. The Act does not apply, *inter alia*, to contracts which are in substance to carry out work or which relate to hire-purchase, or the sale of land or "things in action" such as shares or debts. The law applicable to transactions which involve "sale" of articles under a contract which also includes work or design, is considered at the end of this section.

The Act does not displace the ordinary principles of the law of contract except where they are inconsistent with its provisions. Thus, the formation of contract and the effects of misrepresentation and mistake are the same as for any other contract. The Sale of Goods Act is a codification of the common law, so that its principles can apply outside the limited sphere to which the Act strictly applies. The most important part of the Sale of Goods Act lays down a series of terms which, unless excluded or modified, are to be implied into all contracts to which the Act applies. These consist of four terms relating to description and quality and one relating to title. This legislation was originally contained in the Sale of Goods Act 1893, which was amended and is now consolidated in the Act of 1979, which has itself recently been amended by the Sale and Supply of Goods Act 1994.

Implied terms as to quality

Where there is a contract for sale of goods by description, (which is invariably the case when required for construction work) the goods must correspond with their description. And if the sale is by sample as well as by description, it is not sufficient if the goods correspond only with the sample; they must also correspond with the description (section 13). Even a small deviation from description, (providing it is not *de minimis*), will constitute a breach of the term. Where any variation can be permitted a specific tolerance should therefore be stated. Description may extend not only to the goods but also to their packing (see below).

Where there is a sale by sample, the bulk of the goods must correspond with the sample in quality, and the goods must also be free from defects which would not be apparent on reasonable examination (section 15). The use of a sample does not therefore protect the seller from latent defects in the goods. Sections 13 and 15 apply to all contracts of sale, but section 14 (below) applies only to sales which are made in the normal course of business by a dealer. Thus, in private sales, the only terms to be implied are those relating to correspondence with description and with sample. Dealers are additionally bound by two important terms which relate to fitness for purpose and to quality.

If a dealer is expressly or impliedly told of any particular purpose for which the goods are wanted, they must be reasonably fit for that purpose (section 14(3)). However, such an obligation is not to be implied where the circumstances show that the buyer does not rely, or that it is unreasonable for him to rely, on the seller's skill and judgment. Fitness for purpose may extend to the container as well as the goods. Where the condition applies, the dealer's liability is strict. It is no defence that all reasonable care was taken if the goods are still unfit.

In addition to being fit for their purpose, the goods purchased from a dealer must be of satisfactory quality (section 14(2)). The buyer is under no duty to examine the goods, but if he does so the dealer is not liable for defects which that examination ought to have revealed. Nor is he liable for defects drawn to the buyer's attention before the contract is made. Goods are of satisfactory quality if they meet the standard that a reasonable person would regard as satisfactory, having regard, *inter alia*, to their description and price (section 14(6)).

The type of defect which can render goods of unsatisfactory quality or unfit for their purpose will generally be a more substantial defect than is necessary for a breach of section 13 or 15. However, the fact that goods are substantially defective does not mean there must be a breach of both sections 14(3) and 14(2). In the *Hardwick Game Farm* case,[1] feeding stuff was supplied to the plaintiffs who bred pheasants. It contained a substance which killed the pheasants. When the supplier sought an indemnity from his supplier it was held that there was a breach of section 14(3) as the goods were unfit for their particular purpose. However, the evidence established that the contaminated feeding stuff was acceptable for the manufacture of cattle foods, and consequently it was held that there was no breach of section 14(2) of the Act.

Right to sell

In any sale, unless a different intention is shown, the seller must always have a right to sell the goods at the time of sale (section 12). The right to sell applies usually to the ability to pass title to the goods, but it may also cover other matters which affect the right to deal in the goods. In the case of *Niblett v. Confectioners' Materials*,[2] the contract was for the supply of tins of condensed milk. Some of the cases supplied were labelled "Nissly Brand," which was considered to be an infringement of the "Nestlé" trade mark, and as a result the purchasers were forced to resell the goods unlabelled. Scrutton L.J. held, in the Court of Appeal:

---

[1] *Hardwick Game Farm v. Suffolk Agricultural and Poultry Producers' Association* [1969] 2 A.C. 31.
[2] [1921] 3 K.B. 387.

"The respondents impliedly warranted that they had then a right to
sell (the goods). In fact they could have been restrained by injunction
from selling them, because they were infringing the right of third
persons. If a vendor can be stopped by proceed of law from selling,
he has not the right to sell. Therefore the purchasers . . . have made
out a cause of action for breach of section 12."

The more usual application of the section is where the seller
cannot pass title, for example because he holds the goods on hire
purchase. Where there is a breach of the section, the buyer can
recover the whole price even though he may have used the goods.

Formerly, all the implied conditions described above could, in
principle, be excluded from the contract of sale by an appropriately
worded clause and this was frequently done. The situation now is
that any clause excluding or restricting the operation of sections 13,
14 or 15 of the Act is void in the case of a consumer sale, and in
any other case, unenforceable so far as it is not fair or reasonable.
An exclusion or restriction of section 12 is void. These provisions
are now contained in the Unfair Contract Terms Act 1977, which
also applies the same principles to other types of contract (see
Chapter 6).

Other rights and remedies

Remedies of the buyer under the Sale of Goods Act for breach
of contract depend on whether the term "broken" is a condition or
as a warranty. Breach of a condition gives the buyer a right to
reject the goods and treat the contract as at an end, and to sue for
damages. Breach of a warranty gives only a right to damages, and
the buyer remains liable for the price. The five implied terms set
out above are stated in the Act to be conditions so that a breach of
any one of them, subject to any exclusion clause, gives the buyer a
right to reject the goods. Time of delivery is often made a
"condition" so that the buyer may refuse delivery and cancel the
contract if the goods are not delivered on time. The expression
"time to be of the essence" is often used to establish the right of
rejection. It needs to be emphasised that these principles are
exclusive to contracts of sale.

Contracts for work and materials

Contracts for the sale of goods are essentially simple transactions, compared to construction contracts. The subject matter of
a "sale" is usually ascertained at the date of the contract, and
compliance with the terms of the contract can be ascertained with

reasonable certainty. Thus, it has been possible for the law to prescribe simple direct sanctions, such as the right of rejection and cancellation of the contract. In contracts where the supplier also carries out work as part of the obligation, the situation is more complex. It may not be possible to discover whether the goods conform to the contract until the work has been carried out, so that the supplier has virtually performed his obligations. Where the contract concerns construction work on the purchaser's land, the goods become fixed and simple rejection is no longer possible. In these circumstances it is necessary to revert to common law principles which underlie the Sale of Goods Act. Since the Act codified the common law in regard to particular types of contract, it follows that contracts outside the ambit of the Act continue to be governed by the same principles.

One of the leading cases concerned a contract for making false teeth. In *Samuels v. Davis*[3] the plaintiff alleged that the goods were not reasonably fit for their purpose. It was argued for the dentist, the defendant, that the Sale of Goods Act did not apply, and the defendant's obligation was only to use materials of good quality and reasonable skill and care. The judge at first instance had acquitted the defendant of negligence. In the Court of Appeal, Scott L.J. said:

"In my view, it is a matter of legal indifference whether the contract was one for the sale of goods or one for services to do work and supply materials. In either case, the contract must necessarily, by reason of the relationship between the parties and the purpose for which the contract was entered into, import a term that, given reasonable co-operation by the patient, the dentist would achieve reasonable success in his work."

A similar point arose in relation to construction work in *Young & Marten v. McManus Childs*,[4] where roofing tiles were purchased by a sub-contractor and installed during the construction of a number of houses. The question then arose whether the main contractor was responsible for latent manufacturing defects in the tiles. In the House of Lords, Lord Wilberforce considered the question of the application of warranties in the main contract:

"Before the Sale of Goods Act 1893, the courts had to consider questions of implied warranty under the common law and they did so, both in relation to sales, and to analogous contracts, not strictly or at

---

[3] [1943] 1 K.B. 526.
[4] [1969] 1 A.C. 454.

least not purely sales, in precisely the same way. Their conclusions as to sales were taken into the Act, but the pre-existing principles remained and continued to be applied . . . since the Sale of Goods Act 1893, it has been fully accepted by the courts that suitable warranties, adapted to the nature of the contract, ought to be applied in contracts where there are mixed elements of supply of goods and work to be done."

The Supply of Goods and Services Act 1982 has codified this further area of common law, by analogy with the Sale of Goods Act. The Act creates implied terms, *inter alia*, of quality and fitness for purpose in a variety of contracts which involve the transfer of property in goods. The 1982 Act has not, it appears, brought about any substantive change in those areas which were previously covered by the common law. In regard to quality, therefore, the Acts of 1979 and 1982, and the underlying common law, may be seen as achieving the same result, overlapping in many cases. In addition, where materials are supplied as part of a contract to carry out work on a dwelling, the Defective Premises Act 1972 requires the supply of "proper materials" and that the work be carried out so that the dwelling will be fit for habitation when completed. The purchaser thus has a further alternative in the event of complaint.

Property in goods

It may be important to decide precisely when property in goods passes from seller to buyer if, for example, the goods are damaged or stolen, or one party becomes insolvent before physical delivery of the goods. No property can pass until the goods are ascertained, such as, by separating the number to be sold from a bulk. Once the goods are ascertained or specified the property passes when the parties intend it to pass. But if no intention is expressed or implied, the Sale of Goods Act defines when the property is to pass. In the simplest case where the sale is unconditional and the goods are in a deliverable state, property passes when the contract is made (sections 17 and 18). Unless otherwise agreed, risk passes with the property. Thus, if a contract is made to sell specific goods, delivery to be suspended until payment, the risk passes to the buyer on making the contract. If the goods are damaged or destroyed before delivery, the buyer remains liable for the price.

In building and engineering contracts these problems can occur under supply sub-contracts. The construction industry operates on credit. Suppliers have naturally tried to retain some security over their goods until they are paid for. In *Aluminium Industrie v. Romalpa*[5] the Court of Appeal upheld a clause which prevented the

---

[5] [1976] 1 W.L.R. 676.

passing of property in goods until paid for. The argument in this case was as to the effect of the clause, where the goods had been used in the defendant's manufacturing processes, and then sold on to third parties. The Court of Appeal held that there was no objection to the creation, as between seller and buyer, of a fiduciary relationship which entitled the unpaid seller to claim the proceeds of sale, despite the insolvency of the buyer. The effect of the clause was to place the buyer in the position of an agent. Roskill L.J. said:

> "If an agent lawfully sells his principal's goods, he stands in a fiduciary relationship to his principal and remains accountable to his principal for those goods and their proceeds. A bailee is in like position in relation to his bailor's goods. What, then, is there here to relieve the defendants from their obligation to account to the plaintiffs for those goods of the plaintiffs which they lawfully sell to sub-purchasers?"

Since this decision, many contracts of sale have incorporated so-called *Romalpa* clauses to seek to protect the unpaid seller. The ICE conditions stipulate the opposite effect, namely that all goods when on the site are deemed to be the property of the employer. This provision may run into difficulties when the property in the goods has never passed to the main contractor, since no-one can pass a better title to goods than he has himself. There is, however, a further overriding principle that anything attached to land (or to buildings) must belong to the owner of the land. Thus, once goods or materials are built into the works the unpaid seller will be reduced to the rank of an unsecured creditor, whatever his contract says.

AGENCY

Agency is a broad term describing the relationship between two parties whereby one, the agent, acts on behalf of the other, the principal. Common situations when this arises are when a person is appointed to buy or sell goods, or to conduct business on behalf of the principal. Examples of persons who act as agents are brokers, auctioneers, architects and engineers. An agency may be either special, that is limited to a particular transaction, or it may be general. An agent may represent his principal in many different ways. He may conduct legal proceedings, or even commit a tort on behalf of his principal. This section addresses the ways in which an agent may affect the contractual position of his principal.

When an agent acting on behalf of his principal makes a contract with a third party the usual result is that the agent drops out, leaving the contract enforceable only between the principal and the third party. Agency is thus a substantial exception to the rule of privity of contract. There may be two or three distinct contracts involved in an agency transaction. First, there is the relationship between principal and agent creating the agency. Secondly, there is the contract with the third party which the agent makes on behalf of the principal. There may also be an implied promise by the agent to the third party that he has authority to contract for the principal. This may be called a warranty of authority.

Formation of agency

An agency may arise under a contract, whereby the agent is appointed by his principal to carry out certain duties. Engineers and architects are frequently so appointed to act for promoters of building schemes. The appointment may be to act expressly on behalf of the employer, as when the engineer is appointed to administer the performance of a contract. An agency may also arise where the engineer is appointed to carry out design work. This may involve acting on behalf of the client, for example, dealing with the planning authority, the public health inspector and with other public bodies. An agency may also arise without an express appointment but by virtue of the relationship between the parties in question. A director of a company or a partner of a firm may hold such an agency.

An agency may also arise without actual authority. If the principal acts so as to clothe a person with ostensible authority, the principal will be bound by acts within such authority. An estate agent who is expressly instructed to find a purchaser may have ostensible authority to accept a deposit. If this is so and the agent defaults, the principal will be liable for repayment. An agency may also arise when it becomes an urgent necessity to perform some action on behalf of another person whose instructions cannot be obtained. This may apply where a person is in charge of goods which are in danger of perishing. In such a situation the person in charge may lawfully sell the goods on behalf of the owner. If a person without authority purports to act as agent for an identified principal, that principal may within a reasonable time ratify the agent's act and become bound by it. Ratification of a contract relates back to the date when the agent purported to make it. But the principal must be competent to make the contract both at the date of the agent's act and at the date of ratification.

However the agency is formed, there will be some express authority to act. In addition, the agent will have an implied authority to do such things as are reasonably incidental to his express powers. This authority will also bind the principal. Where there is a limitation upon the authority of the agent it is a matter for the principal to give notice of this as is commonly the case under construction contracts where the consent of the employer is needed, for example, to agree to a variation.

Rights of parties

The position of the third party, that is, the person with whom the agent makes the contract, depends crucially upon whether the third party knows that he is dealing with an agent. If an agent with authority discloses his agency, the third party can in general sue and be sued by only the principal. But if the agent does not disclose his agency, either the agent or the undisclosed principal can sue on the contract. Similarly, the third party, after discovering the agency, may choose whether to sue the principal or the agent on the contract. But his choice is binding, and if a judgment obtained against one is unsatisfied, he cannot afterwards sue the other. These rules will apply to a person dealing with an individual builder who is in fact acting on behalf of a company: either the individual or the company may sue for payment, and the client has the option of suing either of them if he is dissatisfied with the work, but he cannot proceed against both.

Where an agent acts without authority, or in excess of his actual or implied authority, the principal will be bound only if he has created ostensible authority or if he ratifies the contract. Otherwise the agent himself will be liable to the third party, not on the contract which he has purported to make (which is of no effect) but for breach of warranty of authority. The agent is so liable whether he has acted fraudulently or innocently.

An agent properly appointed is entitled to an indemnity against liabilities properly incurred. He has a lien over the principal's goods in his possession for payment of sums due. If the contract of agency makes no express provision for payment, an agent is entitled to a reasonable fee. It is a matter of construction of the contract of agency when payment becomes due. In the case of an estate agent his commission is usually payable out of the purchase money, so that nothing will be due until the property is sold. If, however, the terms required the estate agent only to produce a ready and willing purchaser, the fee would be payable whether or not the sale took place.

An agent must act honestly and obediently, and exercise reasonable skill and care. The agent must generally himself carry out the duties entrusted to him. However, there are circumstances in which an agent may delegate. There may be an express or implied agreement to permit delegation; or it may be necessary for the proper performance of the work. In the construction industry, an architect has no power to delegate his duty without express authority. If an architect agrees to design a building but is unable to perform the structural design work, there are two courses open to him. He may request the client to employ a specialist; or he may, while remaining liable to the client, himself seek advice and assistance. In either event the client will have a remedy for negligent design work: *Moresk Cleaners v. Hicks*.[6]

An agent must not take a secret profit from his work, nor must his own interest conflict with his duty to the principal. In *Salford Corp v. Lever*[7] an agent who arranged coal supplies received secret commission from suppliers. The court held that the employer was entitled to recover damages jointly and severally from the agent and from the suppliers, and in addition to recover from the agent the secret commission. Lord Esher said:

> "Hunter [the agent] had received money from the defendant for the performance of a duty which he was bound to perform without any such payment. Nothing could in law be more fraudulent, dangerous or disgraceful and therefore the law has struck at such conduct in this way. It says that, if an agent takes a bribe from a third person, whether he calls it a commission or by any other name, for the performance of a duty which he is bound to perform for his principal, he must give up to his principal whatever he has by reason of the fraud received beyond his due. It is a separate distinct fraud of the agent."

Termination

A contract of agency may be brought to an end by the parties themselves, or by operation of law. Architects and engineers are normally employed, expressly or impliedly, until the completion of the works, although such an appointment may be limited to separate stages of the work and the terms of engagement usually provide for termination. An agency may at any time be terminated by agreement. An agency contract between individuals will be terminated automatically by the death of either party or by the

---

[6] [1966] 2 Lloyd's Rep. 338.
[7] [1891] 1 Q.B. 168.

bankruptcy of the principal. It may also be terminated by frustration, such as by destruction of the subject matter, or by the contract becoming illegal.

Architects and engineers as agents

An agency between the promoter of a building scheme and his architect or engineer will arise as soon as there is an appointment to carry out design or investigation work. The extent of the agency will initially be limited, but will be enlarged by subsequent instructions, such as to obtain planning consent and then to obtain tenders for the work. The agency will not normally embrace entering into contracts on behalf of the promoter. Building contracts are almost invariably made directly between the contractor and the promoter. The role of the architect or engineer is usually to represent the interests of the promoter during the course of the works, in addition to his duties as independent certifier under the building contract (see Chapter 9).

Architects and engineers are frequently employed under standard conditions of engagement, such as those of the RIBA or the Association of Consulting Engineers (see Chapter 12). Under such conditions, the work is usually divided into stages, and the authority of the client is necessary before each new stage is commenced. The conditions usually deal expressly with particular duties required to be carried out, such as supervision of the works. However, such appointments rarely deal fully with the authority of the agent, and it is necessary to consider what implied or ostensible authority will exist, where not expressly given.

An express duty to certify payments to the contractor will usually carry with it an implied authority to supervise the works. There is no implied authority to vary the terms of the contract nor to warrant the accuracy of information in the contract documents. There is further no implied authority to order variations or extra works. However, standard building contracts invariably give express powers to the architect or engineer. He will have ostensible authority to exercise such powers under the building contract, unless the contractor has been expressly notified of any limitation of authority. Thus, the employer or promoter will not be able to deny the architect's or engineer's authority when sued by the contractor. Where the architect or engineer is an employee of the promoter his ostensible (or actual) authority is likely to be more extensive and may cover negotiation of the contract terms with the builder. This applies particularly to local government officers.[8] This

---

[8] See *Carlton Contractors v. Bexley Corp.* (1962) 60 L.G.R. 331.

may allow the contractor to sue the employer directly on an oral variation order, if the contract requires an order in writing.

When an architect or engineer negotiates with a nominated sub-contractor he is no longer the agent of the employer, who is not a party to the sub-contract. He must therefore exercise caution when conducting such negotiations, especially before the appointment of the main contractor, since he may become personally liable for breach of warranty of authority.

The remuneration due to an architect or engineer under standard conditions of engagement is usually based upon a percentage of the cost of the works but subject to other express agreement. One important function of the conditions of engagement is to define with precision the cost upon which the percentage fee is to be calculated. This method of remuneration is somewhat artificial, and often results in payment being made (apparently) for items not carried out by the person receiving payment. An alternative method is to calculate payment on a time basis; this is often used for partial services. Where there is no express agreement as to the means of payment, the engineer or architect will be entitled to a reasonable fee, which may be calculated either by reference to the standard conditions, or to the time spent.

A question which frequently gives rise to disputes is the ownership of work produced by architects and engineers. The formal documents which are prepared for the purposes of a building project, such as the drawings or specification, become the property of the client; but if there is a dispute about payment of fees, the architect or engineer has a lien over such documents in his possession, against the payment of money due. Documents such as working papers, calculations and correspondence, will not become the property of the employer.

Ownership of the designs produced by an engineer or architect is known as copyright. This remains vested in the designer, and may be transferred or sold like other property. The person who employs an architect or engineer has an implied right to make use of the designs produced in constructing the project. But this does not extend to repeating the design. The case of *Meikle v. Maufe*[9] concerned the design of premises in Tottenham Court Road for Heals. The original building was designed and built in 1912. Using another architect Heals, in 1935, embarked upon extensions based substantially on the original design. It was held that there was no implied right to reproduce the original design in an extension, and the copyright remained vested in the original architect.

---

[9] [1941] 3 All E.R. 144.

INSURANCE

The nature of a contract of insurance is that the insurer undertakes to make payments to or for the benefit of the assured on the happening of some event. The contract may generally be in any form, even oral; but it is usually contained in a document called a policy. The consideration provided by the assured is called the premium.

Insurance and assurance

There are two different types of insurance. Indemnity insurance involves the insurer agreeing to compensate for losses which the assured may suffer in certain events. Non-indemnity insurance provides for the payment of a specified sum on the happening of some event, such as the death of the assured. This is also referred to as assurance. In many ways the two types are governed by the same rules. But there is an essential difference. On an indemnity insurance the insurer pays out only the actual financial loss. The essence of a non-indemnity policy is that the fixed sum should be paid when the event occurs. Common examples of indemnity policies are fire, motor and third party liability insurance. Non-indemnity policies include life and personal accident insurances. Life Assurance policies are often used as a vehicle for investment, with provision for accumulation and repayment of premiums. The only additional benefit likely under an indemnity policy is a "no claims" bonus.

Insurable interest

An essential feature of practically every insurance contract is that the assured must have an insurable interest. This usually means a foreseeable financial loss or liability resulting from the event insured against. But there is no complete definition. A person has an insurable interest in his own life even though his loss will hardly be a financial one.

A person need not own the thing he seeks to insure in order to have an insurable interest. For example a carrier or custodian of goods has a sufficient interest to insure them in his own name. Works under construction may be insured against loss either by the contractor or by the employer, or by both. The parties to an action or arbitration have an insurable interest in the life of the judge or arbitrator, since his death may result in loss of the costs incurred.

But there are cases in which no insurable interest exists. Thus, in *Macaura v. Northern Assurance Co.*[10] the plaintiff, who owned the shares in a timber company, insured the timber in his own name. When the timber was destroyed by fire it was held that he could not recover under the policy. Lord Sumner said of the appellant:

> "He owned almost all the shares in the company, and the company owed him a good deal of money, but neither as creditor nor as shareholder could he insure the company's assets. The debt was not exposed to fire nor were the shares, and the fact that he was virtually the company's only creditor, while the timber was its only asset, seems to me to make no difference. He stood in no legal or equitable relation to the timber at all. He had no concern in the subject insured. His relation was to the company, not to its goods."

Insurance is often effected through an agent or broker. Generally, such a person has no authority to make a binding contract on behalf of the insurer. His duties are limited to issuing and receiving proposals, although a broker may be authorised to issue temporary cover as a separate contract. The broker is, in law, the agent of the assured.

Extent of cover

Because the obligation to pay is dependent upon the happening of an event, it is important for any policy to define the time limits between which it is to apply. If the event occurs outside the time limit, for example, if the proposed assured dies before the policy is effected, there can be no liability.

Cover will usually run from the date of the original policy, or from such later date or time as may be specified in the policy. An accident or indemnity policy may be arranged to run for a specified period. Thus, a specific policy covering building works may be expressed to extend to a date which is the anticipated completion date plus a reasonable margin. At the other extreme, a life assurance policy will usually extend during the life of the assured, although such policies may be taken out for a specific and limited period. Most indemnity policies including professional indemnity, are periodic, usually annual. Although such policies have the appearance of being perpetual, the need for annual renewal means that the insurer annually has the opportunity of increasing the premium or of altering the terms of the policy, or of declining to accept a further renewal.

---

[10] [1925] A.C. 619.

In professional indemnity insurance, it is often important to know in which year a particular claim must be accounted for. In all such policies, the insured is covered not against negligent acts as such, but against claims being made arising from such acts. Consequently, the claim will apply normally during the year in which it is notified to the insurer, irrespective of the fact that the negligent act complained of may have occurred several years earlier. This arrangement is of benefit to the insured person, and to the party making the claim, because they have the advantage of any increased level of cover available under the later policy. There will be a considerable disadvantage, however, to any person who was insured at the time of the event complained of but who ceased to be insured under the policy before the claim was made, for example because he retired from the practice. Such a person will be uninsured against subsequent claims, and must therefore make appropriate arrangements for protection with the remaining partners. A further consequence of annual renewal is that all claims within a particular year must be satisfied out of the indemnity available for that year. Thus, if a professional is insured for £1 million, and three claims arise in the same year for £0.5 million, the liability of the insurer is limited to £1 million, even though there may be no claims at all in the preceeding and succeeding years.

Duty of disclosure

Any insurance contract is said to be *uberrimae fidei* that is, based upon utmost good faith. Thus, the assured must make full disclosure of every material fact known to him. A fact is material if it would influence the judgment of a prudent insurer. The duty of disclosure continues after filling in the proposal form, up to the making of the contract. Non-disclosure of a material fact makes a policy voidable by the insurer. Thus, in *Roselodge v. Castle*[11] diamond merchants had insured their stock without disclosing that their sales manager had a previous conviction for diamond smuggling. It was held that this was a material fact and the policy was therefore voidable. Under a policy which is annually renewable, the duty of disclosure arises upon every renewal. Thus, where a consulting engineer becomes aware of a defect in a structure which could ultimately result in a claim being made, he will be under a duty of disclosure, and failure to bring this to the attention of the insurer may render the next annual policy voidable.

---

[11] [1966] 2 Lloyd's Rep. 113. For a full review of the law on disclosure see *Pan Atlantic v. Pine Top* [1995] 1 A.C. 501.

The policy

Policies such as life insurance tend to be neatly printed on thick paper, relatively easy to follow, and of little interest other than financial. Conversely, indemnity policies, particularly annually renewable ones, tend to exist on many different pieces of paper, sometimes physically attached to a standard policy document and sometimes not. There are often separate endorsements, exclusions and memoranda all of which have to be identified and construed together.

Despite this, there are certain provisions common to most forms of insurance. There must be a definition of the events upon which the insurer agrees to pay, and this may be accompanied by certain exclusions of liability. In an indemnity policy, such as a house insurance policy, the right to payment will be defined by specifying the property or item insured (such as the house and contents) and the risks insured against (such as fire, flood and subsidence). There may be a term requiring notice of an event which may lead to a claim. Some policies, especially motor insurances, contain an "excess clause" requiring the assured to bear the first £x of any claim. Such a term does not, however, prevent the assured himself suing the third party to recover the excess. If goods or property are insured for a sum less than their full value, an "average clause" may be inserted to reduce the sum payable by the proportion of their under-insurance. Where there is more than one indemnity policy covering the same risk there may be provisions which prevent full recovery, or even any recovery, on one or other of the policies.

Exclusions

Exclusion clauses are found in most policies, limiting the risk or circumstances in which the insurer becomes liable. For example, the insurance required in respect of the works under the ICE conditions refer to exclusions known as the "excepted risks".[12] These include "any fault defect error or omission in the design of the works (other than a design provided by the contractor pursuant to his obligations under the contract)". The contractor is not responsible for damage due to such cause, and there will be a corresponding exclusion in the insurance policy taken out to cover his liability. It is important that the wording of the policy follows as

---

[12] See cll. 20–21.

closely as possible the limitation upon the contractor's liability. Where there is an exclusion in respect of damage due to or caused by a specified risk, the exclusion will apply only where the events excluded are to be regarded as the effective or dominant cause of the loss. Where there are two causes one of which is excluded under the policy, the exclusion will not apply unless the event excluded was the effective or dominant cause. In *Wayne Tank v. Employers Liability*[13] the suppliers of equipment to a plasticine factory had been held liable for a fire which destroyed the factory.[14] They now claimed indemnity against their insurer. The insurer, however, relied on an exclusion which provided:

> "the company will not indemnify the insured in respect of liability consequent upon . . . damage caused by the nature or condition of any goods . . . sold or supplied by or on behalf of the insured."

The goods supplied consisted of a pipe intended to carry hot wax, which was unsuitable for that purpose, coupled with a thermostat which did not work. In the result, hot wax escaped and led to the disastrous fire. However, the supplier argued that the cause of the fire was the fact that the factory owner had left the pipe in operation and unattended, before it had been tested. Lord Denning answered the question of causation as follows:

> "I would ask, as a matter of common sense, what was the effective or dominant cause of the fire? To that question I would answer that it was the dangerous installation of a pipe which was likely to melt under heat. It seems to me that the conduct of the man in switching on the heating pipe was just the trigger—the precipitating event— which brought about the disaster. There would have been no trouble if the system had been properly designed and installed."

The Court of Appeal accordingly held that the cause of the loss fell within the exclusion, and the insurer was not liable.

Rights of parties

Upon the happening of an event insured against, the assured has a right to sue for payment under the policy, irrespective of any rights which may exist against a third party. But under an indemnity policy the sum payable is limited to the actual loss, and subject to any excess clause. The right of the assured is a claim

---

[13] [1974] Q.B. 57.
[14] See *Harbutt's Plasticine v. Wayne Tank* [1970] 1 Q.B. 447.

under the contract, and accordingly there can be no claim for losses consequential to the insured risk, unless this loss is itself insured under the policy.

When the insurer pays out on the policy a right of subrogation arises. This is a right to sue, in the name of the assured, any person who could have been sued by the assured in respect of the loss. Thus, a house insurer who has paid out on the policy for a subsidence claim may, in the name of the assured, sue the builder or architect in respect of defective foundations, to recover the sum paid. Where the party insured is legally liable for the loss, the insurer may nevertheless seek, under his right of subrogation, indemnity or contribution from any other party who caused or contributed to the loss.

The attitude of the courts is, therefore, that the existence of insurance policies available to one of the parties, or even to both, is legally irrelevant. The courts once used to avoid even mentioning insurance, on the basis that it might be seen as prejudicial to a party's case if he was known to be insured. This attitude has now changed, and the availability of insurance is now directly relevant in the application of the Unfair Contract Terms Act. In the case of *Smith v. Eric Bush*,[15] a surveyor's negligence case, Lord Griffiths said:

> "There was once a time when it was considered improper even to mention the possible existence of insurance cover in a law suit. Those days are long passed. Everyone knows that all prudent professional men carry insurance, and the availability and cost of insurance must be a relevant factor when considering which of two parties should be required to bear the risk of a loss (under the Unfair Contract Terms Act)."

The assured is under a duty not to prejudice the insurer's right. He must not release a third party from any liability he may be under in respect of the insured loss. Thus, he must not admit the claim, but must preserve the right of the insurer to dispute it. If the insurer pays out a sum on the policy less than the actual loss, and the assured then receives some other payment in respect of the loss, the assured must repay to the insurer anything received in excess of the actual loss.

Third party rights

There is a statutory right by which a third party, who is not a party to the insurance contract, may obtain the benefit of the

---

[15] [1990] 1 A.C. 831.

policy.[16] This applies when the insured person becomes insolvent. If liability is incurred to a third party, either before or after the insolvency, the right against the insurer vests in the third party. The Act imposes duties on the insured person and on the insurer to give information to the third party. Apart from these provisions, the insurer would either escape liability or the insurance money would go to the creditors and not to the injured party. Under a motor insurance policy a third party, having obtained judgment against the assured, may claim against the insurer irrespective of the solvency of the assured.

Insurance in construction projects

There are a variety of provisions and practices in construction work which usually result in there being a considerable variety of policies applying to different aspects of the work, covering different parties and providing different types of cover. Some of these are compulsory, being required by conditions of contract, while others are discretionary and taken out for the protection of individual parties. The result is often that, when a loss occurs, the disputes between the parties turns into a dispute between those who have insured the parties against their loss or liability. This can have the unintended effect that the parties effectively lose control of the dispute, and the decision whether to fight or settle is that of the insurers. A number of alternatives to this situation have been suggested, which are mentioned below.

Construction contracts invariably make a number of express requirements for insurance. Particular cover required under the JCT and ICE contracts is dealt with in Chapters 12 and 13. In general, construction contracts require two different types of cover. First, insurance is required on the works themselves. This is often a policy which is to be taken out in joint names of the contractor and the employer, in terms stated in the particular contract. Thus, the ICE conditions[17] require insurance against loss or damage "from whatsoever cause arising" save for the "excepted risks," which include faulty design, war, radioactivity and like perils. The JCT form[18] conversely requires insurance against specified perils only including fire and storm. The JCT contract provides, alternatively, for the employer to take these risks, in which case the contractor is not required to insure.

---

[16] Third Parties (Rights against Insurers) Act 1930.
[17] cl. 21.
[18] cl. 22.

The other, separate type of cover, is insurance against third party claims. While insurance of the works is insurance of "property," whether the cover is against all risks or specified perils, third party insurance is against "liability." Thus, the ICE conditions[19] require the contractor to insure against damage to any property or person arising out of the works; and the JCT contract[20] makes similar provision. One reason for requiring these two types of cover is that, while the contractor (and the employer) have an insurable interest in the works, they have no such interest in third party property other than through their potential liability for damage to it.

The insurances just mentioned are specific to the contracts in question, although they may be taken out pursuant to standing arrangements with insurers. In addition to these policies, contractors usually maintain a continuing policy covering a variety of matters, called a Contractors' All-Risks (CAR) Policy. The type of cover provided tends to vary, but a CAR policy typically provides some level of cover against liability for design work, for defects in material or workmanship. The usual procedure is for contracts to be noted on the policy, which continues upon annual renewals. The insurances required under construction contracts are usually released at or shortly after completion. The importance of a CAR policy is that it will continue in force, so that claims made, perhaps long after completion, may be covered. In practice, this is often the only cover which the contractor has against latent defects.

The other major insurance cover under construction projects, is the professional indemnity (P.I.) cover taken out by engineers and architects. These are continuing annual policies which cover the professional against legal liability, which will usually arise through negligence. P.I. cover operates on a "claims made" basis. That is, each annual policy covers claims arising during the year of its currency. In addition to terms concerning matters such as notification of claims, there will always be a limit of cover, which may be expressed in terms of each claim or the aggregate of claims during the year. This limit will tend to increase with inflation and with expansion of the professional's business, so that the person making the claim will get the advantage of a higher limit being available in subsequent years. The importance of P.I. cover is that, unlike the contractors' policies under the contract, it will continue (subject to renewals) after completion of the project. Thus, in regard to latent defects, the insurance position is often that the professional's P.I.

---

[19] cl. 23.
[20] cl. 21.

cover is the major insurance available, supplemented by the contractor's CAR policy, if applicable to the claim.

The effect of "liability" insurance is illustrated by the case of *Wimpey Construction v. Poole*,[21] where the contractor undertook a design and construct contract for a new anchored quay wall. The wall suffered partial failure which was found to be due to softening of the clay at the toe. The contractor had a P.I. policy which covered claims arising from "any omission error or negligent act in respect of design or specification of work." The contractor carried out remedial work at his own expense and sought to recover the cost from the insurer, contending that it had carried out the design negligently. The commercial court held that the plaintiff had failed to establish his own negligence, but nevertheless, the failure of the design to make sufficient provision for softening of the clay amounted to an omission or error in respect of that design, and was therefore prima facie covered by the policy. For other reasons, the plaintiff failed to recover the bulk of its loss. The case illustrates the legal contortions that may arise from insurance of liability rather than property.

The availability of insurance has no effect on the ability of a claimant to pursue a claim against a defendant who is held liable. In regard to claims against contractors, recovery of an uninsured claim will depend upon the company's assets. As regards professionals trading as partnerships, the individual partners are liable, and any judgment may be enforced against their personal assets, to the extent of bankruptcy proceedings. For this reason, many professional organisations have turned themselves into limited companies, although this does not rule out the possibility of a tort action being brought against individuals. In addition to the contractor and the professional team, sub-contractors and others involved in disputes may have their own insurance arrangements.

The availability of different insurances essentially covering the same type of loss creates procedural problems, as the cover under different policies will overlap. Also, where cover is based on liability, it is necessary to establish that liability before the insurer becomes bound to pay. A number of solutions have been put forward to deal with these problems. As regards the overlapping cover provided during the currency of the contract, it is possible to take out a "project" insurance policy which is designed to supersede all the different levels of cover otherwise provided, in theory at a lower cost. A more difficult problem is that of latent defects,

---

[21] [1984] 2 Lloyd's Rep. 499.

i.e. those appearing after completion. Here, the problem is even more complicated because employers, particularly developers of commercial buildings, will aim to sell or lease the premises at or soon after completion, often on terms which place liability for latent defects on the purchaser or lessee. These arrangements have in past years led to tort claims against designers and contractors by purchasers or lessees. With the demise of tort claims, warranties have been created giving other rights of action. Such complex arrangements inevitably lead to lengthy and costly litigation, in which the damaged owner receives no compensation until the claims are settled or resolved through the courts.

This unsatisfactory state of affairs was considered in a government committee set up through the National Economic Development Office (NEDO), which in 1988 produced the BUILD (Building Users Insurance against Latent Defects) Report. This recommends a new type of insurance based on the French decennial system whereby the owner takes out a policy effective from completion of the work, which insures the property against latent defects. The report recommends surrender of the right of subrogation so that litigation will not automatically follow a claim. Policies of this type are now available and many have been effected. It is to be expected that the insurer will need to be identified at the outset, and will take an active interest in the design and construction of the building, as well as its subsequent maintenance, in order to protect his liability. BUILD policies are recommended to run for 10 years and to be assignable or otherwise to cover the interests of subsequent owners and occupiers. The major limitation on these policies is that cover is usually limited to major elements such as the structure and weathershield.

SALE OF DWELLINGS

The term "dwelling" is used to cover any form of residential accommodation. The purchase of a dwelling may form the most important economic transaction which many individuals enter into during their lives. It may involve complex problems relating to the title of the property sold, and to the means of raising finance. This section is concerned solely with problems relating to the quality of the building and the rights of parties where there is a dispute. It is further limited to the sale of new or recently built dwellings, where the sale is of the land and building together. When builders are employed to build a house on a person's own land, the rights of the owner will be governed primarily by the building contract.

The term "sale" is used here loosely. The essential feature of the transaction is that the property should be transferred or conveyed from vendor to purchaser. In the case of a house this is usually done by a conveyance by which the title of the land is vested in the purchaser. In the case of flats and maisonettes, and sometimes houses also, the vesting of title may be by a lease, usually for a fixed period, often 99 years. In either case all that needs to be transferred is the physical space in which the building stands (or is to stand, if not completed). The sale automatically transfers with the land everything attached to it, including buildings, paths, walls, trees, etc., and also necessary legal rights. The conveyance or lease, however, usually creates no rights in respect of the building itself.

The conveyance or lease is almost invariably preceded by a contract of sale. This needs to be in writing or at least evidenced in writing. The contract may (in addition to the agreement of sale) contain terms relating to the building, such as a condition that the work has been or will be carried out in accordance with an identified plan or specification, or in a good and workmanlike manner. But contracts of sale are sometimes entirely silent as to the building itself. In the absence of contractual terms, the law was previously expressed by the maxim *caveat emptor*: let the buyer beware. He had no redress if the building proved to be defective.

Substantially, this remains the law in respect of the sale of old houses. Where the building is new or of recent construction, a number of developments in the law have changed the position radically. In *Hancock v. Brazier*[22] the Court of Appeal held a purchaser entitled to damages in respect of defective hardcore which had been incorporated into the foundation of a house before the date of the contract of purchase. Lord Denning summarised the law as follows:

> "When a purchaser buys a house from a builder who contracts to build it, there is a three-fold implication: that the builder will do his work in a good and workmanlike manner; that he will supply good and proper materials; and that it will be reasonably fit for human habitation."

However, this did not protect the purchaser of a completed house, nor subsequent purchasers of newly built houses. An important further measure of protection was introduced by a private body now known as the National House Building Council (NHBC). They publish forms of agreement relating to the quality of the

---

[22] [1966] 1 W.L.R. 1317.

building (see Chapter 11). They also operate a scheme of registration under which builders and developers must undertake to
comply with NHBC Rules and Requirements. This gives a wide
measure of protection to the first purchaser, which is also intended
to protect subsequent purchasers. The scheme is backed by insurance so that purchasers will have a considerable degree of protection in the event of the vendor's insolvency. However, there
remains the possibility of purchasers being unprotected, for example if necessary notices are not given, or because a subsequent
purchaser fails to acquire the right to enforce the agreement. In
such cases the purchaser may have further rights under statute.

Defective Premises Act 1972

In 1972 Parliament passed an Act to impose duties on all
persons taking on work for or in connection with the provision of
dwellings. The Defective Premises Act creates a general duty on
such persons to see that the work is done in a workmanlike or
professional manner, with proper materials so that the dwelling will
be fit for habitation (section 1). The duty applies to builders and to
professional persons such as architects. It may be enforced independently of any contract which may exist, by any person acquiring
an interest in the dwelling. Purchasers' rights under the Act cannot
be excluded by contract (section 6(3)).

The new Act, before it ever came into force, was entirely
overshadowed by the sudden introduction of apparently general
rights under the law of negligence in respect of defective building
works following the decision of the Court of Appeal in *Dutton v.
Bognor Regis UDC*.[23] While some claims were brought under the
Act there was little incentive to do so until the demise of the new
tortious rights, beginning with *Peabody v. Parkinson*[24] and effectively ending with *D. & F. Estates v. Church Commissioners*[25] in
which the Defective Premises Act figures as part of the reasoning
for rejecting the general availability of tort claims in respect of
building defects. Since then, the Act has taken its proper place in
establishing effective and transferable rights in respect of the
quality of construction work for dwellings. The Act creates a
special limitation provision under which a cause of action is
deemed to accrue at the time of completion of the original work or
of any further work done to rectify defects (see Chapter 14). It has

---

[23] [1972] 1 Q.B. 373.
[24] [1985] A.C. 210.
[25] [1989] A.C. 177.

been suggested that the incidence of quality disputes would be greatly reduced if the Act were extended to cover commercial buildings as well.

PRIVATE FINANCE INITIATIVE

In 1992 the Government announced its support for a new policy known as the Private Finance Initiative. This involved relaxation of previous finance policy, encouragement of public-private joint ventures and promotion of opportunities for private sector financing. There is no definition of PFI, which has now extended well beyond construction projects, into the provision of services formerly provided through public finance in many different fields. An early and substantial example of PFI is the cross-channel rail link.

PFI projects usually involve the creation of a special purpose vehicle (SPV) company which is intended to undertake the primary contractual obligation, financed through equity and loans in whatever proportions the promoters may decide. The involvement of government or public authorities is usually limited to the provision of land, with operating agreements under which the project is usually to revert back to public ownership (as in the case of the channel tunnel) but may involve outright sale. PFI is currently utilised for the provision of roads, prisons, hospitals and other capital projects and services. The essence of PFI projects is that they involve long-term operation agreements coupled with construction contracts in which the terms are modified to fit the wider roles being undertaken by the parties. For example, contractors are likely to have a financial interest in the project, and to undertake substantially enhanced risks under the construction contract. They may also be involved in the associated services contracts. Payment provisions will be related to the contractor's overall interest in the project. The design will also play an important role in the overall viability of the project, and its provision is likely to be integrated with the arrangements for financing and constructing the capital works. Certain construction contracts entered into under the Private Finance Initiative are excluded from the operation of the Housing Grants, etc., Act 1996[26] and are thus not required to conform to the payment provisions under the Act, nor to include the right to adjudication.[27]

---

[26] The Constuction Contracts (England and Wales) Exclusion Order 1998 (S.I. 1998 No. 648).
[27] HGCRA, s. 198.

In the developing world much construction work has been financed by the World Bank which has favoured standard procurement methods using the FIDIC form of contract. In more recent years PFI has become widely used in a variety of forms depending on the particular project. The procurement methods employed are variously known as Build Operate Transfer (BOT), Build Own Operate Transfer (BOOT) and latterly, Design Build Finance Operate (DBFO). Projects vary greatly in their financial and administrative detail, but all involve the provision of capital works financed through external private sources. The promoters are granted leases or licenses to provide and operate the capital works, with the objective of recouping their investment and profit, the works ultimately being transferred to the government or other promoter of the scheme. Such projects have included power stations, hydro-electric schemes and all forms of building and construction throughout the developing world. Typically, the design and construction will be split between the major contractors, who also contribute to the finance through a joint venture agreement.

CHAPTER 8

DOCUMENTS

IN most cases an oral contract is as good as a written contract in the eyes of the law. However, there are obvious practical differences. When parties to an oral contract are in dispute they may disagree over the terms or even as to whether a contract was concluded. Building and engineering contracts are usually put into some recorded form. But many problems can arise, for example, as to whether the documents represent the whole agreement, as to the status of various documents and as to their true meaning. In terms of construing documents, the problem is usually that they have been drafted for a particular purpose or in particular circumstances and the events which have in fact occurred are not those that were foreseen.

In addition to their construction, written documents can give rise to other problems. One of the parties may claim that a document does not record what was agreed. If this is so he may, in certain circumstances, obtain rectification of the contract through the courts or in an arbitration. If the parties agree that they intended something different from the written agreement, or if they change their intentions, they may themselves alter the contract. It may then be necessary to determine the legal effect of the alterations. These problems are discussed in this chapter.

INTERPRETATION

As a general rule a written document is interpreted as the sole declaration of the parties' intention and it is from the words used that the intention must be discovered. It is therefore important to ensure that what is written truly records what the parties have agreed. One way to do this is to use words and phrases which have acquired accepted meanings through precedent. These may make a contract sound archaic but they are more likely to cover an unexpected situation. This is one advantage of using a standard form of contract. A contract will generally be construed as a whole so that no words can have an absolute meaning out of context. But the meaning of similar words in another document is often a guide to construction, and previous decisions of the courts on the meaning of the standard forms of contract are treated as binding precedents.

Evidence admissible

The general rule that intention is to be inferred from the words alone has several exceptions, when extrinsic evidence (that is, evidence outside the document) is admissible to interpret the terms. Thus, evidence may be admitted to show the meaning of technical terms or to establish a special trade usage, *i.e.* that a particular word or phrase has a special meaning and not its ordinary meaning. The principal exception to the general rule is in the admission of evidence to prove surrounding circumstances. The precise extent of this exception may be a matter of dispute, since the "circumstances" relied on by one side may be much wider than the other side is prepared to admit. In the case of *Prenn v. Simmonds*[1] the House of Lords considered the amount of evidence admissible to construe a share option, the exercise of which was dependent upon the available profits. The issue was whether "profits" meant the separate profits of one company or the group profits. In giving judgment, Lord Wilberforce observed that there had been prolonged negotiations between solicitors leading ultimately to the ambiguous clause. The judgment continued:

"The reason for not admitting evidence of these exchanges is not a technical one or even mainly one of convenience . . . it is simply that such evidence is unhelpful. By the nature of things, where negotiations are difficult, the parties' positions with each passing letter, are changing and until a final agreement, though converging, still divergent. It is only the final document which records a concensus. . . . It may be said that previous documents may be looked at to explain the aims of the parties. In a limited sense this is true: the commercial or business object of the transaction, objectively ascertained, may be a surrounding fact. . . . But beyond that it may be difficult to go: it may be a matter of degree or judgement how far one interpretation or another gives effect to a common intention: the parties indeed may

---

[1] [1971] 1 W.L.R. 1381.

be pursuing that intention with differing emphasis and hoping to achieve it to an extent which may differ, and in different ways. The words may, and often do, represent a formula which means different things to each side, yet may be accepted because that is the only way to get 'agreement' and in the hope that disputes will not arise. The only course then can be to try to establish the 'natural' meaning. . . . In my opinion, then, evidence of negotiations, or of the parties intentions, and a fortiori of Dr Simmonds' intentions, ought not to be received, and evidence should be restricted to evidence of the factual background known to the parties at or before the date of the contract, including evidence of the 'genesis' and objectively the 'aim' of the transaction."

The principles by which contractual documents are construed by the courts were reviewed by Lord Hoffmann in *ICS v. West Bromwich BS*,[2] where the effect of *Prenn v. Simmonds* was described as follows:

"The result has been, subject to one important exception, to assimilate the way in which such documents are interpreted by judges to the common sense principles by which any serious utterance would be interpreted in ordinary life. Almost all the old intellectual baggage of 'legal' interpretation has been discarded."

The principles of construction were then summarised as follows:

"(1) Interpretation is the ascertainment of the meaning which the document would convey to a reasonable person having all the background knowledge which would reasonably have been available.

(2) The phrase 'matrix of fact', if anything, understates what the background may include.

(3) The law excludes from the admissible background the previous negotiations of the parties.

(4) The meaning which a document would convey to a reasonable man is not the same thing as the meaning of its words.

(5) The rule that words should be given their 'natural and ordinary meaning' reflects the common sense proposition that we do not easily accept that people have made linguistic mistakes, particularly in formal documents."

The application of these principles to ambiguities is often difficult. The safer course is to define any terms which may give rise to dispute. This is often done by incorporating a "definitions" clause such as that often found at the beginning of the standard forms of contract.

Sometimes the body or operative part of the document is preceded by a recital relating what has led up to executing the

---

[2] [1998] 1 W.L.R. 896

document. For example, the JCT forms of contract commence with a number of recitals beginning "Whereas . . ." which give a brief description of the works with the name of the architect and a list of contract drawings. In the absence of doubt as to construction, the body of the document alone is effective. But where there is an ambiguity in the body, the recital, if clear, may give the true meaning. Recitals can also be useful in setting out the surrounding circumstances so as to provide an agreed background for the construction of any phrases which may later appear ambiguous.

Where a contract is partly in a printed standard form and partly in terms specially written, the latter will usually prevail in the event of an inconsistency, on the basis that they represent the parties' true intention, rather than a document which was prepared by others (see Chapter 9). Thus, provisions of a standard form may be overridden by an inserted clause, or by a contrary provision in the specification or bill of quantities. This is, however, subject to the terms of the contract itself.[3]

Maxims of interpretation

Where doubt as to the precise meaning of a document remains after allowing for such extrinsic evidence as may be admissible, and after giving due weight to the different parts of the document, there are a number of principles or *maxims* of interpretation which may assist in arriving at a definite or at least a more probable meaning. They are often quoted in Latin but, for the most part, an English rendering is given here.

(1) The law prefers a reasonable to an unreasonable meaning. This is part of a wider legal principle by which many things are judged against an objective standard of reasonableness. Thus, if a document can be read as having a sensible meaning or an absurd meaning, it will more readily be taken to have the sensible meaning. Further, if there is either a lawful or unlawful meaning, the lawful meaning will usually be adopted.

(2) An erroneous description can be given effect as it should have been stated, provided it is clear what was meant. This maxim can be used to correct an obvious error in a document. It may apply for example to the statement of a price in pounds when pence was obviously meant, and vice

---

[3] See JCT, cl. 2.2.1 and ICE, cl. 5.

versa. If there is genuine doubt, however, the contract may have to be enforced as written.

(3) Express mention of some things will exclude others of the same class not mentioned. This may assist where it is not clear what is to be included in a list of items. Thus, a contract to sell a house and a factory with the fixtures of the house will be taken to exclude the fixtures of the factory. Further, a contract to supply and lay bricks and to supply paving slabs would not include laying the paving slabs. Uncertainty of this sort would normally be resolved, in a construction contract, by reference to the standard method of measurement, but even that may contain ambiguities.

(4) The meaning of a doubtful word may be ascertained from the words associated with it. For example, the term "general contractors" might include almost any commercial activity; but in the context "engineers and general contractors" it must be limited to the field of engineering. This is part of a wider rule that words are to be construed in their context, which may include looking at the whole document and the admissible surrounding circumstances.

(5) Where a series of words comprises a class and is followed by general words, the general words cover only things of that class. This is known as the *ejusdem generis* rule, and an example is probably simpler than a statement of the rule. Thus, the words "iron, steel, brass, lead and other materials" could include copper since the class is one of metals; but stone or wood could not be included. However, if the words had been ". . . and other materials of whatsoever kind" they would preclude the operation of the rule and include any other materials. Further, a list reading "steel, bricks, plywood and other materials" forms no particular class, so that again the rule is excluded.

Contra proferentum

The words of a document are to be interpreted against the person proffering it. This is perhaps the best known maxim of construction, but it is equally capable of being misunderstood. The simple notion that any uncertainty in a contract is to be resolved against the party who drafted it is wrong and can lead non legally qualified arbitrators seriously astray. As regards ordinary contractual provisions, *contra proferentum* has limited application.

Whoever is responsible for drafting a contract, it is ordinarily to be construed by balancing all the competing arguments as to its construction and ascertaining the meaning which most fairly represents the presumed mutual intention of the parties. The maxim can be applied in the case of an ambiguity (see below), *i.e.* where the contract can have two possible meanings. In this event it is permissible (after considering all other means of the construction) to construe the document against the person who drafted the contract (the *proferens*).[4]

There is a separate application of the maxim in relation to exclusion and similar clauses, which are to be construed against the party seeking to rely on them. In such a case, any uncertainty of construction will be resolved in favour of the party against whom the clause is being applied. As an example, there are a series of cases in which exclusion clauses have been held not to cover negligence where the clause could be construed as covering other events. The same principle is applied to an indemnity clause, which is the obverse of an exemption clause. In the case of *Smith v. South Wales Switchgear*[5] the plaintiff motor manufacturer employed the defendant, an electrical company, to carry out maintenance work upon the plaintiff's standard conditions of contract, which provided that the defendant should indemnify the plaintiff against "any liability, loss or claim or proceedings whatsoever under statute or common law in respect of personal injury to, or death of any person whomsoever. . . ." One of the defendant's employees suffered injury as a result of the plaintiff's negligence and breach of statutory duty. The plaintiff claimed to be indemnified under its standard conditions. The House of Lords held that the clause did not afford indemnity against the plaintiff's own negligence because there was no express provision, and the clause could not be construed as covering negligence by the plaintiff's own servants. Lord Dilhorne said:

> "When considering the meaning of such a clause one must, I think, regard it as even more inherently improbable that one party should agree to discharge the liability of the other for acts for which he is responsible. In my opinion, it is the case that the imposition by the proferens on the other party of liability to indemnify him against the consequences of his own negligence must be imposed by very clear words. It cannot be said, in my opinion, that it has been in the present case."

This very strict approach to exemption clauses has been modified by a number of factors. First, the law concerning fundamental

---

[4] See *Chitty on Contracts* (27th ed.), Vol. 1, paras 12-071, 14-009.
[5] [1978] 1 W.L.R. 165.

breach, much of which involved construing exclusion clauses so as
not to cover supposed flagrant breaches of contract, has been
transformed by the House of Lords' decision in *Photo Production v.
Securicor*.[6] Secondly, the advent of a series of statutes culminating
in the Unfair Contract Terms Act 1977, which permit the courts to
override "unfair" exempting provisions, has removed the need for
the application of strained methods of interpretation. Thirdly, the
courts have recognised the need to apply less exacting standards
where the parties have entered into an arrangement which limits,
but does not entirely exclude, the right of the injured party to
compensation. Thus, in *Ailsa Craig Fishing v. Securicor*[7] the House
of Lords held that a security company was entitled to rely on a
clause which clearly limited its liability to £1000 even though it was
admitted that the loss had been caused by their negligence. Lord
Wilberforce held:

> "Whether a clause limiting liability is effective or not is a question of
> construction of that clause in the context of the contract as a whole. If
> it is to exclude liability for negligence, it must be most clearly and
> unambiguously expressed and in such a contract as this must be
> construed contra proferentum. I do not think that there is any doubt
> so far. But I venture to add one further qualification, or at least
> clarification: one must not strive to create ambiguities by strained
> construction, as I think the appellants have striven to do. The
> relevant words must be given, if possible, their natural, plain mean-
> ing. Clauses of limitation are not regarded by the courts with the
> same hostility as clauses of exclusion: this is because they must be
> related to other contractual terms, in particular to the risks to which
> the defending party may be exposed, the remuneration which he
> receives and possibly also the opportunity of the other party to
> insure."

Another example of the application of the *contra proferentum
maxim* in building and engineering contracts, is in the interpretation of the extension of time clauses. These are to be regarded as
benefitting the employer, who is, therefore, seen as the *proferens*,
because the clause protects his right to recover liquidated
damages.[8]

Mandatory or permissive

Many contract clauses which set out procedures to be adopted
employ the words "may" or "shall" or other equivalent words. It is

---

[6] [1980] A.C. 827.
[7] [1983] 1 W.L.R. 964.
[8] *Peak v. McKinney* (1969) 1 B.L.R. 111 and see Chap. 9.

frequent to find them used inconsistently and sometimes it is necessary to conclude that one in fact means the other. Once the meaning of a clause is ascertained, however, it will be clear whether the provision is mandatory (sometimes called "directory"), or whether it is permissive. The difference is important in relation, for example, to the service of notices under a contract which are intended to achieve a particular objective. If this is the exercise of a determination clause, the question will be of vital importance. Clause 28.1.1 of the JCT conditions entitles the contractor to determine his employment for non-payment of a certificate conditional upon prior notice having been given "by registered post or recorded delivery".[9] These words are mandatory and the clause will not be operated unless the prescribed notice is given. Conversely a clause reading "notice may be served by post or by actual delivery" would not rule out other equally expeditious means such as delivery by (legible) fax or perhaps E-mail.

Building contracts tend to make excessive use of "shall", sometimes leading to unnecessary dispute as to whether one party is bound to do something which common sense would suggest to be optional. For example clause 51 (1) of the ICE conditions states that "the Engineer shall order any variation . . . that is in his opinion necessary for the completion of the work". While this clause can be given a rational meaning, it is to be noted that the JCT form uses "may" (clause 13.2).

Ambiguity

This is a term which is often used loosely: it is frequently taken to be synonymous with doubt or uncertainty. In law, however, an ambiguity is a provision which has two (or more) possible meanings, which cannot be resolved by application of the normal rules of construction. Much of the law concerning ambiguity relates to wills and trusts, but the same principles apply to any type of written instrument which has to be construed by the courts. Where construction cannot produce an answer, evidence will be admitted going beyond that which is ordinarily admitted to establish surrounding circumstances (see above). An example of this principle occurred in a case concerning a will in which the testator left £100 "to my grand-nephew Robert". There was no such grand-nephew, but there were four of other names. Evidence was therefore admitted which showed that the testator in fact thought that one of

---

[9] See Chap. 12—amendment 11 also permits notice "by actual delivery".

his grand-nephews, Richard, was called Robert, and he was held entitled to the bequest.[10]

Clause 5 of the ICE conditions refers to "ambiguities or discrepancies" which are to be explained or adjusted by the engineer. Strictly, these two concepts are different: the former implying two (or more) meanings which cannot be resolved, but the latter not necessarily being uncapable of resolution by means of construction. The purpose of this provision, clearly, is to permit the work to proceed by resolving "uncertainty" but in regard to the question of payment the difference could be relevant.

Building and engineering contracts, particularly the latter, are not noted for their clarity and consistency of drafting. Problems often arise because "the contract" is contained in several long and complex documents, often written by different persons at different times. In addition, contract documents are often prepared from standard drafts which are adapted to the particular circumstances. These processes can easily lead to inconsistencies appearing between different parts of the contract. In drafting documents, the contents remain only as clear as the thought which went into them. Where uncertainties arise, it may be impossible to ascribe a definite meaning. If not, the parties may be forced to bring the issue before the court or before an arbitrator, as appropriate, for resolution.

DRAFTING

The opposite process to interpretation is drafting. Logically it precedes interpreting, but if properly done there should be no room for doubt and therefore little which needs to be interpreted. The rules of interpretation must be kept clearly in mind when putting together a document. The draftsman must constantly ask himself what his words and sentences mean, but with this added fillip: they will ultimately be read by others, who can be relied on to search out any loopholes he may leave.

There are no rules or *maxims* of drafting. The writer must use his common sense and proceed in an orderly way. He may be assisted by the following considerations, the order of which is not significant.

(1) What is the object of the document? It may be of assistance to set out in a concise form what is to be achieved by the

---

[10] *Re Ofner* [1909] 1 Ch. 60.

exercise. This may involve selecting key words or phrases to be incorporated in the full draft. The object may also be expressed in diagrammatic or symbolic form. For example, a variation of price clause can usually be written as a simple algebraic formula. The many lines of prose needed to express the formula in words are a purely mechanical exercise.

(2) Is the document dependent on other documents? If it is to stand by itself, it will need to contain or incorporate all necessary material; for example, if the document is to constitute an agreement it must encompass the terms agreed and show the assent of the parties. If the document forms part of another, such as an additional term of a contract, it must be drawn so as to effect all necessary alterations or to prevail over any inconsistency. Examples of the care and thoroughness needed to make effective amendments can be seen in statutes which amend earlier Acts, often by way of detailed schedules.

(3) What is the most appropriate form of document? A formal contract may seem appropriate, containing recitals, articles of agreement, standard conditions, special conditions and other incorporated material. But other ends call for simpler means. If notice is required to be given under a contract, one way is to draw up a formal document stating "WHEREAS . . . NOW WE HEREBY, in exercise of the said. . . ." However, it may be equally effective (and clearer) to write: "Dear Sir, We give you notice under clause. . . ." The appropriate form of document is that which will achieve the object with certainty and efficiency.

(4) What form of drafting is called for? Where the brief is to achieve a stated object the draftsman usually has no difficulty and can use formal language including words such as "notwithstanding" to arrive at certainty. However, there may be objection to this approach. First, the document may be the subject of negotiation and compromise. The other party may not accept language which seems heavily weighted against him (consider the drafting of the standard forms of contract). Secondly, the document may need to be expressed in simple, direct language to fulfil its purpose. A form of contract which places duties on a supervising officer does not fulfil its purpose if the supervising officer cannot

understand what is required of him without taking legal advice.[11]

(5) What formal requirements are necessary? A contract should incorporate evidence of the parties' agreement to the terms set out. Formal signature and dating are unnecessary but desirable in the interests of clarity. The most important practical consideration is the identity of the contracting party. Where one party uses a trading name, the true identity of the proprietor should be discovered; and where groups of companies are in evidence, their relationship should be ascertained. In case of doubt or suspicion, a clause prohibiting assignment or sub-contracting may be added to the contract.

(6) Good drafting is an amalgam of clarity, style and choice of appropriate language. Its object is to achieve a result, rather than to hold the interest of the reader. Short sentences are clearer than long ones. They can be more easily adapted if provisos or further clauses need to be added. Where there is doubt in the draftsman's mind there are devices which may assist. Thus, where a clause is to be added to an existing document of less than perfect clarity, the draftsman may add words such as: "For the avoidance of doubt it is agreed that . . ." and add for good measure "notwithstanding anything to the contrary."

(7) Care should be taken to use consistent language where a consistent meaning is intended. This may produce a stilted impression, especially when coupled with overuse of "the said. . . .", but greater clarity may result. The legal draftsman has one great advantage over the ordinary user of the English language. He can say with conviction "When I use a word it means just what I choose it to mean, neither more nor less".[12] Thus, documents may expressly define (often in a separate clause) particular words used. This can be a considerable aid to brevity and clarity, provided the defined meaning is adhered to. If it is not, confusion will result. Where the draftsman is in doubt he may take refuge in the formula ". . . save where the context otherwise requires".[13] A good example of the difficulty of assigning a definition is

---

[11] Consider the JCT form, Chap. 12.
[12] Humpty Dumpty in Lewis Carroll, *Through the Looking Glass*.
[13] See ICE conditions, cl. 1(1).

in use of the word "completion". Most forms of contract permit completion to occur in a variety of circumstances such that it is not easy to define the term without the risk of confusion.

Difficulties, both of drafting and interpreting, lead to the frequent use of precedents, whose meaning and effect is reasonably certain. This applies both to individual clauses and to whole documents (such as contracts, leases, notices, pleadings, etc.). This has become even more prevalent since the advent of the word processor. The problem is then to ensure that the various standard and special parts are consistent.

ALTERATION OF TERMS

Rectification

If, by a mistake, a document does not record the true agreement between the parties, the courts have power to rectify or alter the document so as to give effect to the true agreement. There can be no rectification of a mistake in the transaction, but only of the way in which the transaction was put into writing. Rectification is an equitable remedy, and it is therefore not available as of right, but is a matter of discretion. One consequence of this is that rectification will not be granted if relief can be obtained by other means. The court itself will correct an obvious mistake such as a clerical slip or even an erroneous "not," without recourse to formal rectification. Rectification is not a *panacea* for badly written contracts, and it is a remedy which is only rarely granted in practice.

In a claim for rectification it must be established that the document was intended to carry out the parties' prior agreement, and not to vary it. If the opposing party claims that the prior agreement was intended to be varied by the subsequent document, a heavier burden of proof falls upon the claimant. However, it is not necessary to prove that a prior concluded agreement was reached before drawing up the document. It is sufficient to show that the parties had a continuing common intention, and that the written contract failed to conform to that intention. Usually the mistake to be rectified is one of fact, but it may also be as to the legal effect of the words used.

Generally the mistake must be common to both parties, but in a few situations a unilateral mistake may be rectified. These situations include the case where one party is mistaken but the other is fraudulent, and also where one party is mistaken and the other party knows of his mistake. In *Roberts v. Leicestershire C.C.,* [14] when a building contractor had tendered to build a school in 18 months, the employer, after accepting the tender, inserted a period of 30 months into the formal contract without the contractor's knowledge. It was held that the employer knew of the contractor's mistaken belief as to the term, and that the contractor was entitled to have the contract rectified by insertion of a completion period of 18 months.

Voluntary variation

This refers only to variations to the terms of a contract and not to variations made pursuant to an express power in the contract, such as the usual provisions for the variation of the work found in building and engineering contracts. An agreement to vary a contract is like any other contract in that it requires either to be for consideration or under seal in order to be binding. A variation may take the form of an alteration of some of the terms of the contract, or its replacement by a new contract, or even its complete discharge. If the original contract is one which is not required to be evidenced in writing, it may be varied by an oral agreement even if the original is in writing or under seal. An example of a variation is where each side surrenders some outstanding obligation. Each surrender constitutes consideration and the agreement will be binding. Thus, if the employer (without express power) wishes to omit a piece of work and the contractor agrees to the omission, the variation is binding and no action will lie for breach of contract by either party.

Where one party has completely performed his obligations under a contract any variation or release of the obligations of the other party will be binding only if made under seal or if the party being released provides some new consideration, since he has no rights under the contract to give up. The consideration may take any form, and what is commonly surrendered is a potential claim. Thus, where a contractor has completed his work and agrees to accept a sum less than the full amount claimed in return for the surrender by the employer of a claim for defects in the work, the agreement is binding provided that the claims on both sides are bona fide, even though in fact not sustainable. Such an arrangement is called an "accord and satisfaction".

---

[14] [1961] Ch. 555.

Waiver and estoppel

When one party only agrees not to insist on some right under a contract, the other party gives no consideration. Nevertheless, if the other party has acted on the agreement the court may treat it as a binding waiver. Such a waiver remains effective until reasonable notice of withdrawal has been given. Waiver is a principle which parties in litigation frequently seek to rely on. Its effect may be that, where a party has not insisted on his legal rights, he may be unable to claim the benefit retrospectively. In the case of *Richards v. Oppenheim*[15] the defendent ordered a Rolls Royce from the plaintiff, to be completed by a certain date. It was not finished on time, but the defendant continued to press for delivery. Eventually the defendant stated that if the car was not completed by a specified date he would not accept it; the car was not finished in the time. It was held that the defendant had waived the original completion date. Denning L.J. giving judgment in the Court of Appeal went on to hold:

> "It would be most unreasonable if, having been lenient and having waived the initial expressed time, he should thereby have prevented himself from ever thereafter insisting on reasonably quick delivery. In my judgement, he was entitled to give a reasonable notice making time of the essence of the matter. Adequate protection to the suppliers is given by the requirement that the notice should be reasonable."

There are, however, certain rights which, once waived, cannot subsequently be relied on. This applies, for example, to the right to written notice of a claim within a specified period. If oral notice is received within the period and is acted on, the recipient cannot, after expiry of the period, insist on written notice.

A simpler but distinct doctrine which may effectively vary the terms of a contract is estoppel by convention. Where both parties to a transaction act upon an agreed assumption, both may subsequently be precluded from denying the truth of the assumption. This applies both to questions of fact and to the true construction of a contract. In the *Vistafjord* case[16] an agreement for the charter of a cruise ship was negotiated by agents. Both the agents and owners believed that commission was payable under a previous agreement but no such right existed on the true construction of the agreement. It was held that there was an estoppel by convention

---

[15] [1950] 1 K.B. 616.
[16] *Norwegian American v. Paul Mundy* [1988] 2 Lloyd's Rep. 343.

binding on the owners by which they were bound to pay commission which both sides assumed to be payable. This form of estoppel is to be distinguished from estoppel by representation, by which a party may be prevented from denying the existence of a fact which has been the subject of representation. Estoppel by representation is a rule of evidence but its effect can be the same as a rule of law.

CHAPTER 9

CONSTRUCTION CONTRACTS

THE essence of a construction contract is that a contractor agrees to supply work and materials for the erection of a defined building or other works for the benefit of the employer. The detailed design of the work to be carried out is often supplied by or on behalf of the employer, but may also be supplied in whole or in part by the contractor. In legal terms there is no difference between a building and an engineering contract, and the term Construction Contract is adopted to cover both. For the first time under English law, the Housing Grants, Construction and Regeneration Act 1996, Pt II includes a definition of "construction contract" (see below). This is solely for the purpose of identifying types of contract to which the Act applies, and does not apply. Many of the excluded activities fall within the normal definition of a construction contract, and will be subject to the general principles discussed in this chapter.

Almost invariably there will be other parties involved in a construction project in addition to the contractor and the employer. There may be an architect or engineer who provides the design and supervises the work; and there are likely to be sub-contractors employed to carry out parts of the work. The status and capacity of these parties is considered in Chapter 4. This chapter deals with those particular areas of the common law which help to define the rights and duties of the parties and which regulate the performance of construction contracts.

The number of statutory provisions which directly affect construction contracts, as opposed to construction operations, is not great. The number of decided cases which apply to construction contracts has grown very considerably over past two decades. Since the advent of systematic reporting of construction cases (in the *Building Law Reports, Construction Law Reports* and elsewhere) there has accumulated a large number of decisions on the standard forms and on other principles of construction law. However, there remain areas in which there is no direct authority. In such situations assistance may be obtained from the standard textbooks, which are often consulted by, and sometimes expressly approved by the courts in deciding new legal points. Reference is also made frequently to foreign decisions, from the Commonwealth and the United States of America, where direct authority may be found on areas not yet decided under English law.

In Chapter 7 a number of special types of contracts are considered. Each of these contracts has its own particular features; for example, a sale of goods is governed by extensive statutory provisions. The special nature of construction contracts arises from the form which most contracts take and from features such as the role assigned to the architect or engineer and the provisions for payment as the work proceeds. These matters are dealt with in this chapter. The following chapter covers factors outside the contract itself which affect the parties' rights. The particular provisions of common forms of building and engineering contracts are considered in Chapters 11, 12 and 13.

New statutory definition

As noted above, the Housing Grants, etc., Act provides in sections 104 and 105 an extensive but far from comprehensive definition of "construction contract". Thus, drilling for oil or gas, tunnelling generally, plant or steelwork for nuclear processing, power generation, water or effluent treatment or chemical, oil, gas, steel or food and drink production are excluded; as are the supply (excluding installation) of components, materials, plant and machinery generally. A contract with a residential occupier is also excluded. The Act does, however, include matters not ordinarily considered subject to construction contracts, such an agreement to do architectural, design or surveying work, or an agreement to provide advice on building, engineering, interior or exterior decoration or the laying-out of landscape in relation to construction operations. Further, by additional regulations, particular types of contract are excluded, such as PFI contracts and highway and sewerage works for adoption. Thus, it is necessary to consider carefully whether particular operations are within the Act. The subject matter of some contracts will be partly within the Act. In this case section 104(5) provides *"where agreement relates to construction operations and other matters, this part applies to it only so far as relates to construction operations"*. This provision may create difficulty in relation to the resolution of disputes by adjudication (see Chapter 2) or in regard to the right of suspension.[1]

New forms of contract

While the majority of construction work in the United Kingdom and abroad is carried out under conventional arrangements with a main contractor, sub-contractors and a professional team, a number of radical alternatives have emerged in recent years. The "prime cost" contract (see below) has developed into the management contract, in which there is a main contractor ostensibly responsible for the whole of the work. However his legal liability is substantially restricted in regard to the performance of sub-contractors, who normally perform the entirety of the physical work. The role of the main contractor then becomes that of a manager, charged with organising and co-ordinating the work of the sub-contractors. A major feature of management contracts is that the design is evolved as the work proceeds, with the main contractor participating in or advising on design decisions. This arrangement offers the opportunity of commencing work at an early stage, without waiting for the design to be finalised, and it also permits more financial control, to the extent that this depends on design choices.

A further, logical development from management contract is the "construction management" contract (sometimes called a project management contract) in which the party filling the role of management contractor does not enter into contracts with those who carry out different elements of the work, but supplies only management and other professional services. The work is carried out under a series of direct contracts. The construction management contract contains obligations limited to managing and co-ordinating these individual direct contracts. These new forms of contract involve certain common drafting difficulties in the definition of the obligations being taken on, the provision of workable sanctions and the residual responsibility of the contractor. However, many of the forms of contract utilised tend to be individual, or in-house forms offered by one contractor. It is, therefore, difficult to identify detailed principles that apply to these new forms of contract. Further development will inevitably occur. Mention should also be made of the Latham Report[2] which reviewed contemporary problems of the United Kingdom con-

---

[1] s. 112 and see below.
[2] *Constructing the Team* (1994).

struction industry. The JCT and ICE forms of contract came in for particular (although not detailed) criticism, and the report identified the need for a single standard form in which "adversarial attitudes" would be discouraged. The report identified the New. Engineering Contract (subsequently renamed the New Construction Contract) as the most likely contender. This form is said to promote good management and thereby to avoid confrontation and dispute. The latest version (second edition) contains newly drafted dispute machinery based on adjudication, also recommended by Latham. Use of the new form, since its first appearance in 1991, has been limited and its success is difficult to assess.

Performance and Payment

In a construction contract, the contractor undertakes to carry out the works, including the provision of all things necessary for completion. The employer's side of the bargain is usually the payment of money. Problems may arise in deciding when the contractor's obligation is discharged, what amount of money is payable and when. In each case the answer depends first on construction of the contract, since the parties may make whatever contractual arrangements they choose. There are, however, some general principles which may amplify the parties' intentions.

Where the contract is to carry out and complete a specific item of work, the general rule is that only complete performance can discharge the contractor's obligation and no payment is due until the work is substantially complete. In *Sumpter v. Hedges*[3] a builder contracted to erect two houses and stables on the defendant's land for a lump sum, but abandoned the contract part-completed. It was held that in the absence of entitlement under the contract, the builder was not entitled to further payment for the unfinished work, despite the fact that the employer retained the benefit: A.L. Smith L.J. giving judgment said:

> "The learned Judge had found as a fact that he abandoned the contract. Under such circumstances, what is the building owner to do? He cannot keep the buildings on his land in an unfinished state forever. The law is that, where there is a contract to do work for a lump sum, until the work is completed the price of it cannot be recovered. Therefore the plaintiff could not recover on the original

---

[3] [1898] 1 Q.B. 673.

contract. It is suggested, however, that the plaintiff was entitled to recover for the work he did on a quantum meruit but, in order that that may be so, there must be evidence of a fresh contract to pay for the work already done."

The contractor in such a situation is not, however, always without a remedy. He may recover if he can show that completion was prevented by the employer, or that a fresh agreement to pay for the partially completed work is to be implied. In the above case the builder did succeed in recovering payment for his materials which had been used by the employer, these not being attached and having remained his property (see also Chapter 5).

The contract price

Construction contracts usually state a price for which the work is to be completed. This is invariably subject to modification as the work proceeds on account of ordered variations, allowable price fluctuations, re-valuation of prime cost or provisional sums, claims, and other matters. Where the price of the original contract work remains fixed, the contract may be called a "lump sum" contract. But if the original contract work is based on quantities which are to be recalculated when the work is done, the contract is called a "re-measurement" contract. This is so when there is an express right to have the work re-measured; or where the bills are stated to be provisional or approximate. The JCT form of contract is a lump sum contract, whether or not it is based on quantities (see Chapter 12). The ICE form creates a re-measurement contract; this is emphasised by the fact that the stated price of the work is referred to as the "tender total" (see Chapter 13). A further term some-times used is "fixed price." This is generally taken to mean a contract where the sum payable is not adjustable by reason of price increases (fluctuations). The price may, however, be adjustable on many other grounds. Where the employer wishes to know the exact price of the work in advance, none of the common forms of contract are appropriate. While it is possible, an "invariable price" contract would be uneconomic and difficult to draft.

Stage payments

In most construction contracts of any substance there are express provisions for interim or stage payments to be made as the work proceeds. The usual provision is for the contractor to be paid the value of the estimated quantities of work done and materials supplied less a retention which the employer holds as security for completion of the works. In such cases the rule of payment on substantial completion (see above) may still apply to each payment, but subject also to the provision as to certificates. Even where there is no provision for interim payments there may be, in the absence of express provision to the contrary, an implied term for reasonable interim payments as the work proceeds.

Contracts within the definition contained in the Housing Grants etc., Act must provide for stage or other periodic payments, unless the duration of the work is to be less than 45 days.[4] The contract must also provide a mechanism for determining what payments become due and when, and provide a final date for payment. Further, the contract must provide for the giving of notice not later that 5 days after the date on which a payment becomes due, specifying the amount of the payment made or proposed to be made.[5] If the contract does not comply with these requirements, the Scheme for Construction Contracts Part II is to apply. This provides for calculation of instalments by reference to the value of work done and materials supplied, less previous payments. Sections 111 and 112 of the Act contain important rights regarding set-off (see chapter 2) and the right to suspend performance (see below under Extension of Time). Section 113 prohibits "pay when paid" clauses except where the person from whom the payer is receiving payment is insolvent (see chapter 10).

Stage payments based on measurement can be regarded as unnecessarily complex, requiring more or less detailed measurement, usually on a monthly basis. There is usually nothing to prevent the contractor artificially adjusting the rates and prices under the bill to achieve inflated early payments, which also has the effect of reducing the contractor's incentive to complete. It has been suggested (by Latham and others) that a fairer and more efficient system is to agree lump sum instalments in advance, dependent only upon the contractor's rate of progress. Such sums are referred to as "milestone" payments. They are found routinely in construction contracts in the United States of America, where bills of quantities are rarely employed.

No price agreed

In most substantial contracts, the primary sums payable are provided for in detail. There are, however, many situations in

---

[4] s. 109.
[5] s. 110.

which sums are to be paid where the contract terms are inapplicable. In an extreme case this may arise from the absence of a contract, or from a contract making no provision for payment. In such case, the contractor is entitled to a reasonable sum (see Chapter 5). However, it is frequent to find provisions under the standard forms, where the pricing mechanisms provided lead, after eliminating various provisions which do not apply (for example application of contract rates or analogous rates) to the conclusion that the contractor is entitled to payment of a sum based on "a fair valuation"[6] or which is "reasonable and proper".[7] There are many similar provisions in the forms. In each such case it is for the court, or arbitrator, to decide a reasonable sum based on (1) any materials which the contract may require to be taken into account, and (2) such other material as is placed before the tribunal. There are no rules as to what is admissible and what is not. A fair or reasonable price may be based on rates quoted by other contractors for similar work; or on time and materials plus other allowances. A reasonable sum must include both profit and overheads. A claim limited to "cost" will generally include overheads but not profit.[8]

CONTRACT DOCUMENTS

A common feature of most construction contracts is the incorporation of a variety of different types of document. These are not limited to documents expressed in words: drawings appear in most contracts and have to be interpreted and given legal meaning and significance. Typically, a construction contract of any importance will contain a set of conditions of contract, a specification, a bill of quantities, a set of drawings, and other documents of varying sorts. There may also be a separate "agreement" in which the parties formally bind themselves to execute the work. The question necessarily arises, how these documents fit together, which (if any) are to have precedence, and what is to happen if they conflict.

A further related question, is the definition of "the works" to be performed. Is this governed by all the contract documents, or some of them only, or is there an independent definition? These questions need to be addressed to decide, for example, whether

---

[6] ICE, cl. 52(1).
[7] ICE, cl. 52(2).
[8] ICE, cl. 1(5).

particular work constitutes a variation, and whether completion has been achieved. Every contract is different, and the definition of what is to be performed depends upon its particular provisions. However, there are two distinctly different approaches to the question. The first, and simplest, is to make all contract documents of equal weight and significance. This is the solution adopted in the ICE Conditions of Contract (clause 5). The problem that is then thrown up is, what happens in the event of discrepancies? The ICE conditions provide that these are to be "explained and adjusted" by the engineer, but it is not clear whether, in doing so, the engineer will vary the contract. Another solution sometimes found, is to provide that the contract documents shall have an order of precedence, *i.e.* a conflicting requirement in two documents is to be resolved in favour of that having the higher priority.

The second solution is adopted in the JCT forms, which typically provide that the quality and quantity of the work to be carried out is that contained in the contract bills (or in the case of a contract without quantities, in the specification).[9] The effect of these provisions is that the bill (or specification) is given a limited function, and may not override the contract conditions. This type of provision has given rise to some unusual results. For example, a sectional completion provision written into the bills is likely to be ineffective if there is no corresponding amendment to the conditions of contract and the appendix. The principle was applied in *English Industrial Estates v. Wimpey*,[10] where, under a JCT form of contract, the contract bills provided for the employer to take possession and occupy parts of a factory being constructed by the defendant. But there bad been no relevant amendment to the conditions of contract, which laid down a procedure for the employer to take possession of completed parts, and to become responsible for such parts. The employer in fact took over part of the works which were then destroyed by fire. The question arose, who was responsible? The Court of Appeal (relying on clauses equivalent to those given above) held that, despite the provisions in the bill, the contractor remained responsible for the parts taken over, because the procedure under the conditions had not been followed. Stephenson L.J. said:

> "It follows from the literal interpretation of clause 12[11] that the court must disregard—or even reverse—the ordinary and sensible rules of

---

[9] See cll. 14.1 and 2.2.1.
[10] [1973] 1 Lloyd's Rep. 118.
[11] Now cl. 2.2.1.

construction and that the first of the documents (the conditions of contract) . . . expressly prevents the court from looking at the second of those documents (the bills) to see what the first of them means. But that is because the second document is . . . a hybrid document and part of it deals with matters which should have been incorporated in the first."

Contract conditions

Most contracts incorporate a set of conditions whose primary purpose is to lay down procedures of general application to a variety of types of work. It is often convenient to use a set of standard conditions, such as those dealt with in more detail in Chapters 11, 12 and 13. There is no rule as to what should be included in conditions of contract, but most sets of conditions follow a standard pattern. Typically, conditions deal with:

(1) general obligations to perform the works;

(2) provisions for instructions, including variations;

(3) valuation and payment;

(4) liabilities and insurance;

(5) provisions for quality and inspection;

(6) completion, delay and extensions of time;

(7) role and powers of the certifier or project manager;

(8) disputes.

One of the main objects of the conditions of contract is to facilitate the efficient control and administration of the work, while at the same time providing certainty so that, for example, queries as to the nature of the work to be done are dealt with timeously. One of the recurrent problems under United Kingdom forms of contract is the extent to which the works are fully specified at the outset and the assumptions of the contract in this regard.

Frequently, there are additional conditions variously described as "special conditions" or "conditions of particular application". Such conditions will generally be construed on an equal footing with "general" conditions, but there is a rule of construction that greater weight should be given to conditions which have been particularly drafted against those which are of a standard nature. The principle is sometimes expressed as "type" prevailing over "print" and it can also be applied to handwriting prevailing over typescript. Lord Denning, giving a dissenting judgment in the *English Industrial Estates* case, above, described the principle thus:

> "In construing this contract we should have regard to provisions C and D (of the bills). They were carefully drafted and inserted in type in the bills of quantities. They were put in specially so as to enable the contractors to make their calculations. It was on the basis of these that the contractors made their tender and the employers accepted it. They were incorporated into the formal contract just as much as the conditions in the RIBA form. In contrast, conditions 12 and 16 were not specially inserted at all. They were two printed conditions in the middle of 23 pages of small print. It was in quite general terms. On settled principles they should have taken second place to the special insertion."

However, the other two Lords Justices did not agree that, by this means, the bills could be allowed to override the conditions. Had it not been for the express provision that the conditions were to prevail, however, the typed bill would have decided the meaning of the contract.

Specification

This is the document which describes the work to be carried out, often in great technical detail. There are, however, many ways in which work may be specified without such detail. For example one of the standard specifications used in many parts of the construction industry may be incorporated. There may also be reference to appropriate British standards or codes of practice. Alternatively, the specification may describe the performance required, leaving the details to the contractor. However, in this case if a JCT form of contract is used, a performance specification is in danger of being held ineffective in seeking to override the conditions of contract (see above). This was the case in *Mowlem v. British Insulated Callenders*,[12] where the bills required "waterproof concrete". This provision was held insufficient to impose a design liability on the contractor. In a JCT contract without quantities, the specification is the overriding description of the quality and quantity of the work required.

The specification may make requirements for the method of working to be adopted. In such a case, where it has full contractual effect, a change in the specified method may require a variation order, with the contractual consequences that that entails (see

---

[12] (1977) 3 Con. L.R. 64.

below). In a design-and-build contract, the specification acquires particular significance, because it must set out the employer's requirements to which the contractor's design, or detailed design, must comply. There are often also "contractor's proposals" as submitted with the tender, and these will also need to be incorporated as part of the specification (see below).

Bill of quantities

These documents originated historically as non-contractual measurements, taken off drawings to assist tenderers in quoting lump sum prices. The practice developed for tendering contractors to retain a quantity surveyor to draw up a bill which all tenderers could use as the common basis for their pricing. Bills subsequently acquired another use, namely for assessing interim payments by approximate measure under a lump sum contract. This is today the primary use of bills under a JCT form of contract. Under engineering contracts, conversely, a different practice developed of using the rates quoted, but recalculating (or remeasuring) the actual quantities of work carried out for the purpose of the final payment. A refinement of this process, effectively limited to civil engineering contracts, is the provision for adjustment to the quoted rates, where the actual quantity of any item of work of itself makes the quoted rate unreasonable or inapplicable.[13]

A further refinement in the use of bills of quantities is standardisation of the descriptions of work and of what those descriptions are deemed to include. The applicable rules are set out in a separate document known as a standard method of measurement, separate versions of which exist for different types of work. For JCT contracts, the RICS method is generally used; while ICE contracts use either the ICE method or, for some government contracts, the method of measurement for road and bridge works. These documents are not part of the contract, but it is provided that the bill of quantities is "deemed to have been prepared in accordance with" the appropriate standard method; and any error or omission in relation to the standard method becomes a "deemed variation" (see Chapters 12 and 13).

The substantive effect of the RICS and ICE standard methods may be quite different, reflecting the contrasts between building and civil engineering work. Civil engineering bills tend to be much shorter and often include many complex and difficult operations

---

[13] ICE, cl. 56(2).

rolled up into one simply described item. For example, construction of a tunnel may be measured as a single item, per metre
of length. Under the RICS method, although the JCT form of
contract does not permit or recognise claims for unforeseen ground
conditions, the method of measurement requires separate items to
be provided for "excavating in rock" and "excavating in running
silt or running sand." Thus, if such material is encountered, and
has not been billed, the contractor is entitled to payment for such
work as an extra. Under both methods, the normal way of
describing work items is in terms of the work or the product, for
example concrete or brickwork. An alternative way of producing
bills, sometimes used in civil engineering work, is to use "method
related" items, which are not to be paid by measured quantity, but
by fixed charge or time-related charge.

Drawings

All construction contracts have some drawings. It is usually
necessary to distinguish between those which have been incorporated into the contract (the contract drawings) and those which
follow, which may be amendments of the contract drawings or
further details necessary for the construction of the work. For
example, in a contract for the construction of a building with a
reinforced concrete frame, the contract drawings may show the
dimensions of the structural frame, and the bill of quantities will
record the quantities of concrete and reinforcing steel. The details
of the design, including drawings showing the placement of reinforcing bars, and bending schedules showing the shapes of individual bars, will usually be issued at a later stage, when the contractor
is approaching the point at which these details are required. In
practice, this type of detailing will often lead to large numbers of
drawings coming into existence after the work has started. This
situation sometimes leads to contractors claiming that the details
are more complex than had been envisaged or that they are not
issued in sufficient time, in either case giving rise to claims for
additional payment. Ideally, all drawings and details would be
issued at the date of the contract (except in the case of design-and-
build, or management contracts).

Drawings often contain notes and other written material, which
is to be construed as part of the drawing. Difficult questions of
construction can sometimes arise from such notes where they
conflict with provisions elsewhere in the bill or in the specification.
A particular problem for designers in large complex buildings is the
interaction between the different elements of the design, for

example, the services and the structure. Modern building services, particularly heating, ventilating and air-conditioning (HVAC), often require the provision of physically large ducts, which may interact with the structure and with other services and components. Design work is often carried out by different specialist teams; for example, HVAC design may be undertaken in part by a prospective nominated sub-contractor. The problems of integration may not always appear until the work is put in hand. These problems are typical of those which give rise to construction disputes.

Method statement

Recently a practice has grown up, particularly under civil engineering contracts, of requiring contractors to specify their intended method of construction. This was originally regarded as affording protection to the employer, by ensuring that he would have the benefit of the contractor's particular method (which could otherwise be changed at the contractor's option). However, in two recent cases, this principle has been applied in reverse, by contractors contending that a specified method has become impossible, leading to a requirement for the engineer to order a variation, so that the employer is required to pay the additional cost of changing the method. In the case of *Yorkshire Water Authority v. McAlpine*[14] the contractor gave a method statement, which was approved by the engineer, and incorporated into the contract. The method statement provided for pipe jacking to be carried out working upstream. The contractor maintained that the work was impossible within the meaning of clause 13 of the ICE conditions. It was held that, on the assumption that the work was impossible, the contractor was entitled to a variation order to carry out the work by some other method. This case was followed by the Court of Appeal in *Holland Dredging v. Dredging and Construction Co.*,[15] where the plaintiffs were dredging sub-contractors to the defendant, the main contractor, for the construction of a sea outfall pipeline under conditions of contract substantially the same as the ICE fifth edition. The sub-contract incorporated a method statement which defined the area from which material could be excavated for backfilling. There proved to be insufficient material within the specified area, and the plaintiffs incurred extra cost in winning additional material from elsewhere. The Court of Appeal held that the method statement was to be given full weight as a contract document. Purchas L. J. concluded:

---

[14] (1985) 32 B.L.R. 114.
[15] (1987) 37 B.L.R. 1.

"The occurrence of the shortfall and the consequential necessity to look elsewhere for material necessary to backfill the trench to the pre-existing levels disclosed an omission in the specification and bill, unless it is to be accepted that the sub-contract is impossible to complete on its terms as agreed."

Employers should, therefore, be clear as to the effect of a method statement if they wish to have such a document incorporated into a contract. While these cases were decided under the ICE conditions, the same principles could apply under a JCT or other lump sum contract on the basis that the bills and other description of the work restrict the means by which the contractor may carry out the work.

Other documents

The contract itself is likely to be formed by signing a specially prepared "form of agreement" or "articles of agreement" which may be provided with the printed form of contract, or drawn up specially. This gives the opportunity of having the contract executed under seal, thereby increasing the limitation period. Another function of a formal contract is to list the contract documents themselves, and this gives the opportunity of including any other documents to be incorporated. Additional documents may include the contractor's tender and correspondence in which the parties have negotiated the final agreement (this may qualify the tender or the conditions of contract). Many contracts contain "conditions of tendering" or like documents which set out matters which may become relevant to contractual disputes. These are usually not incorporated, but each contract depends on its own terms.

A document frequently referred to, but usually not incorporated, is the site investigation or other data concerning soil conditions. These are usually provided for the contractor's information, and frequently contain a statement disclaiming responsibility or requiring the contractor to form his own judgement. Such documents will still be relevant to a claim for unforeseen ground conditions and it is unnecessary for this purpose for the investigation to be a full contract document. Even when incorporated, it usually adds little to the contractor's protection: a typical ground investigation consists of factual statements which are highly specific to the particular probes or tests that have been carried out, and any interpretation that may be offered will be no more than a statement of opinion.

The works

This expression appears in most construction contracts as part of the stated obligation undertaken by the contractor. It is always a matter of construing the contract to discover the meaning of the expression, and whether the term contains the whole of the contractor's obligation. For example, the contractor may be obliged to carry out "the works" as defined, but also liable for their performance, in which case the contractor will be bound to carry out any other work necessary to comply with such requirements. Each contract depends on its own terms.

Under civil engineering contracts, there is a potentially important distinction between "permanent works" and "temporary works," both expressions being included within the overall term "works." The precise distinction between the two expressions is often unclear, and there may be an overlap, for example, in relation to steel sheet piling intended to facilitate the construction process, which then becomes incorporated and left in as part of the permanent works. The importance of the distinction relates to the placing of responsibilities for different categories of work. Also, under standard methods of measurement, temporary works are not usually included or priced in the bill, unless they are designed by the engineer or are otherwise of sufficient importance to justify inclusion in the bill. Temporary works are not included at all in building contracts, it being assumed that the contractor will carry out all such necessary works within the prices quoted for the measured work.

VARIATIONS

One of the common features of construction contracts is that the design of the work contracted for may require variation as the work proceeds. The magnitude of variations tends to be greater in engineering works, reflecting the greater element of the unknown in such operations. But it is still a rare event for even the smallest of building jobs to be completed exactly according to the original contract provisions. Strictly the contractor is not bound, without express provision, to execute more than the contract work; and the employer will be in breach of contract if he omits a part of the work included in the contract without a contractual provision enabling him to do so. Construction contracts, therefore, provide that the employer (or his agent) may require alterations, additions or omissions to the contract work and bind the contractor to carry them out.[16]

---

[16] See JCT, cl. 13; ICE, cl. 51.

When there has been a departure from the work specified in the contract, it is necessary to decide whether there is, in law, a variation under the contract; if there is a variation, whether the contractor is entitled to be paid extra; and if so, the amount of the extra payment. These questions are considered below.

Contractual variations

Contractors sometimes make claims on the basis that the contract work has cost more than was anticipated. This is not a variation and the contractor is entitled to no extra payment unless he can bring a claim under the contract for additional payment. When the contractor has undertaken to carry out and complete the work for a stated price he is bound to do so, however difficult or expensive it may prove to be. Thus, in the old case *Bottoms v. Mayor of York*[17] the contractor undertook to build sewerage works in unknown ground which turned out to be marshy. He abandoned the works when the engineer refused to authorise additional payment. It was held that since there was no express warranty as to the nature of the site, the contractor was entitled to no additional payment. Lord Esher said:

> "In such circumstances, contractors sometimes seek to rely The corporation insisted upon his observing the contract and going on with it. But he resisted and refused to have anything more to do with it. If that be true, he brought himself into a very difficult position, and was not able to enforce any payment whatever. . . . I take it that the real reason why he has come to this misfortune, indeed, is that he would go and tender when there was no guarantee given to him as to the kind of soil, and when there was no information given to him as to what the soil was—when there was no contract entered into by the people who asked him to tender as to what the nature of the soil was, and that he either too eagerly or too carelessly tendered and entered into the contract without any such guarantee or representation on their part, and without due examination and enquiry by himself. That is what has produced the difficulty."

In such circumstances, contractors sometimes seek to rely on any instruction or direction which may be given by the engineer or architect.[18] However, an instruction relating to an existing obligation will not entitle the contractor to additional payment.[19] In the case illustrated, the contractor may well have had a claim under the

---

[17] (1892) *Hudson's Building Contracts* (4th ed.), Vol. II, p. 208.
[18] *Simplex Concrete Piles v. St Pancras B.C.* (1958) 14 B.L.R. 80.
[19] *Howard de Walden v. Costain* (1991) 55 B.L.R. 124.

modern version of the ICE Conditions for unforeseen conditions. In the absence of such a claim, the contractor must bear the loss.

The contract work

Extra work for which the contractor is prima facie entitled to be paid must constitute something additional to the work contracted for. It is, therefore, necessary to construe the contract to ascertain whether work claimed as extra is covered by the contract. The shorter and simpler the description of the work to be carried out, the more difficult it will be for the contractor to contend that work is extra. Thus, in a contract to construct "a three bedroomed house," the scope for variations may be limited to extra bedrooms. In *Sharpe v. San Paulo Railway Co.*[20] the contractor had agreed to build a railway in Brazil between fixed termini for a stated sum. Redesign became necessary as a result of difficulties in completing the work. There was no extra to the contract, and the contractor was entitled to no additional payment. James L.J. said:

> "The (plaintiff) says that the original specification was not sufficient to make a complete railway and that it became obvious that something more would be required to be done in order to make the line. But their business, and what they had contracted to do for a lump sum, was to make the line from terminus to terminus complete, and both these items seem to me to be on the face of them entirely included in the contract. They are not in any sense of the word extra works."

Where the contract is to carry out itemised works without an overall obligation to deliver a finished product, the contractor may then contend that any items omitted from the contract description are extras. In each case, this depends upon the contract documents reasonably interpreted. Many items will be taken as necessarily included even though not specifically mentioned. In *Williams v. Fitzmaurice,*[21] where the contract was to build a house for a fixed sum, the specification omitted to mention the floorboards, and the contractor claimed the boards were an extra. It was held that they must be taken to be included. Pollock C.B. said:

> "It is clearly to be inferred from the language of the specification that the plaintiff was to do the flooring, for he was to provide the whole of the material necessary for the completion of the work; and unless it

---

[20] (1873) L.R. 8 Ch. App. 597.
[21] (1858) 3 H. & N. 844.

can be supposed that a house is habitable without any flooring, it must be inferred that the flooring was to be supplied by him. In my opinion the flooring of a house cannot be considered an extra any more than the doors or windows."

Where the contractor is obliged to carry out the whole project, he may nevertheless be entitled to payment for additional items under the terms of the contract, as automatic or "deemed" variations. This is usually the case where there is a bill of quantities drawn up by reference to a standard method of measurement. Items which should have been included in the bill, so that they would have been priced by the contractor, are to be paid for as extra work.[22] In addition, under the ICE conditions, increases or decreases in the actual quantities of work required are to be treated, for the purpose of payment, as variations (clause 56(2)).

Cost-plus contracts

In addition to these traditional forms of contract, there are many alternative forms where the concept of a "variation" may be of less importance. The simplest type is an agreement to pay the contractor the cost of the work (usually by some specified means of calculation) plus a further sum which may be called profit, or overheads, or a fee. The statement of quantities or a contract sum will be of little significance. An example of this type of contract is the JCT Fixed Fee Form of prime cost contract. An alternative form is the "target" contract under which the contractor is paid the cost of the work together with an additional sum which varies according to how close the final cost is to the pre-stated "target." This is intended to give the contractor an incentive to adhere to a particular figure, there being no such incentive under the "fixed fee" form. These contracts, however, tend to become more difficult to operate when made subject to a "cap", beyond which the contractor is to accept all financial liability.

Payment for extras

If the contractor carries out work which is an extra to the contract, he will be able to recover payment for that work only if he can show that the employer is bound under contract to pay. The mere doing of extra work, or doing work in a way different from that specified, does not of itself bind the employer to pay for

---

[22] See JCT, cll. 14.1, 2.2 and ICE, cl. 55(2).

extras. If the building contract provides for the ordering of and payment for extras, the contractor may claim payment under the contract, provided that any condition precedent to payment is satisfied. Most contracts provide that a written order is necessary. However, there may be an implied promise to pay if an appropriate order is refused. In *Molloy v. Leibe*[23] a building contract provided that no payment for extras would be made without a written order, it was held that there was an implied promise to pay for the works if they were extras. Lord Macnaughten held:

> "As Molloy insisted on the works being done, in spite of what the contractor told him, the umpire naturally inferred . . . that the employer impliedly promised that the works would be paid for either as included in the contract price or, if he were wrong in his view, by extra payment to be assessed by the architect. It is difficult to see how the umpire could have drawn any other inference from the facts as found by him without attributing dishonesty to Molloy."

As an alternative, the question whether a variation order should be given for work which the employer insists on having done, will usually be within the powers of an arbitrator appointed under the contract, so that the requirement for an order in writing is a formality only.

If a promise is made to pay for extra works, that promise may be enforceable as a separate contract whether or not the extras are claimable under the construction contract. This will not avail the contractor if the "extras" are in fact no more than he was bound to do under the building contract, since the promise is then unsupported by consideration. But the court took a different view in *William v. Roffey*.[24]

Most contracts which make provision for extras also lay down means of valuation and these will determine what the contractor is entitled to be paid. In the absence of such provision, extra work will be paid for at the contract rates or at reasonable rates. In addition, it is common for building contracts to provide for some additional payment (or "claim") if the ordering of variations causes expense beyond the payments allowed. The JCT form allows recovery of loss or expense not recoverable elsewhere in the contract (clause 26). The ICE form permits a refixing of the rates for any items of work in addition to those varied (clause 52(2)). The International (FIDIC) form allows the contractor to recover additional payment if variations exceed 15 per cent, of the net contract sum.

---

[23] (1910) 102 L.T. 616.
[24] [1991] 1 Q.B. 1; see Chap. 6.

Contracts do not usually place any limit on the permissible extent of variations. The usual provision that no variation is to vitiate (or invalidate) the contract makes it difficult to imply any limit. However, there must always be some limit to what may be added to a contract. If work exceeding such limit is ordered, the contractor may be entitled to be paid on a *quantum meruit* basis (see Chapter 5). The question of such entitlement arose in the case of *Thorn v. London Corporation.*[25] In that case, the Lord Chancellor said:

> "Either the additional and varied work which was thus occasioned is the kind of additional and varied work contemplated by the contract, or it is not. If it is the kind of additional or varied work contemplated by the contract, he must be paid for it, and will be paid for it, according to the prices regulated by the contract. If, on the other hand, it was additional or varied work, so peculiar, so unexpected, and so different from what any person reckoned or calculated upon, that it is not within the contract at all; then, it appears to me, one of two courses might have been open to him; he might have said: I entirely refuse to go on with the contract—*Non haec in foedera veni*: I never intended to construct this work upon this new and unexpected footing. Or he might have said, I will go on with this, but this is not the kind of extra work contemplated by the contract, and if I do it, I must be paid a *quantum meruit* for it."

CERTIFICATES

A common feature of construction contracts is a provision for an independent third party to issue certificates signifying particular events and usually embodying administrative decisions. The events range from sums becoming due to one party to extensions of time and other matters. A certificate is merely a manifestation of the parties' private agreement and its effect is no more than the parties have agreed it to be. The duty of issuing certificates is usually given to the architect or engineer under the contract, either as a personal appointment or as a firm. In this section such a person is referred to as the certifier. In modern building contracts the role of the certifier is invariably to act impartially between the employer and the contractor. This is distinct from the other role of the engineer or architect as the employer's agent, when he must act as agent in the best interests of his principal. The role of the certifier was clarified in the case of *Sutcliffe v. Thackrah*,[26] where Lord Reid said:

---

[25] (1876) 1 App. Cas. 120; see Chap. 6.
[26] [1974] A.C. 727.

"The building owner and the contractor make their contract on the understanding that in all such matters the architect will act in a fair and unbiased manner and it must therefore be implicit in the owner's contract with the architect that he shall not only exercise due care and skill but also reach such decisions fairly, holding the balance between his client and the contractor."

The function of a certificate, is usually to record factual events only. But this frequently involves the certifier forming a judgment or giving an opinion, for example as to the value of work performed or whether it complies with the terms of the contract. In some cases he can impose his own standard, such as where particular items of work are required to be to the "approval" or "satisfaction" of the engineer. These phrases can create considerable difficulty in their application. Depending on the context, the engineer's approval may be to no more effect than that the work complies with the express terms of the contract, but more often it will have the effect of imposing some further unstated requirement. This can lead to dispute as to whether material which has been "approved" can subsequently be rejected for latent defects. Generally such approval does not preclude subsequent rejection.[27]

A certificate may be conclusive as to what it purports to certify. However, most modern construction contracts draw a firm distinction between "interim" and "final" certificates, with the latter only being given qualified finality.[28] What is more difficult is the status of a certificate which is "interim," but where the contract permits no review other than by arbitration at completion of the works. This may apply, for example to an "interim" extension of time.[29] Generally, the certifier will be entitled to exercise the function once only and will therefore be *functus officio* thereafter.

A requirement for any certificate is that it must be properly made in order to have effect as provided in the contract. Thus, a certificate which is not in the correct form or which is given by the wrong person is of no effect. Subject to this the courts will uphold the parties' agreement as to the effect of a certificate. Thus under the JCT form the House of Lords have held that the courts are bound by a final certificate which is to be conclusive evidence that the work has been properly carried out. In *Kaye v. Hosier & Dickinson*[30] an architect gave his final certificate under a JCT contract during the course of court proceedings concerning defects.

---

[27] But see *Rotherham v. Haslam* (1996) 78 B.L.R. 1, CA.
[28] See JCT, cl. 30.9. The ICE final certificate involves no finality.
[29] See ICE, cl. 44.
[30] [1972] 1 W.L.R. 146.

The certificate was held to have the effect of preventing the employer from continuing to contend that the work had been executed defectively. Lord Pearson said:

> "The architect's function is not primarily or essentially an arbitral function. The works have to be carried out to his satisfaction, and accordingly he must give or withhold his expression of satisfaction. He may notify defects and require them to be made good. He has to issue certificates showing how much money is owing. Incidentally, his certificates and instructions may resolve some controversial points, and he has to act fairly, but he is not primarily or characteristically adjudicating on disputes. If in a contract such as this the parties agree that the architect's final certificate shall be conclusive evidence of certain matters, I do not think that there is any invasion of the court's jurisdiction or any affront to its dignity. The court's function in a civil case is to adjudicate between the parties, and if they have agreed that a certain certificate shall be conclusive evidence the court can admit the evidence and treat it as conclusive."

The fact that the law recognises and requires a fair and unbiased decision from the certifier, who is in other instances bound to act only in the employer's interest, is remarkable. This is particularly so when the certifier is an employee of the building owner, as is often the case in local authority contracts. Unfortunately, despite the professional independence of many engineers and architects acting in this role over many decades, there has grown up over recent years concern as to whether this role should continue. Within the United Kingdom a number of contracts have been produced without independent certifiers, with persons appointed to the role of "employers' representative" and "adjudicator". At the same time, particularly in foreign contracts based on the FIDIC form, employers have sometimes sought to appoint themselves as the certifier, effectively creating a right for one party to change the terms of the contract.[31] The future role of the certifier is therefore uncertain. It is to be noted that outside the United Kingdom, for example in the United States of America, many substantial contracts operate without an appointed certifier.

Types of certificate

Certificates may be of three types. First, interim or progress certificates are those which are issued periodically during the course of the work to signify quantities of work carried out and interim payments due to the contractor. The most usual contrac-

---

[31] See *Balfour Beatty v. DLRL* (1996) 78 B.L.R. 42.

tual provision is for a monthly valuation and certificate. Such payments form a vital part of the economics of contracting, sometimes referred to as the "life-blood" of the industry. An interim certificate, properly given, creates a debt due from the employer. In a series of cases, starting with *Dawnays v. Minter*,[32] the Court of Appeal held that an interim certificate was payable without set-off, save for liquidated or established claims. This case was disapproved by the House of Lords in *Gilbert–Ash v. Modern Engineering*,[33] which held that the general right of set-off was available against sums certified in favour of the contractor or a sub-contractor. In this case, the main contractor had refused payment due to a sub-contractor upon an architect's certificate, relying on a cross-claim. Lord Dilhorne held:

> "Consideration of the terms of the main contract leads me to the following conclusions: There is nothing in it to justify the conclusion that it excludes the contractor's right to counter-claim and set-off under the common law and in equity. . . . An interim certificate does not create a debt of a special nature. It is a certificate of the value of work properly executed and it is only for the work properly executed, less any deduction that may properly be made, that the employer has to pay the contractor and the contractor to pay the portion attributed to the sub-contractor. . . . I see no ground for holding that . . . the contractor cannot seek to deduct from the amount claimed from him the amounts bona fide claimed by him from the sub-contractor. Even if the subcontract does not give, as it does, an express right to make such deduction, I can see nothing in it to exclude the contractor's common law and equitable rights to set off and counter-claim."

Thus, where there has been delay or defective work, the employer may generally withhold the amount of his cross-claim from certified sums due to the contractor, and the contractor may similarly withhold from a sub-contractor. The unrestrained exercise of rights of set-off, however, has given rise to serious economic strains. This has led to the introduction of a new process referred to as "adjudication" by which such cross-claims are subject to rapid and summary review.[34] The Latham Report has recommended statutory support for the inclusion of such a scheme in all construction contracts.[35] Even without adjudication, the new Arbitration Act 1996 contains measures enabling rapid interim review of the withholding of sums due.[36]

---

[32] [1971] 1 W.L.R. 1205.
[33] [1974] A.C. 689.
[34] JCT sub-contract, NSC/4, cl. 24.
[35] Currently contained in the Housing Grants, Construction and Regeneration Act 1996.
[36] s. 39—see Chap. 3.

A different aspect of "finality" arose in the case of *Lubenham v. South Pembroke D.C.*[37] where an architect had wrongly deducted liquidated damages from an interim certificate. The error was patent but the employer refused to pay the difference and the contractor terminated his employment on the grounds of non-payment. The court treated the certificate as effective even though demonstrably wrong. May L.J. delivering the judgment of the court said:

> "We can for our part see no sufficient reason for differentiating as suggested between certificates which contain patent errors and those which contain latent errors. Whatever the cause of the undervaluation, the proper remedy available to the contractor is, in our opinion, to request the architect to make the appropriate adjustment in another certificate, or if he declines to do so, to take the dispute to arbitration under clause 35. In default of arbitration or a new certificate the conditions themselves give the contractor no right to sue for the higher sum. In other words we think that under this form of contract the issue of a certificate is always a condition precedent to the right of the contractor to be paid."

It is clear that such a certificate may be challenged, but only by invoking the appropriate contract machinery, and until revised, the certificate must stand.

The second type of certificate is the final certificate which may be issued after completion of the works. A final certificate may fulfil either or both of two functions. It may state the sum finally due to or from the contractor; and it may certify approval of the works. The final certificate issued under the JCT form (clause 30.9) fulfils both functions. Under the ICE conditions, the final certificate is merely a document of account (clause 60(4)) and the maintenance certificate signifies final completion of the work (clause 61).

The third type of certificate is that which records some event for the purposes of the contract. Examples of this type are certificates of substantial completion[38] or practical completion[39] of the works; and a certificate of non-completion where the work is delayed.[40] An extension of time given by the architect or engineer, although not so called, has the effect of a certificate.

Recovery without a certificate

There is a substantial amount of case law concerning the recovery of money under construction contracts where no certifi-

---

[37] (1986) 33 B.L.R. 39.
[38] ICE, cl. 48.
[39] JCT, cl. 17.
[40] JCT, cl. 24.

cate has been given. This situation arises only when, on the construction of the contract, the certificate is a condition precedent to recovery. Most of the cases concern allegations that the certifier has acted improperly, for example, by colluding with the client. In such circumstances, the courts have, in a number of cases, held the certifier to be disqualified so that the contractor was entitled to recover payment in the absence of a certificate. Thus, if the leading case of *Panamena Europa v. Leyland*[41] the certifier was alleged to have considered extraneous matters before coming to his decision. The certifier insisted upon being satisfied that the work had been done economically, but it was held that the contract limited his function to deciding whether the work was satisfactory. The contractor was accordingly entitled to recover payment without the certificate. In giving judgment Lord Thankerton held:

> "[The surveyor] declined to proceed with the matter unless he was provided with the information to which on his erroneous view of the contract he held himself entitled; in this view [the employer] concurred and this position was maintained up to and after the issue of the writ. This means that an illegitimate condition precedent to any consideration of the granting of the certificate was insisted on by [the surveyor] and by [the employer]. It is almost unnecessary to cite authority to establish that such conduct on [the employer's] part absolved [the contractor] from the necessity of obtaining such a certificate and that [the contractor] is entitled to recover the amount claimed in the action. . . . If [the employer] had taken the contrary view of their surveyor's function under clause 7, it would have been their duty to appoint another surveyor to discharge that function."

However, under most modern construction contracts the actions and decisions of the certifier are expressly open to challenge in arbitration. A typical arbitration clause empowers the arbitrator to "open up, review and revise any certificate, decision, opinion"[42] of the architect. Instead of seeking to persuade the court that the certifier should be regarded as disqualified, the contractor may simply seek a second opinion from an arbitrator. Most modern contracts permit arbitration, at least on the withholding of certificates, to proceed while the work is continuing, so that questions of disqualification will rarely arise. Furthermore, provisions of the Arbitration Act 1996 will permit a rapid provisional review of any withholding of payment (see Chapter 3). These measures are likely to lead to a major change of balance in the relationship between the parties to a construction contract and the certifier.

---

[41] [1947] A.C. 428.
[42] JCT 80, cl. 41.4.

Immunity of certifier

It has long been the law that judges and those performing judicial functions, including arbitrators, are generally immune from actions in negligence. The immunity of arbitrators except where acting in bad faith, is confirmed under the Arbitration Act 1996, s. 29. In an earlier case it was held by the Court of Appeal[43] that such immunity applied to an architect giving a final certificate. This was, however, overruled by the House of Lords in the leading case of *Sutcliffe v. Thackrah.*[44] In this case the architect gave an interim certificate, including work not properly done. The builder, having been overpaid for the work, became insolvent so that the employer could not recover the loss. The architect contended that, even though negligent, he was immune from action. In holding the architect liable for negligence, the court found no inconsistency in owing a duty to the employer to act with due care and skill, while being under a duty to hold the balance fairly between his client (the employer) and the contractor. The rule giving immunity applied only where there was a dispute which called for a judicial decision. Such immunity is not confined to formal arbitration proceedings, and might include a person acting as adjudicator. An adjudicator appointed under the Housing Grants, etc., Act 1996 is entitled to contractual immunity.[45]

ENGINEERS AND ARCHITECTS

A particular feature of construction contracts is the position in law of the engineer or architect. This varies according to the function being performed. It is important for him, and for those affected by his decisions, to know his status. The engineer or architect may perform functions as the agent of, or as an independent contractor engaged by the employer, or as an impartial certifier. He may also do things which incur a duty under the law of tort to other persons. The position of the certifier (see above) and the relationship between principal and agent generally (see Chapter 7) are considered elsewhere. In this section specific duties and liabilities to the employer and to others are considered. An additional factor to be considered in some modern construction contracts is the imposi-

---

[43] *Chambers v. Goldthorpe* [1901] 1 K.B. 624.
[44] [1974] A.C. 727.
[45] s.108(4).

tion by the employer of constraints upon the exercise of the functions of the certifier. This may take the form of placing limits upon the power to certify, or requiring impartial decisions to be confirmed by the employer. The impact of such requirements remains to be considered by the courts.

Duties to the employer

The scope of the work normally performed by the engineer or architect may be divided broadly into pre-contract duties and duties which arise under or by virtue of the construction contract. In the pre-contract stage, the duty is to prepare skilful and economic designs for the works, acting as an independent contractor for the employer (unless the designer happens to be the employee of the building owner). When the work is in progress, the duties arising under or by virtue of the contract are to supervise and administer the carrying out of the works in the best interests of the employer. Such functions will generally be performed as the agent of the employer. In each case the duty is owed in contract and the common law requires such duties to be exercised with reasonable skill and care. Whether particular conduct will incur liability for any resulting loss depends primarily upon established practice, that is, whether the conduct falls within the range to be regarded as acceptable. The required standard was laid down in *Bolam v. Friern Hospital*,[46] which was a medical negligence case decided by a jury (like many building cases in the USA). The summing up in the case is still regarded as the classic exposition of the appropriate standard of case. McNair J. described the standard as follows:

> "Where you get a situation which involves the use of some special skill or competence, then the test as to whether there has been negligence is not the test of the man on top of the Clapham Omnibus, because he has not got this special skill. The test is the standard of the ordinary skilled man exercising and professing to have that special skill. A man need not possess the highest expert skill; it is well established law that it is sufficient if he exercises the ordinary skill of an ordinary competent man exercising that particular art."

This is ultimately a question (in the absence of a jury) for the judge to decide. But in practice unless the failure is gross and obvious it is usually necessary to call a person practising in the same technical

---

[46] [1957] 1 W.L.R. 582.

field to give expert evidence as to the breach of duty complained of.[47]

There may be occasions on which the court will find that the parties intended a different standard of duty. Where the engineer or architect is employed by a contractor who is himself liable for the fitness for purpose of the works, the court may impose a higher duty. In *Greaves Contractors v. Baynham Meikle*[48] an engineer was employed in such circumstances to design the structure of a building known to be subject to vibrating loads. The floors were not adequately designed to resist the vibrations. The court accepted that the engineer had not failed to exercise reasonable skill and care, but found there to be an implied term of his engagement that the building would be fit for its purpose. The engineer was therefore held liable. Lord Denning held:

> "The law does not usually imply a warranty that (a professional man) will achieve the desired result, but only a term that he will use reasonable care and skill. The surgeon does not warrant that he will cure the patient. Nor does the solicitor warrant that he will win the case. But when a dentist agrees to make a set of false teeth for a patient, there is an implied warranty that they will fit his gums. What then is the position when an architect or an engineer is employed to design a house or a bridge? Is he under an implied warranty that, if the work is carried out to his design, it will be reasonably fit for the purpose? Or is he only under a duty to use reasonable care and skill? This question may require to be answered some day as a matter of law. But in the present case I do not think we need answer it. For the evidence shows that both parties were of one mind on the matter. Their common intention was that the engineer should design a warehouse which would be fit for the purpose for which it was required. That common intention gives rise to a term implied in fact."

This case has not, however, been followed, and a number of other reported cases have restated the ordinary duty to be one of reasonable skill and care.[49]

Design duties

What constitutes reasonable skill and care in the design of work depends upon the circumstances of each case. The duty may normally be discharged by following established practice, but there is no rule that doing what others do cannot give rise to liability.

---

[47] See *Lusty v. Finsbury Securities* (1991) 58 B.L.R. 66.
[48] [1975] 1 W.L.R. 1095.
[49] *George Hawkins v. Chrysler* (1986) 38 B.L.R. 36.

There may be situations where there is no established practice, such as where a new construction technique is used. In such cases the duty of reasonable skill and care may be discharged by taking the best advice available and by warning the employer of any risks involved. In an old case, *Turner v. Garland & Christopher*[50] the employer had instructed his architect to use a new patent concrete roofing which proved to be a failure. It was held that where an untried process was used, failure might still be consistent with reasonable skill. This case was also tried with a jury. Earle J. said, in summing up:

> "The plaintiff will merit your verdict if the defendant was found to be wanting in the competent skill of an ordinary architect. If he possesses competent skill and was guilty of gross negligence, although of competent skill, he might become liable. If of competent skill, he had paid careful attention to what he undertook, he would not be liable. You should bear in mind that if the building is of an ordinary description in which he had had abundance of experience, and it proved a failure, this is an evidence of want of skill or attention. But if out of ordinary course, and you employ him about a novel thing, about which he has had little experience, if it has not had the test of experience, failure may be consistent with skill. The history of all great improvements shows failure of those who embark on them; this may account for the defect of roof."

However, when a novel design is to be undertaken, or where tried and traditional methods are to be superseded by cheaper processes, the risks involved must be brought to the attention of the employer who is to bear such risk. It is the employer who must decide the course to adopt, and he must be given all necessary information to enable him to reach a proper and considered decision.

When dealing with the liabilities of the parties and with insurances, the standard forms of contract often use the criterion of whether or not loss or damage is caused by the design of the works.[51] Such a criterion does not necessarily coincide with that of whether the designer has exercised sufficient skill and care. In *Queensland Railways v. Manufacturers Insurance*[52] a river bridge failed during erection because the piers were subjected to forces beyond those which could be predicted by existing knowledge. It was held that the failure was in fact due to faulty design, even though there might be no fault attributable to the designer.

---

[50] 1853; *Hudson's Building Contracts* (4th ed.), Vol. II, p. 1.
[51] ICE, cl. 20 and JCT, cl. 21.
[52] [1969] 1 Lloyd's Rep. 214.

Delegation of design

There may be many situations where design work is undertaken by persons other than the engineer or architect named in the building contract. Problems may then arise as to who can be sued for a design defect. If the employer directly employs a consultant or even the contractor to do design work, he will have a remedy for design defects, depending upon the terms of the contract. Where the engineer or architect himself delegates design work, there is no contract between the employer and the designer. As a general rule the engineer or architect will remain liable for the design unless the employer concurs in a delegation of responsibility. Thus, in *Moresk Cleaners v. Hicks*,[53] an architect delegated the design of a reinforced concrete structure to the contractor. The design proved to be defective and it was held that the architect was liable. The official referee in his judgment said:

> "If the defendant was not able, because this form of reinforced concrete was a comparatively new form of construction, to design it himself, he had three courses open to him. One was to say: 'This is not my field.' The second was to go to the client, the building owner, and say: 'This reinforced concrete is out of my line. I would like you to employ a structural engineer to deal with this aspect of the matter.' Or he can, while retaining responsibility for the design himself seek the advice and assistance of a structural engineer, paying for his service out of his own pocket but having at any rate the satisfaction of knowing that if he acts upon that advice and it turns out to be wrong, the person whom he employed to give the advice will owe the same duty to him as he, the architect, owes to the building owner."

However delegation may be permissible in the case of specialist processes. In the case of *Merton LBC v. Lowe*[54] an architect was held to have a design responsibility in respect of a proprietory plaster system used for a swimming pool ceiling, but was found not to be in breach because he was entitled to rely on the specialist manufacturer's expertise, where details of the design were not revealed. Waller L.J. held:

> "It was submitted (by the plaintiffs) that the fact that Pyrok (the subcontractor) maintained secrecy was immaterial, and reliance was placed on the case of *Moresk Cleaners v. Hicks*. I entirely agree with the judgment in that case. There the architect had literally handed over to another the whole task of design. The architect could not

---

[53] [1966] 2 Lloyd's Rep. 338.
[54] (1982) 18 B.L.R. 130.

escape responsibility for the work which he was supposed to do by handing it over to another. This case was different. Pyroc were nominated sub-contractors employed for a specialist task of making a ceiling with their own proprietory material. It was the defendant's duty to use reasonable care as architects. In view of successful work done elsewhere, they decided that to employ Pyrok was reasonable. No witness called suggested that it was not at the beginning."

The architects were, however, held liable under their general design responsibility, for failing to take adequate steps to remedy the design deficiency which subsequently became apparent.

Direct warranty

If the design work is done by a nominated sub-contractor, the employer may protect himself by obtaining a direct (or collateral) warranty from the sub-contractor. This constitutes a separate contract under which the sub-contractor warrants his work, usually in consideration of his nomination by the employer. The warranty must be given before the sub-contract is entered into, otherwise there is no consideration and the warranty may be unenforceable. The principle was first applied to a construction contract in *Shanklin Pier v. Detel Products*.[55] In this case a supplier stated to the employer that his paint had a life of seven to 10 years. The particular paint was specified by the employer and duly used by the contractor. The paint in fact lasted for about three months. It was held that the statement made concerning the quality of the paint constituted a warranty so that the employer was entitled to damages from the supplier for breach. Payment of a fee will be equally effective as consideration to support any such warranty.

Standard forms of warranty have been issued by the RIBA and now the JCT (see Chapter 11). The general subject of warranties has become more important since the retrenchment which has occurred in the law of negligence. Developers and purchasers who hitherto relied on tort claims against those with whom they had no contract are now, in general, unable to advance such claims. Lawyers have responded by producing a new generation of warranties intended to replace such rights by creating directly enforceable contractual obligations. No new principle of law is involved in these documents, but the following points should be considered when preparing or giving a warranty:

(1) Who is to give the warranty? Frequently warranties are requested from a variety of parties involved in a con-

---

[55] [1951] 2 K.B. 854.

struction project, sometimes from all parties. Where professional organisations take on work as a partnership or limited company, care should be taken not to allow individuals to sign warranties, so losing any corporate protection.

(2) Terms of the warranty: the warranty may be made to cover any part of the work, not limited to that which is the responsibility of the person giving the warranty. The party signing should be careful to understand what he is taking on.

(3) Standard of duty: warranties requested from professionals sometimes require a promise that the work will be reasonably fit for its purpose, going beyond the ordinary duty of reasonable skill and care.

(4) Limitation period: warranties may be expressed so as to extend the ordinary limitation period in contract, either by requiring the warranty to run for a stated period or by requiring an indemnity.

(5) Assignment of rights: warranties frequently provide that the recipient (usually the developer) may assign the warranty or rights in it to third parties who acquire an interest in the building or works. This is a substantial enlargement of potential responsibility and the terms should be checked carefully.

(6) Contribution rights: a party who gives a warranty has a right to expect that those who may share responsibility for any loss will be capable of being sued for contribution, on the basis that they are also shown to be liable. The most convenient way of ensuring this is to provide that the warranty is not to become effective unless and until other (identified) parties involved in the project have also given similar warranties.

(7) Indemnity insurance: professionals in particular must check with their insurers whether they are covered in respect of the wider responsibility created by the warranty; many insurers will not cover duties beyond reasonable skill and care.

(8) Fee: there is no reason why the person giving the warranty should not charge a realistic fee, although some developers will offer design commissions only on condition that the professional will agree to give warranties at no additional cost.

(9) Standard forms: a number of these have evolved, issued by professional bodies and by insurers. There are, however, many ad hoc forms in circulation, often drafted for individual projects or clients.

The effect of a contractual warranty on a claim in tort was considered in two cases. In *Greater Nottingham Co-op. v. Cementation*[56] the Court of Appeal held that the taking of a warranty from a sub-contractor covering design but not performance of the work precluded a duty in tort relating to the performance of the work, on the footing the parties had had the opportunity to create a direct duty and had not done so. Conversely, however, in *Warwick University v. McAlpine*[57] Garland J. held that the failure to place any direct duty in contract on a sub-contractor did not prevent the existence of a duty of care in tort. The reduction in the ambit of tort claims makes it unlikely that there will be further developments in this area of the law (see further Chapter 5).

Supervision and quality assurance

The purpose of supervision is to ensure that the works are carried out by the contractor in accordance with the requirements of the construction contract; and the engineer or architect must provide reasonable supervision for this purpose. The amount of supervision required depends on the nature of the works. The building of a house may require visits every two weeks, while engineering operations may require constant attention from a resident staff. The duty of supervision was considered in *East Ham v. Bernard Sunley*,[58] a case concerning the meaning of "reasonable examination" under the JCT form of contract. Lord Upjohn said:

"As is well known, the architect is not permanently on the site but appears at intervals, it may be of a week or a fortnight, and he has, of course, to inspect the progress of the work. When he arrives on the site there may be many very important matters with which he has to deal: the work may be getting behind-hand through labour troubles; some of the suppliers of materials or the sub-contractors may be lagging; there may be physical trouble on the site itself, such as, for example, finding an unexpected amount of underground water. All these are matters which may call for important decisions by the architect. He may in such circumstances think that he knows the builder sufficiently well and can trust him to carry out a good job;

---

[56] [1989] Q.B. 71.
[57] (1988) 42 B.L.R. 1.
[58] [1966] A.C. 406.

that it is more important that he should deal with urgent matters on the site than that he should make a minute inspection on the site to see that the builder is complying with the specifications laid down by him. . . . It by no means follows that, in failing to discover a defect which a reasonable examination would have disclosed, in fact the architect was necessarily thereby in breach of his duty to the building owner so as to be liable in an action for negligence. It may well be that the omission of the architect to find the defects was due to no more that an error of judgement, or was a deliberately calculated risk which, in all the circumstances of the case, was reasonable and proper."

Whatever the frequency of inspections, they must be sufficient to check important items, especially those which will be covered up by later work. In the old Scottish case of *Jameson v. Simon*[59] the architect had made weekly visits to a house under construction, but failed to inspect the bottoming of the floors, which was defective. He was held liable to the employer for the failure. The Lord Justice Clerk said:

"There may, of course, be many things which the architect cannot be expected to observe while they are being done-minute matters that nothing that daily or even hourly watching could keep a check upon. But as regards so substantial and important a matter as the bottoming of a cement floor of considerable area, such as this is shown by the plans to have been, I cannot hold that he is not chargeable with negligence if he fails before the bottoming is hid from view by the cement to make sure that unsuitable rubbish of a kind that will rot when covered up with wet cement has not been thrown in in quantities as bottoming contrary to the specification."

The whole subject of supervision of building and construction works has been the subject of numerous reports and proposals aimed at introducing procedures based on Quality Assurance and self-supervision. Such arrangements are now regularly required, but in addition to existing contractual obligations. The problem which still confronts the drafters of contracts is to devise appropriate means whereby external supervision and its attendant sanctions is replaced by quality control procedures, involving entirely different sanctions.[60] Such procedures are well known in the offshore oil industry but require further development for land-based construction.

Administration

The term is here used compendiously to describe the various functions which the engineer or architect may or must perform

---

[59] (1899) 1 F. 1211.
[60] See J.N. Barber, *Quality Management in Construction*, S.P. 84 CIRIA (1992).

under a construction contract. Administration is part of the duty normally included under the umbrella description "supervision." But it includes many matters not related directly to superintendence on the site. The most important of these are: issuing certificates and ordering variations, which are considered above; and issuing instructions and drawings. In each case the scope of the particular powers or duties depends on the express or implied terms of the contract.

In the absence of express terms[61] the contractor is entitled to have instructions and drawings supplied in reasonable time. Whether the express or implied obligation has been complied with is often difficult to determine, for example when instructions are needed to enable a sub-contract to be placed so as to permit timely completion. This is often a source of contention between contractor and employer. Under the standard forms the contractor is usually required to give notice of any instruction which he considers to be necessary.

During the course of carrying out construction work the engineer or architect may sometimes issue instructions permitting a deviation from the contract to assist the contractor, when strictly the employer is entitled to rely upon the contract. In such a case the instruction should be carefully distinguished from a true variation order. In *Simplex v. St. Pancras B.C.*,[62] the contractor undertook to install piles of specified capacity. This proved impracticable and the contractor offered alternative, differently priced, schemes. The architect accepted one of these "in accordance with quotations submitted." It was held that although the contractor would have been liable for the failure of the first scheme, the architect's acceptance of the alternative amounted to a variation. The contractor was therefore entitled to be paid the price of the alternative scheme and not the (lower) price originally tendered. Edmund-Davies J. held:

> "The architect's letter of 30th July contained an instruction involving a variation in the design or quality (or both design and quality) of the works which the plaintiffs were being instructed to perform, and I have already indicated my view that he did so in circumstances in which he was accepting on the employers' behalf that they would be responsible for the extra cost involved. Such an action fell, in my judgement, within the 'absolute discretion' vested in him by clause 1 and was motivated by his great desire 'to get the job moving' as he put it, and regardless of the legal position of the plaintiffs under their

---

[61] See ICE, cl. 7 and JCT, cl. 5.4.
[62] (1958) 14 B.L.R. 80.

contract. It was an action which led to the plaintiffs doing something different from that which they were obliged to do under their contract, and it was an action which involved the defendants in responsibility for the extra expense which it entailed."

This case has been criticised and was not followed in *Howard de Walden v. Costain.*[63] The issue is also dealt with expressly under the ICE conditions, sixth edition, where it is provided as follows:

"51(3) . . . The value (if any) of all such variations shall be taken into account in ascertaining the amount of the Contract Price except to the extent that such variation is necessitated by the Contractor's default."

This provision would not necessarily reverse the effect of the *Simplex* case. The best course is for any instruction for alternative work made necessary by default of the contractor, to be given expressly on the terms that it will not involve additional payment. The instruction should also deal with the impact of any change on timing of the work, since the varied work may still constitute a variation, even though no payment is due.

Quantity surveyors

The duties of a quantity surveyor include taking off quantities from drawings, preparing bills of quantities and measuring the works in progress. The quantity surveyor is named in the JCT forms of contract, where he is engaged by the employer. In the ICE form such duties are placed on the engineer, but in practice are usually carried out by quantity surveyors employed by the engineer. On most construction works of any substance there will, in addition to those employed by or on behalf of the employer, be quantity surveyors employed by the contractor. In smaller construction contracts, a surveyor alone may be appointed as certifier, and may also be given the functions of supervising and administering the contract. The training of quantity surveyors includes contract administration and legal studies. Consequently, they frequently become involved in claims and disputes. A number of firms have been established specialising in these areas. Quantity surveyors as such are substantially unknown outside the United Kingdom, and the Commonwealth. In the United State of America, for example, bills of quantities are rarely encountered and interim valuations are generally based on pre-agreed milestone payments.

---

[63] (1991) 55 B.L.R. 124.

Other liabilities

The duties of engineers and architects to the building owner arise by virtue of their employment under contract. Acts performed for or on behalf of the employer may at the same time give rise to duties and liabilities to other persons. This may arise by virtue of the position as agent for the employer, when there may be personal liability on a contract or liability for acting without authority. In addition, architects and engineers are subject to the Defective Premises Act 1972. Under section 1, they owe a duty to any present or future owner of a dwelling to see that their work is done in a professional manner (see Chapter 7).

The question of liability in tort is subject to the recent retrenchment (see Chapter 14). There are, however, many circumstances in which professionals may still be under tortious liability in relation to construction work. In *Clay v. Crump*[64] an architect had failed to examine a dangerous wall and allowed it to remain in the belief that it was safe. He was held liable in negligence to a workman who was injured when the wall collapsed. The contractor who employed the man and the demolition contractor were also liable. In this case Ormerod L.J. held:

> "It may be that there was negligence in some degree on the part both of the demolition contractors and builders. If there was such negligence, it may be that it was a contributory cause of the accident. It cannot however, in my judgement, absolve the architect from a share in the blame. To hold otherwise would be to hold that an architect, or indeed anyone in a similar position, could behave negligently by delegating to others duties he was under an obligation to perform and escape liability by the plea that the injuries caused were caused by the negligence of that other person and not of himself. I do not accept that as being the true position in law."

While this case is of interest as regards the type of duty which may be owed, there will be difficulty in recovering loss not resulting from physical damage to person or property, unless the case is brought within the principles of *Hedley Byrne v. Heller*[65] (see Chapter 14).

A more recent and disturbing case concerning tortious liability is *Eckersley and Others v. Binnie & Partners,*[66] the *Abbeystead* case. Owing to the undetected presence of methane gas in an underground pumphouse, an explosion occurred, killing or injuring many

---

[64] [1964] 1 Q.B. 533.
[65] [1964] A.C. 465.
[66] (1988) 18 Con. L.R. 1.

visitors who had been invited to a view by the Water Authority. The trial judge found fault on behalf of the designer (Binnie), the contractor and the Water Authority and the loss was apportioned between them. The Court of Appeal found no liability on the part of the contractor and the authority, but by a majority held Binnie alone liable. The trial judge suggested that the designer might be under a continuing duty, after completion of the project, to advise on new information which might indicate a danger. Bingham L.J. while not prepared to rule out any possibility of such a continuing duty, said:

> "What is plain is that if any such duty at all is to be imposed, the nature, scope and limits of such a duty require to be very carefully and cautiously defined. The development of the law on this point, if it ever occurs, will be gradual and analogical. But this is not a suitable case in which to launch or embark on the process of development because no facts have been found to support a conclusion that ordinarily competent engineers in the position of (Binnie) would . . . have been alerted to any risk of which they were reasonably unaware at the time of handover."

The duty owed in regard to the carrying out of the works will be more limited. In the case of *Oldschool v. Gleeson*[67] it was contended that consulting engineers were liable to contractors who suffered loss through the collapse of a party wall. Sir William Stabb, the senior Official Referee, held:

> "The duty of care of an architect or of a consulting engineer in no way extends into the area of how the work is carried out. Not only has he no duty to instruct the builder how to do the work or what safety precautions to take but he has no right to do so, nor is he under any duty to the builder to detect faults during the progress of the work. The architect, in that respect, may be in breach of his duty to his client, the building owner, but this does not excuse the builder for faulty work.
>
> I take the view that the duty of care which an architect or a consulting engineer owes to a third party is limited by the assumption that the contractor who executes the works acts at all times as a competent contractor. The contractor cannot seek to pass the blame for incompetent work onto the consulting engineer on the grounds that he failed to intervene to prevent it.
>
> . . . The responsibility of the consulting engineer is for the design of the engineering components of the works and his supervisory responsibility is to his client to ensure that the works are carried out in accordance with that design. But if, as was suggested, here, the design was so faulty that a competent contractor in the course of

---

[67] (1976) 4 B.L.R. 103.

executing the works could not have avoided the resulting damage, then on principle it seems to me that the consulting engineer responsible for that design should bear the loss."

Design Contracts

In traditional construction contract practice there is a more or less rigid distinction between design and construction. Design is the task of the engineer or architect and is taken to be excluded from the contractor's function. This distinction is entirely removed in certain modern forms of contract, sometimes described as "package" or "turnkey" contracts. Before dealing with the particular difficulties of these forms, it is necessary to examine the extent to which the contractor's traditional responsibility does exclude design.

First, the word "design" has no precise meaning in building contracts. It certainly encompasses the planning of the form of the finished works. The ICE conditions draw a distinction between design of permanent and temporary works (clause 8(2)), the latter normally being the contractor's responsibility. Under the JCT forms, temporary works are entirely the contractor's responsibility unless otherwise provided for. In regard to the permanent works, no contract can lay down every detail of the "design," for example, the precise positioning of screws or the mixing of mortar. Each such operation involves an element of design, which is left to the contractor. In simpler forms of contract, this design element may be extensive and important. In addition to a term of good workmanship, there will generally be an implied term that the work and materials will be reasonably fit for their purpose, to the extent they are not fully specified. The contractor should thus be responsible for elements of "design" left to him.[68]

This principle may be limited by the form of the contract. Both the ICE and JCT forms entitle the contractor expressly to be given instructions necessary to complete the works.[69] The ICE conditions require any design responsibility for the permanent works to be expressly stated (clauses 8(2), 58(3)). Under the JCT form, the contractor's obligation is limited to the work shown in the contract drawings and bills (clause 2.1). Thus, in *Mowlem v. BICC*[70] the bills

---

[68] *Young & Marten v. McManus Childs* [1969] 1 A.C. 454; *Cammell Laird v. Manganese Bronze* [1934] A.C. 402; but see *Rotherham v. Haslam* (1996) 78 B.L.R. 1, CA.

[69] ICE, cl. 7(1); JCT, cl. 5.4.

[70] (1977) 3 Con. L.R. 64.

stipulated "waterproof concrete," leaving the means of achieving the result unspecified and unpriced. This was held insufficient to make the contractor responsible when the concrete (otherwise constructed in accordance with the contract) leaked. Sir William Stabb held:

> "I should require the clearest possible contractual condition before I should feel driven to find a contractor liable for a fault in the design, design being a matter which a structural engineer is alone qualified to carry out and which he is paid to undertake, and over which the contractor has no control. I agree that the construction for which (counsel for the employer) contends places the contractor in an impossible position. He cannot alter the faulty design without being in breach of contract, for this fault in the design is not, in my view, a discrepancy or divergence between the contract drawings and/or the bill of quantities, and yet if he complies with the design he would still be in breach. I decline to hold that the specification in the bill of quantities makes the contractor liable for the mistakes of the engineer and, in so far as they may purport to do so, I think that it is ineffective by reason of clause 12(1) of the [JCT form of] contract."

Drafting a design contract

Where it is intended to make the contractor fully responsible for the design, the standard forms require substantial amendment, beginning with the clauses mentioned above. A simple design-and-build contract could be written in the form "Build a house with six bedrooms." Difficulties will arise when the owner seeks to elaborate the contract to retain control over the appearance, lay-out and cost of the work. A method of overcoming these problems is to invite tenderers to submit their own designs with a lump sum price. Further difficulties arise if, having selected a design, the employer wishes to vary it. Package deal contracts usually allow the contractor to object to a variation which affects his design responsibility.

The procedure usually adopted in design contracts for determining the design is for the tender documents to contain particulars of the "Employers' Requirements" and for the contractor's tender to include "Contractor's Proposals". These two documents then become incorporated into the contract which makes provision for generation of the necessary details of the work, in general by the contractor exclusively. The contractor must always retain some level of discretion and choice in regard to unspecified details. The contractor is entitled by these means to achieve economies in the work. Employers sometimes seek to circumvent this by the provision of considerable detail in the employer's requirements. There will, however, always be areas of choice for the contractor.

A design contract should include provision expressly making the contractor responsible for the design. It should make clear that responsibility is for the adequacy (and not mere provision) of the design. In principle, the responsibility should be on the basis of fitness for purpose, and this can often be achieved conveniently by incorporating performance requirements, for example for mechanical plant. In some contracts, the obligations of the designer will be found to be limited to reasonable skill and care. This creates potential difficulty where liability for individual components may be on a higher basis. In the absence of express or clear provision the courts have resolved doubts in favour of the building owner where it was clear that the design had been carried out by the contractor, or his sub-contractors. This was the case in *Newham L.B. v. Taylor Woodrow*,[71] where the contractor disputed liability arising from the partial collapse of a tower block known as Ronan Point. A modified JCT form of contract had been used, which was nevertheless held sufficient to put the contractor under an absolute responsibility to comply with Building Regulations. The contractor was therefore liable despite being absolved of negligence. Conversely, in *Independent Broadcasting Authority v. EMI and BICC*,[72] the sub-contractor who designed the Emley Moor T.V. mast (BICC) was held liable in negligence for its collapse. The House of Lords, however, expressed their view on the basis there was no negligence. They held that the main contract included design responsibility, although this was to be carried out by the sub-contractor alone. Lord Fraser went on to say:

"If the terms of the contract alone had left room for doubt about that, I think that in a contract of this nature a condition would have been implied to the effect that EMI had accepted some responsibility for the quality of the mast, including its design, and possibly also for its fitness for the purpose for which it was intended. . . . It is now well recognised that in a building contract for work and materials a term is normally implied that the main contractor will accept responsibility to his employer for materials provided by nominated sub-contractors. The reason for the presumption is the practical convenience of having a chain of contractual liability from the employer to the main contractor and from the main contractor to the sub-contractor-see *Young & Marten Ltd v. McManus Childs Ltd.*[73] . . . In the present case it is accepted by BICC that, if EMI are liable in damages to IBA for the design of the mast, then BICC will be liable in turn to EMI. Accordingly, the principle that was applied in *Young & Marten Ltd* in

---

[71] (1981) 19 B.L.R. 99.
[72] (1980) 14 B.L.R. 1.
[73] [1969] 1 A.C. 454.

respect of materials, ought in my opinion to be applied here in respect of the complete structure, including its design. Although EMI had no specialist knowledge of mast design, and although IBA knew that and did not rely on their skill to any extent for the design, I see nothing unreasonable in holding that EMI are responsible to IBA for the design seeing that they can in turn recover from BICC who did the actual designing. On the other hand it would seem to be very improbable that IBA would have entered into a contract of this magnitude and this degree of risk without providing for some right of recourse against the principal contractor or the sub-contractors for defects of design."

A number of standard design and build contracts are now available including those published by the JCT and by the ICE. They all possess similar characteristics in terms of the design mechanism while retaining essential features from the parent form of contract. Both of these forms involve difficulties over the degree of liability which the contractor accepts in respect of the design (see Chapter 11).

COMPLETION

This covers the time period within which the work must be carried out and the consequences of delay; what is necessary to achieve completion of the work; and the effect of maintenance or defects liability clauses. Generally building contracts do not require the contractor to carry out individual items of work at particular times, and a programme of work is rarely made a term of the contract. Consequently the contractor will not be regarded as in breach by reason of delay during the course of the work unless there is a failure to comply with a term requiring "due diligence". Similarly the contractor is not necessarily in breach by reason only of defective work, provided he can complete in accordance with the contract.[74] Conversely when defects appear within the maintenance period, although the contractor has the right and duty to make good, he is nevertheless in breach so that the employer may sue for damages, through being deprived of use of the works. In this event, the damages recoverable will not be limited by any provision for liquidated damages, which are recoverable for delay in achieving completion.

---

[74] *Kaye v. Hosier & Dickinson* [1972] 1 W.L.R. 146; *Lintest v. Roberts* (1980) 13 B.L.R. 38.

Time for completion and damages

The time within which the work is to be performed is a matter of economic importance both to employer and contractor. In most contracts dates will be specified for the start and completion of the work. The contractor is bound to do the work within the period set, and will be liable in damages if he fails to complete, subject to entitlement to extensions of time. The contractor also has an entitlement to carry out the work. Thus, if the employer prevents completion, for example by failing to give possession of the site, he will be liable in damages to the contractor. Where no time period is specified, the same principles apply, save that the contractor is obliged and entitled to complete within a reasonable time.

Damages recoverable by the employer for delay are usually limited to "liquidated damages".[75] When the delay is caused partly by the employer's default, it has been held that no liquidated damages may be recovered unless the contract allows an extension of time to be granted on the ground of the default, and such extension is in fact granted. Thus, in *Peak v. McKinney*[76] building works were suspended after the discovery of defective piles, for which the contractor was responsible. The employer caused further delay before work restarted. The contract did not provide for an extension of time for the employer's default. It was held that no liquidated damages could be recovered for any of the delay. Salmon L.J. held:

> "The liquidated damages and extension of time clauses in printed forms of contract must be construed strictly contra proferentum. If the employer wishes to recover liquidated damages for failure by the contractors to complete on time in spite of the fact that some of the delay is due to the employers' own fault or breach of contract, then the extension of time clause should provide, expressly or by necessary inference, for an extension on account of such a fault or breach on the part of the employer."

The grounds of this decision were that liquidated damages may be recovered only from a date fixed under the contract. If no date can be fixed, time is "at large." For this purpose the liquidated damages and extension of time clauses are regarded as being for the employer's benefit and are construed against him, so that general words cannot be relied on. However, this appears to ignore the fact that pre-determined damages are as likely to benefit the

---

[75] See JCT, cl. 24; ICE, cl. 47.
[76] (1971) 69 L.G.R. 1; (1970) 1 B.L.R. 111.

contractor as the employer. If damages for delay are not "liqui-dated," the employer may sue for his actual loss.

Extension of time and acceleration

Construction contracts usually provide for extensions of time to be granted by the architect or engineer on a variety of specified grounds.[77] Where the ground of extension would otherwise be the contractor's risk, the extension is purely a concession, such as for inclement weather. Where the extension is based on some act or default of the employer, for example, the ordering of variations or giving late instructions, the contractor may be entitled also to extra payment. For this reason, the contractor may seek to attribute the actual delay to grounds carrying reimbursement in respect of the period granted.[78] While the employer may not be entitled to rely on general words to protect his right to liquidated damages, such words will benefit the contractor. Thus under the ICE conditions, clause 44(1), the contractor is entitled to an extension on the ground of "other special circumstances of any kind whatsoever."

It is a matter of construction of the extension of time clause whether the contractor must establish that the event relied on will cause the works to be delayed beyond the completion date under the contract. In this event, a further problem arises if the contrac-tor has programmed to finish early.[79] Alternatively, where the clause entitles the contractor to be granted an extension by reference to "the delay that has been suffered by the contractor as a result of the alleged cause"[80] it may be unnecessary for the contractor to demonstrate any net delaying effect to the contract. The operation of these clauses will be important where a bonus is to be paid for achieving early completion.

Claims for acceleration may arise where the employer wishes to have the work completed earlier than the projected completion date. Provisions to this effect are included in a number of contracts[81] but usually on the basis of agreement with the contrac-tor. Alternatively, the contractor may contend that as a result of express or implied instructions, the engineer or architect has ordered acceleration measures. This may arise where the contract is held subsequently to be entitled to an extension of time, but at

---

[77] JCT form, cl. 25; ICE form, cl. 44.
[78] JCT form, cll. 25 and 26.
[79] See *Glenlion Construction v. Guinness Trust* (1987) 39 B.L.R. 89.
[80] ICE, cl. 44(2).
[81] ICE, cl. 46(3); GC/Works/1 Ed. 3, cl. 38.

the time of the delay the extension is denied and the contractor is required to adhere to the original completion date. This is sometimes referred to as "constructive" acceleration. There is no direct English authority on such claims but they have been recognised in the United States of America[82] and in the Commonwealth.[83]

Right to suspend work

At common law, there is no right of suspension outside the terms of the particular construction contract. The contractor will be justified in suspending work where a ground of delay exists which gives rise to an extension of time; and suspension may be ordered under most forms of contract by the engineer or architect. There is normally no right to suspend work for non-payment or other breach by the employer. For a construction contract falling within the definition in the Housing Grants, etc., Act 1996, section 112 provides that where a sum due is not paid by the final date for payment and no effective notice to withhold payment has been given (see stage payments above), the contractor has the right to suspend performance until full payment is made, subject to first giving seven days notice and stating the grounds. Any such period of suspension is to be disregarded in computing any contractual time limit. It is thought that the right must apply to the whole of the works, even though non-payment relates to part. Difficulties may arise where there is a dispute as to the contractor's right to payment which is subsequently resolved in the employer's favour, either by statutory adjudication (see Chapter 2) or by arbitration.

Meaning of completion

Generally, full and complete performance is required to discharge contractual obligations. However in construction contracts the purpose of signifying completion is not to release the contractor, but to permit the employer to take possession of the works and to allow the contractor to leave the site. Contracts, therefore, use terms such as practical completion[84] and substantial completion.[85] While such terms do not permit the contractor to achieve comple-

---

[82] See *Norair Engineering v. U.S,*. 666 F. 546 (1981).
[83] *Morrison-Knudsen v. B.C. Hydro & Power* (1978) 85 D.L.R. 3d. 186; 7 Const. L.J. 227; *Perini Corp. v. Commonwealth of Australia* (1969) 12 B.L.R. 82.
[84] JCT, cl. 17.1.
[85] ICE, cl. 48(1).

tion without finishing the whole of the work (save for permitted exceptions), it is thought they allow completion to be certified despite the existence of non-material departures from the contract.

Completion is not affected by the existence of latent defects.[86] If defects are discovered after apparent completion (whether during or after the maintenance period) the employer is entitled to sue for damages, including loss of use.

Defects clauses

These oblige the contractor to rectify faults appearing within a specified period, often six or 12 months following completion. They may also oblige the contractor to maintain the works and to put right defects not due to his default, the latter at the employer's expense.[87] When a default is due to the contractor's failure to comply with the contract, he is in breach. The maintenance clause permits the contractor to mitigate the effect of the breach by carrying out rectification himself. A provision which entitles the employer to have defects rectified within a specified period does not absolve the contractor from liability for defects appearing after expiry of the period. Clear words are required to make a maintenance clause operate also as an exclusion clause. Such words are found in the standard form of contract for mechanical and electrical works, MF/1, where the contractor is expressly absolved of further responsibility after carrying out the rectification of defects.[88]

---

[86] *Jarvis v. Westminster C.C.* [1969] 1 W.L.R. 1448.
[87] Compare JCT, cl. 17 and ICE, cl. 49.
[88] Cl. 36: see Chap. 11.

CHAPTER 10

VICARIOUS PERFORMANCE AND INSOLVENCY

CHAPTER 9 deals with the operation of construction contracts and the parties' rights arising from them. This chapter covers a number of matters outside the contract itself which may affect the parties' rights. Vicarious performance refers to the carrying out of contractual obligations by a person not party to the contract. This may be either by sub-contract or assignment. Sub-contracts are found in most construction work, since very few contractors have the resources to carry out the whole of a project themselves. Particular problems arise when the sub-contractor is nominated. Assignment is the means by which a party may transfer to another the whole or part of his rights or duties under a contract. When the contractor assigns part of his obligation to perform the work, the effect is similar to a sub-contract save that the assignee is in direct contract with the employer. Finally, insolvency and bonds deals with the rights of the parties when one of them becomes unable to perform the contract by reason of financial difficulties.

SUB-CONTRACTS

In the traditional system of contracting in the construction industry the whole of the work is initially let to the main contractor. Subject to the provisions of the main contract, the contractor is entitled to sub-let portions of the work, save where the contract is let by reason of some special skill or quality of the contractor. Thus, a contract for specialist site-investigation work is likely to be one which may not be sublet without consent of the employer. However, performance by a sub-contractor constitutes vicarious performance on behalf of the contractor, who remains fully responsible for the work, save where the main contract provides otherwise.

The standard forms usually contain terms restricting the right of sub-letting. The JCT form prohibits sub-letting without the architect's consent, which is not to be unreasonably withheld (clause 19). The withholding of consent may be challenged, if considered to be unreasonable. Under the ICE form sub-letting of parts of the work is permitted without the need for consent, but sub-letting the whole requires consent (clause 4). If sub-letting occurs without any necessary consent, this may give the employer no right beyond nominal damages. But a refusal to resume performance by the contractor may constitute repudiation. The better course is to provide express remedies in the contract, such as determination. The employer may waive his right to object, and would be held to do so by making payment for the sub-contracted work with knowledge of the sub-contract.

Engineering sub-contractors tend often to be specialists who carry out limited parts of the works, such as piling. In building, it is not uncommon for the majority of the work on a substantial contract to be carried out by a large number of trade sub-contractors. As an alternative to the traditional main contract, under a management contract the whole of the work is sub-contracted, the contractor's role becoming that of a manager. Payment is usually made on the basis of prime cost together with a fee which is retained by the contractor. One advantage claimed for the management contract is flexibility in development of the design.

Rights and obligations

A sub-contract creates no privity of contract between the sub-contractor and the employer. The sub-contractor can sue only the main contractor for the price of the sub-contracted work. The advantage to the employer is that while the work may actually be performed by various specialists, the main contractor alone remains responsible for the whole operation and, perhaps most important, for the co-ordination of his own work and that of sub-contractors. Sums included in the main contract and designated as prime cost (P.C.) usually represent work intended to be sub-let. This may be called prime cost work, since the employer must pay the actual cost. Prime cost work often represents an important part of the project which has not been designed in any detail, and which is to be sub-let to a nominated specialist who will furnish the design. Prime cost work is the subject of special provisions in the standard forms of contract, which are dealt with further below. A provisional sum usually represents work of uncertain scope which may be sub-contracted.

A common feature of main contracts is for the employer to retain the right to make direct payments to a nominated sub-contractor.[1] These provisions recognise that payments due to a nominated sub-contractor are generally fixed by the architect or engineer, so that the contractor acts merely as a channel for payment. However, neither nomination nor direct payment to sub-contractors gives a sub-contractor the right to sue the employer and no privity of contract is thereby created. The only exception to this principle is where the employer obtains a collateral warranty from the sub-contractor. A warranty as such gives the sub-contractor no right to sue for payment, but the RIBA forms of warranty oblige the employer to operate the direct payment provisions. Such rights will generally be subject to any cross-claim which the employer may have against the contractor.

When making a sub-contract, the main contractor will usually wish to ensure, so far as is possible, that the arrangements are "back to back", that is that the sub-contractor automatically becomes liable for any act or omission which causes breach of the main contract. This is sometimes attempted by seeking to incorporate the whole of the main contract into the sub-contract. But this may create problems of interpretation and may not achieve the effect intended. A more useful device, adopted by the standard forms of sub-contract, is to require the sub-contractor so to perform that the contractor is not caused to be in breach of the main contract. This device is likely to be effective as regards the standard or quality of work. But as regards time, it is much more difficult to devise obligations which will ensure that the sub-contractor performs to a programme entirely conformable to the needs of the main contract. Any programme fixed in advance is vulnerable to the need to make adjustments before or even during the performance of the sub-contract work. A device which is likely to be effective is to require the sub-contractor to start only as and when instructed to do[2] but it may be difficult to secure the sub-contractor's agreement.

Payment to sub-contractors

A sub-contract falls within the definition of "construction contract" under the Housing Grants, etc., Act 1996 and the sub-

---

[1] See JCT, cl. 35.13 and ICE, cl. 59(7) and see below.
[2] *Kitson's Sheet Metal v. Matthew Hall* (1989) 47 B.L.R. 82.

contractor is, as against the main contractor, entitled to the benefit of all the payment provisions contained in sections 109 to 111, including the right to suspend performance for non-payment without effective notice under section 112 (see Chapter 9). In addition, sub-contracts of all kinds have for many years contained provisions commonly known as "pay-when-paid" clauses, by which the main contractor could withhold payment from the sub-contractor until the sums in question were received from the employer. Depending on the terms of the contract, such clauses could take effect as meaning "pay-if-paid". The same provisions are frequently found in sub-sub-contracts and contracts of supply.

Under the Housing Grants, etc., Act 1996, s. 113, such clauses are rendered ineffective unless the condition of payment is that the payer (or any other person payment by whom is under the contract (directly or indirectly) a condition of payment by the payer) is insolvent. Thus, it is permissible for the main contractor (or sub-contractor in the case of a sub-sub-contract) to provide that payment may be withheld where the provider of the funding becomes insolvent; otherwise any withholding of payment must be made in accordance with clause 111 of the Housing Grants, etc., Act 1996 by service of an effective notice. The validity of any such withholding may then immediately be challenged by adjudication under section 108 of the Act.

Nomination

The usual procedure for letting a nominated sub-contract is for the architect or engineer to obtain competitive quotations for the work in question direct from prospective sub-contractors, and then, under powers contained in the main contract, to instruct the contractor to place an order with the chosen tenderer. The terms of the sub-contract will largely be settled before the nomination, usually by incorporating a standard form, but both the contractor and sub-contractor may seek to amend or alter the terms. The main contract usually gives the contractor some protection by entitling him to object to a nomination which does not contain certain beneficial terms.[3] The JCT form requires the sub-contractor's tender to be subject to form NSC/T.

Liability for defects

One of the problems relating to sub-contracts is that of determining the obligations of the main contractor in respect of

---

[3] See ICE, cl. 59(1).

materials and work of nominated sub-contractors, and thus determining the rights of the employer in the event of default by a nominated sub-contractor. There is generally no problem in regard to the express description of the work. This will become incorporated in the main contract by virtue of the nomination. The problem arises when the sub-contractor's work, while complying with the express terms of main and sub-contract, is not of good quality or not fit for its purpose.

Generally, the main contractor will be responsible both for the quality and for the fitness of materials used, unless there are circumstances which are such as to negative the existence of either term. Where there is a nomination, the materials will be specifically chosen by the architect or engineer, so that the main contractor will not be responsible for their fitness or suitability. In *Young & Marten v. McManus Childs*,[4] the House of Lords held that the choice of tiles made by only one manufacturer excluded the warranty of fitness but was not sufficient to exclude the implication of a warranty of quality in the main contract, so that the main contractor was liable for latent defects in the tiles. Lord Pearce expressed his view thus:

> "It is frequent for builders to fit baths, sanitary equipment, central heating and the like, encouraging their clients to choose from the wholesalers' display rooms the bath or sanitary fittings which they prefer. It would, I think, surprise the average householder if it were suggested that simply by exercising a choice he had lost all right of recourse in respect of quality of the fittings against the builder who normally has a better knowledge of these matters. Of course, if a builder warned him against a particular fitting or manufacturer and he persisted in his choice, he would obviously be doing so at his own risk. And a builder can always make it clear that he is not prepared to take responsibility for a particular kind of fitting or material."

In the case of *Gloucestershire County Council v. Richardson*,[5] the main contractor discovered, during the course of construction works, defects in pre-cast concrete columns provided by a nominated supplier. The form of contract (RIBA form, 1939 edition) did not permit the contractor to object to the terms of the nominated supply contract, which was subject to a limitation of liability. It was held that the main contractor was not liable for the quality (or the fitness) of the components, and was not therefore in breach of contract. Lord Wilberforce held:

---

[4] [1969] 1 A.C. 454.
[5] [1969] 1 A.C. 480.

"The design, materials, specification, quality and price were fixed between the employer and the sub-supplier without any reference to the contractor: and so far from being expected to secure conditions or warranties from the sub-supplier, he had imposed upon him special conditions which severely restricted the extent of his remedy. Moreover, as reference to the main contract shows, he had no right to object to the nominated supplier . . . In these circumstances, so far from there being a good reason to imply in the contract . . . a condition or warranty binding the contractor in respect of latently defective goods, the indications drawn from the conduct of the contracting parties are strongly against any such thing."[6]

A further problem arises if a nominated sub-contractor repudiates. Is the contractor himself obliged to complete the work at his own expense, so far as this exceeds the agreed sub-contract price, or must the employer pay the additional cost of finding an alternative sub-contractor? In *N.W. Metropolitan Hospital Board v. Bickerton*[7] it was held that under the 1963 JCT form the architect was obliged to re-nominate, so that the employer must bear the loss. The decision was based on the fact that the contract contemplated that prime cost work would be carried out only by a nominated sub-contractor. The decision may therefore apply to other forms of main contract such as the ICE fourth edition. The ICE fifth and sixth editions contain express provisions dealing with parties' rights upon the default of a nominated sub-contractor, so that there is little application for the *Bickerton* principle. The 1980 JCT form now deals expressly with re-nomination in respect of sub-contractors but not nominated suppliers (see Chapter 12).

A combination of the above problems occurs if a nominated sub-contractor repudiates leaving defects in partly completed work. If the facts are similar to the *Young & Marten* case above, the main contractor would be liable for the quality of the sub-contractor's work. In *Fairclough v. Rhuddlan B.C.*,[8] the main contractor was prima facie liable for the work of the sub-contractor, but was held not to be under any responsibility for defects discovered after the sub-contractor had repudiated, relying on the principle established in the *Bickerton* case, that the main contractor had neither the right nor the duty to carry out work within a P.C. sum.

Liability for delay

If a nominated sub-contractor causes delay without repudiating, the contractor generally remains liable and may pass on the

---

[6] See also *IBA v. EMI* (1980) 14 B.L.R. 1 and Chap. 9.
[7] [1970] 1 W.L.R. 607.
[8] (1985) 30 B.L.R. 26.

employer's claim to the sub-contractor in default. However, the JCT form provides expressly for an extension of time on the grounds of delay by a nominated sub-contractor.[9] This effectively deprives the employer of remedy, save under any direct warranty.

Where the nominated sub-contractor repudiates, it appeared to follow from the *Bickerton* case that the employer would remain responsible for the delay in providing a replacement sub-contractor. However, it was held by the House of Lords in *Percy Bilton v. GLC*[10] that such delay was not the responsibility of the employer. Lord Frazer held:

> "Withdrawal of a nominated sub-contractor is not caused by the fault of the employer, nor is it covered by any of the express provisions of clause 23 . . . accordingly, withdrawal falls under the general rule and the main contractor takes the risk of any delay directly caused thereby."

Where a nominated sub-contractor has achieved apparent completion and defects are thereafter discovered in the work, the question arises, under the JCT forms of contract, whether this can constitute "delay on the part of nominated sub-contractors". These facts occurred in the case of *Jarvis v. Westminster Corporation*,[11] where defects were discovered in bored piles after the sub-contractor had withdrawn from the site. The necessary remedial work resulted in substantial delays to the main contract. Lord Dilhorne held:

> "A practical completion certificate can be issued when owing to latent defects the works do not fulfil the contract requirements and . . . under the contract works can be completed despite the presence of such defects. Completion under the contract is not postponed until defects which became apparent only after the work had been finished had been remedied . . . I conclude that the [sub-contractor] had completed the sub-contract works to the reasonable satisfaction of the architects and the [main contractor] and so were not guilty of delay."

Accordingly, as in the *Percy Bilton* case, the subsequent delay was not covered by the extension of time clause, and the main contractor remained responsible for the delay.

ASSIGNMENT

An assignment is a transfer, recognised by the law, of a right or obligation of one person to another. Most rights and obligations

---

[9] 1963 ed., cl. 23(g); JCT 80, cl. 25.3.7.
[10] [1982] 1 W.L.R. 794.
[11] [1969] 1 W.L.R. 1448.

are capable of assignment. This may be achieved in a number of ways. Assignments are sometimes brought about by operation of law. This section is concerned primarily with assignment of rights and obligations under building contracts, but the principles involved cover many other things.

An assignment, in common with other legal transactions, is distinct from a contract to make an assignment. An assignment does not generally require consideration. But a contract to assign, in order to be enforceable, must comply with the same requirements as any other contract, including the need for consideration. An assignment of a right or obligation arising under a contract is a further exception to the doctrine of privity in that rights or burdens are conferred upon persons who are not party to the contract (see Chapter 5).

Assignments not permitted

Building contracts and sub-contracts often contain terms restricting or prohibiting assignments[12]: Such terms have the effect of making any purported assignment invalid as against the other party to the contract.[13] However the right to prevent an assignment may be lost by waiver. Thus, if the contractor assigns his right to receive payment, the employer will waive his right to object if payment is made to the assignee.

There are some rights which may not be assigned. It is a fundamental principle of English law that a "bare" right to sue for damages cannot be assigned. Thus, a party who has suffered personal injury must pursue the claim for compensation himself: he cannot sell that right to another. However, in commercial transactions, there may be a good reason for transferring a right of action. The law on this topic was reviewed by the House of Lords in *Trendtex v. Credit Suisse*,[14] where a bank sought to uphold the validity of an assignment of a claim for damages arising out of a transaction which had been financed by the bank. Lord Roskill stated the law as follows:

> "The Court should look at the totality of the transaction. If the assignment is of a property right or interest and the cause of action is ancillary to that right or interest, or if the assignee has a genuine commercial interest in taking the assignment and in enforcing it for his own benefit, I see no reason why the assignment should be struck

---

[12] JCT, cl. 19; ICE, cl. 3.
[13] *Linden Gardens v. Lanesta Sludge* [1994] 1 A.C. 85.
[14] [1982] A.C. 679.

down as an assignment of a bare cause of action or as savouring of maintenance."

In the result, it may now be possible for commercial claims under construction disputes to be transferred. For example, it may be possible for a sub-contractor who has suffered loss to take over the rights of the main contractor to enforce claims under the main contract, including his own claim.

Methods of assignment

A legal assignment is one which complies with section 136 of the Law of Property Act 1925. This requires that the assignment is in writing and is absolute, and that notice in writing is given to the other party. No particular form is needed and the document need not be under seal. An assignment which is conditional, for example, until a loan is repaid, is not absolute. The assignment takes effect and becomes enforceable against the other party only on receipt of notice.

A transfer which does not comply with the requirements of a legal assignment, such as one made orally, may be enforceable as an equitable assignment. But an equitable assignment will require to be evidenced in writing if it relates to an interest in land.

In either type of assignment it is necessary to draw a distinction between a benefit and a burden. In a construction contract the benefit to the contractor is the right to be paid and the burden is the obligation to do the work. A benefit may be assigned irrespective of the wishes of the other party (subject to rights under the contract). The burden may be assigned only with the consent of the other party. Thus the contractor may not assign the obligation to carry out the work without the employer's consent; and the employer may not assign the duty to make payments without the contractor's consent.

Apart from the above methods, the assignment of certain rights are governed by statutory provision. These include transfers of shares and debentures in companies, and assignment of life insurance policies. In addition to transfers brought about by act of the parties some assignments take place by operation of law. Thus on death, the rights of the deceased person vest generally in his personal representative. Upon bankruptcy, the rights of the bankrupt vest in the official receiver and, upon appointment, in his trustee in bankruptcy.

Effects of assignment

Upon a valid assignment, the assignor loses his rights in the things assigned. The assignee acquires the right to sue in his own name. However, if the assignment is equitable, the assignor may need to be made a party to the action. The right acquired is subject to any rights of the other party against the assignor, including the right of set off. Thus if a contractor assigns money payable under a certificate, the employer may set off against the assignee any claim for defects or delay. He cannot counterclaim, but the set-off may reduce or extinguish the debt.

Assignment does not generally discharge the party assigning from his own contractual obligation. Thus when a lease is assigned, the landlord is entitled to look to the assignee or to the original tenant for payment of rent. Similarly if the contractor assigns the obligation to carry out work, he may still be liable to the employer for breaches, such as defective work. From the contractor's point of view, a more satisfactory arrangement is that there should be a substitution of the new contractor. This is referred to as a novation. Where both the benefit and the burden of a contract are assigned, the latter requiring express consent, this may operate as a novation. Where a new contractor is substituted in the course of a construction contract, difficulties may arise as to existing matters which may later give rise to disputes, as for example, where the work is behind programme or there are grounds for a claim. These matters are best dealt with by express agreement.

The effect of purported assignments in breach of contract was considered by the House of Lords in the consolidated appeals in *Linden Gardens v. Lanesta Sludge* and *St Martin's Property v. McAlpine*.[15] It was held that there was no reason for holding a contractual prohibition on assignment to be contrary to public policy and that purported assignments in breach of clause 17 (1) of the 1963 JCT form of contract[16] were invalid. The result, in each case, was that the purported assignee who had suffered damage was not entitled to enforce rights under the contract. However, the House held that the original contracting party could recover damages on behalf of the party suffering loss, thereby rejecting the submission that the party otherwise in breach should escape liability. Lord Browne-Wilkinson held:

> "McAlpine had specifically contracted that the rights of action under the building contract could not without McAlpine's consent be transferred to third parties who became owners or occupiers and might suffer loss. In such a case it seems to me proper, as in the case of the carriage of goods by land, to treat the parties as having entered

---

[15] [1994] 1 A.C. 85.
[16] Now cl. 19.1, JCT 80.

into the contract on the footing that [the original employer] would be entitled to enforce contractual rights for the benefit of those who suffered from defective performance but who, under the terms of the contract, could not acquire any right to hold McAlpine liable for breach. It is clearly a case in which the rule provides "a remedy where no other would be available to a person sustaining loss which, under a rational legal system, ought to be compensated by the person who has caused it."

Lord Griffiths was in favour of expressing the principle even more widely (see Chapter 5).

Assignment of warranties

This topic has acquired currency through the creation of contractual warranties intended to replace claims previously available in tort. The assignees are those subsequently acquiring an interest in construction works. Thus, parties involved in carrying out a construction project, particularly the designers and other professionals, may be asked to give forms of warranty (see Chapter 9) with the intention that these may be assigned to subsequent owners or leases. In accordance with the principles discussed above, the benefit of such a warranty is generally assignable without need of express provision or consent. However, an assignment can do no more than to transfer rights available to the assignor; it is not capable of creating new rights in favour of an assignee. Thus, while the client can in theory assign the right to have a building adequately designed, it is unclear what right would be transferred to sue for damages in the event of breach. If the developer/assignor has sold the building or created a full-repairing lease, then his right would be to nominal damages only. A further difficulty would arise where a building is sold or leased to a number of different purchasers. The assignment to each of a right to have the building properly designed creates a number of problems as to what enforceable right (if any) has been transferred to each purchaser.

These problems are capable of resolution, in principle, by the terms of the document to be assigned, particularly if created by deed (so as to avoid problems of consideration). Thus, the nature and extent of rights to be transferred may be defined, as well as the damage that may be claimed in the event of a breach. A solution which avoids these problems is to draft the document in the form of a novation, whereby the assignee third party may take over the full contractual rights of the developer, as though named as an original party to the transaction.

The effect of a contract expressly made with the intention that its benefit be assigned to another party was considered by the Court of

Appeal in *Darlington v. Wiltshier Northern*.[17] This concerned the familiar method of financing local authority development works by use of a bank (Morgan Grenfell) as the primary employer. But unlike the *Linden Gardens* case (see above) the contract package provided expressly for assignments. After completion the development (a recreational centre) was alleged to be subject to serious defects. Darlington then took an assignment of rights under the original contract and brought proceedings in their own name. The Court of Appeal held the council entitled to recover the damages, following the *Linden Gardens* case. They also held that the bank could, before any assignment, have recovered damages as a constructive trustee for the council (see Chapter 5), providing another route by which the council could recoup its losses. An appeal to the House of Lords was compromised before hearing. The result of *Darlington* and *Linden Gardens* is that it is now much less likely that the courts will uphold a "no loss" argument where contractual rights and damages have become separated.

INSOLVENCY

Insolvency is not a term of art, but means, in practical terms, inability to pay debts. The effects of this depend on whether the debtor is an individual or a company, but in either case the consequences are severe, both for the debtor and for the creditor who is unpaid. The laws of bankruptcy and of winding up companies provide for the realisation and distribution of assets, with certain debts having priority for payment. In construction contracts the insolvency of one party will usually bring the work to an end. The law on both bankruptcy and winding up are contained mainly in the Insolvency Act 1986 and the Insolvency Rules.

The fact that an individual or a company is insolvent does not mean that there will be a bankruptcy or winding-up. This depends on the action of the creditors (and of the debtor). The creditors may simply defer the enforcement of their rights; or they may agree to a formal arrangement by which the debtor attempts to pay off or reduce the debts. This shows that the concept of insolvency is rather uncertain. Insolvency is often brought about not by the loss of assets but by the loss of credit facilities or, particularly in the building industry, by adverse cash-flow. This reflects the fact that

---

[17] [1995] 1 W.L.R. 68.

construction companies often have a large cash turnover against comparatively small assets.

Bankruptcy proceedings

Bankruptcy applies to individuals and not to incorporated companies. It is a process under which possession of the debtor's property is taken for the benefit of his creditors. The debtor obtains release from his debts and liabilities, but is subject to certain restrictions.

Bankruptcy proceedings are set in train by a bankruptcy petition. The procedure to be followed is set out in Part IX of the Insolvency Act 1986. In the most usual case of a petition by a creditor, it must be shown that there is no reasonable prospect of the debt being paid. The debt must also exceed a minimum "bankruptcy level", currently £750. The debtor may himself present a petition. Where the court makes a Bankruptcy Order, the assets and affairs of the bankrupt pass, usually, to the official receiver initially and then to a trustee in bankruptcy, for the purpose of realising and distributing the assets of the bankrupt person. Essentially, all his assets including future earnings are taken for the benefit of creditors. The bankrupt person is allowed to retain tools, vehicles and equipment necessary for use in his employment or business, and basic household equipment and provisions needed by his family. These provisions are harsh, but it should be remembered that the statutory bankruptcy process has evolved as the alternative to imprisonment for civil debt (graphically described in a number of the novels of Charles Dickens). The bankrupt person can usually obtain discharge after two years, but this may be extended in various circumstances.

The Insolvency Act of 1986 contains new provisions designed to encourage individuals who are in financial difficulties to make binding arrangements with creditors in order to stave off bankruptcy proceedings. In such a case, the individual may apply for an "Interim Order" from the court to obtain a short period of protection against bankruptcy. The effect of such an order is that no bankruptcy petition, or other proceedings or execution, may be commenced against the debtor without leave of the court.

Where a Bankruptcy Order is made and the process of realising and distributing assets commences, the trustee has an important statutory power[18] to disclaim "onerous property". This is defined as

---

[18] Insolvency Act 1986, s. 315.

including (1) any unprofitable contract, or (2) any other property that may give rise to liability to pay money or perform any other onerous act. This provision will operate to the serious disadvantage of any person in contract with the bankrupt, since the trustee is enabled to retain and realise any benefit which may accrue under a contract, but effectively cannot be held liable.

Assets available to trustee

The assets available for realisation include all property belonging to the bankrupt at the date of the bankruptcy, which vests automatically in the trustee. Property acquired after the commencement of bankruptcy may, with certain exceptions, be claimed by the trustee. Where debts or claims are available, the trustee is empowered to enforce these by bringing claims, which may be settled in the trustee's discretion. The defendant in such proceedings is not wholly without protection. Even though the bankrupt cannot be held liable for cross-claims, they may still be raised as a defence of set-off. Further, the trustee may be required to give security for costs, so that the defendant has some protection in the event that the action fails.

The trustee cannot override charges on property so that, as in the case of the winding-up of companies, the secured creditors will be entitled to realise their debts, leaving the trustee to realise and distribute the remaining assets to unsecured creditors.

Winding-up companies

Winding-up procedure is dealt with in Chapter 4. While bankruptcy is always conducted through the courts, voluntary winding-up may be initiated by the company itself resolving to wind-up. The company may then appoint its own liquidator. If the company is insolvent, the voluntary winding-up will be controlled by the creditors who may have their own liquidator appointed. Winding-up which is not voluntary is conducted by the court, in most cases being on the ground of insolvency and at the suit of the creditors. In winding-up by the court, the official receiver initially acts as liquidator, and continues unless replaced by order of the court.

The liquidator's duty is to collect in the assets and apply them in discharge of the company's liabilities. Unlike the trustee in bankruptcy, the assets do not vest in the liquidator, unless by specific order of the court. The liquidator has powers to carry on the company's business, to sell the assets and to compromise claims. A Winding-up Order operates to discharge all the directors of the company, so that only the liquidator retains the power to act for the company.

As an alternative to winding-up by the court, the Insolvency Act 1986 contains an interim procedure which empowers the court to make an "Administration Order". The purpose of this is to protect the company from the ordinary consequences of insolvency, where there is a possibility that such a temporary arrangement may save the company or at least improve the realisation of its assets. This procedure has the same intention as the Interim Order in relation to personal bankruptcy proceedings (see above). An administrator of the company is appointed who has wide powers to control the company, investigate its affairs and to make proposals for consideration by the creditors. The administration will lead either to a scheme allowing the company to continue or to winding-up.

Assets available to liquidator

The liquidator must collect all assets belonging to the company which includes contributions due from members on shares which are not fully paid-up. Since the winding-up is followed by dissolution, no assets are retained by the company (unlike a bankrupt individual). A liquidator is given power to disclaim unprofitable contracts.[19] As in bankruptcy, an important limitation on the company's assets arises when they are used as security. A mortgagee or debenture holder may exercise his security, and may prove in the winding-up for any unsatisfied balance, including unpaid interest.

Receivers

A receiver is a person appointed to collect and preserve property. The courts have wide jurisdiction to appoint receivers, such as in a pending action to protect the subject matter of the dispute. A receiver may also be appointed by a party under an instrument, for example, a mortgage or debenture. The receiver's duty in this case is to take possession of the assets mortgaged or charged in order to protect and realise the security. Such an appointment is usually indicative of the debtor's insolvency or financial difficulty. The appointment and exercise of the receiver's powers may have an important effect on the debtor's ability to perform a subsisting contract.

---

[19] Insolvency Act 1986, s. 178.

Receivers appointed under mortgages or charges may exercise statutory powers.[20] But these are invariably enlarged by the terms of the instrument itself. This may provide for powers of management and for the receiver to be the agent of the debtor company. The instrument under which the receiver is appointed will also specify when the right to appoint arises. When the company is being wound-up, a receiver may be appointed either before or after the appointment of a liquidator.

When a liquidator is appointed after a receiver, any powers as agent of the company will be terminated, since the liquidator takes precedence in managing and winding-up the company. But in other matters concerning the right to the company's assets, the liquidator and the receiver may be in conflict.

The Insolvency Act 1986, following recommendations of the Review Committee on Insolvency Law and Practice, introduced the new title of "administrative receiver". This is a person appointed as receiver or manager of the whole or substantially the whole of the company's property. Such an appointment will occur under a "floating charge" by which the whole or substantially the whole of the company's assets are charged. The effect of such an appointment is to place the receiver virtually in control of the company. The procedure is designed to create powers and duties enabling the administrative receiver to obtain information and make reports about the affairs of the company, to be made available also to unsecured creditors.

Insolvency under a construction contract

If the contractor becomes insolvent, the first effect is usually that the work is brought to a stop by the inability to continue financing the work. If the employer is the insolvent party, his inability to meet interim payments will stop the work. In either case, this produces a potentially serious financial loss for the other party, which will not be satisfied by the insolvent party. Such loss may be reduced or even avoided by an employer appropriating retention money or the contractor's plant and goods or by enforcing a bond. A contractor is likely to fare less well. His work and materials, whether paid. for or not, pass to the employer and to the employer's trustee or liquidator on insolvency, when they become attached to the land. Thereafter they cannot be removed.

The parties' rights are usually regulated by provisions of the contract which operate upon various events indicative of

---

[20] Law of Property Act 1925, ss. 101, 109.

insolvency. When one party intimates that he cannot continue with a contract by reason of insolvency, he will repudiate the contract and the other party has no real choice but to accept. However, standard forms of contract usually provide that the innocent party may terminate the contractor's employment without ending the contract, so that advantage may be taken of contractual terms applying after such termination, for example, as to the rights in goods and plant, and claims.

The ICE form does not give the contractor any right of determination under the contract for insolvency, so that he must rely on common law repudiation. The employer is given the right to terminate the contractor's employment in the event, *inter alia*, of the contractor becoming bankrupt or having a receiving order or administration order made against him or going into liquidation. The employer then has the right to complete the contract by other contractors and to claim or set off the additional cost of completion (clause 63). To fortify these rights the contractor's plant and unfixed goods and materials are deemed to be the property of the employer when brought to the site, so that these are available as security in the event of determination (clause 63).

The JCT form gives rights of determination to both employer and contractor. The contractor may determine for non-payment of certificates, whether or not due to insolvency (clause 28). The contractor may then claim his loss from the employer. But such claim is not secured since goods and materials, when paid for, become the employer's property (clause 16). The employer has similar rights if the contractor, *inter alia*, becomes bankrupt or has a Winding-Up Order made or a receiver of his business appointed. In such cases the contractor's employment is automatically ended, subject to reinstatement. The employer may claim the additional costs of completion from the contractor (clause 27), but the only security for such claim is the retention of money and any performance bond (see below).

All the above provisions take effect subject to the laws of insolvency. The principles which may conflict with such contractual rights are:

(1) provisions which vest the debtor's property, upon insolvency, in a particular creditor may be void; and

(2) the statutory right of disclaimer of a trustee in bankruptcy or liquidator cannot be excluded.

The impact of insolvency law is illustrated by the difficult case of *British Eagle v. Air France*[21] in which the House of Lords held the International Air Transport Association (IATA) scheme to be subject to the laws of insolvency. The scheme provided for settlement of debts and credits between airlines and precluded direct claims between them. Upon the insolvency of British Eagle it was held that claims could be brought direct against other airlines. The case has given rise to concern generally about the ability of contractual provisions to survive insolvency.

Clause 53 of the ICE form states that the contractor's property is deemed to vest in the employer when brought to site. While such a provision may be enforceable on insolvency,[22] it was held in *Re Cosslett (Contractors)Ltd*[23] that the clause was ineffective to pass title in the contractor's plant. The effect of the clause depends also on property in goods and materials having passed to the contractor (see Chapter 7). Determination following insolvency is not contrary to the insolvency laws and it is thought that a determination correctly carried out under the above clauses will be effective against a liquidator. In *Re Cosslett (Contractors) Ltd* (above) it was also held that the employer had the right to use the contractor's plant after determination on the ground of insolvency, even though property in the plant had not passed to the employer. But there is doubt as to the validity of a determination where the trustee or liquidator seeks to exercise his right to complete the contract and the employer may not be entitled to recover his loss where the trustee or liquidator exercises his right of disclaimer. Insolvency of the main contractor may raise issues as to whether the employer may make a direct payment to a sub-contractor of a sum otherwise due to the insolvent main contractor. In *Re Tout & Finch*[24] it was held that the equivalent direct payment provision under the RIBA form[25] could be operated where the conditions of the clause were satisfied, namely that the contractor had failed to provide proof that a previous payment in respect of the sub-contractor had been discharged. The right of direct payment did not become available upon the contractor's insolvency but by reason of his previous default.

---

[21] [1975] 1 W.L.R. 758.
[22] *Re Walker, ex p. Barter* (1884) 26 Ch. D. 510.
[23] [1998] Ch. 495.
[24] [1954] 1 W.L.R. 178.
[25] Now cl. 35.13.5, JCT 80.

BONDS

When the contractor fails to complete the contract, whether by reason of his own default or the employer's determination, and the employer is unable to recover his loss from the contractor, the employer may have some further protection if the contractor has provided a bond.

A bond is an undertaking by a surety to make payment upon the contractor's default. The usual form of bond guarantees the contractor's performance of the contract with an undertaking to be bound in a specified sum until (and unless) such performance is achieved. Upon the contractor's failure to perform in full, the employer is entitled to call on the surety (or bondholder) to make good the loss, up to the maximum amount of the bond. Since a bond is a contract of guarantee, it requires to be evidenced in writing. Further, since the employer gives no consideration (save that the contractor must include the cost of the bond in his price for the work) the bond must be made by deed.

Subject to the terms of the bond, a surety may be discharged from liability by a material alteration in the contractor's obligation which has been guaranteed, such as extra works being ordered or an extension of the contract period being granted. The employer is also under a duty to mitigate his loss, since otherwise it may be said the loss is not caused by the contractor's non-performance. The ICE form of contract incorporates a form of bond which provides:

> "no alteration in the terms of the contract . . . or in the extent or nature of the works . . . and no allowance of time . . . nor any forebearance or forgiveness . . . on the part of the employer or the said engineer shall in any way release the surety from any liability under the above-written bond."

These provisions are designed to overcome the above difficulties. The ICE conditions (clause 10) and tender provide that the contractor may be required to obtain a bond in a specified sum, not exceeding 10 per cent of the tender total. The JCT form does not provide expressly for a bond, but this may be·incorporated into the tender documents. The provision of a bond may be made a condition precedent to the execution of the contract or to the contractor's right to payment.

A bond of the form under discussion is in the nature of a guarantee. The bondsman is not an insurer, and consequently there is no automatic duty of disclosure (see Chapter 7). The terms of the bond may, however, require that notice be given of relevant events. A further matter which should be dealt with in the bond is its intended duration. This may be, for example, until the end of the maintenance period or for some further stipulated period. It is a matter of some difficulty for contractors if bonds do not provide for release, since banks may be unwilling to continue to provide further bonds while those given earlier remain in force.

The law on bonds took an unexpected turn in the case of *Trafalgar House v. General Surety*.[26] A groundworks sub-contractor had given a bond for 10 per cent of the value of the sub-contract. The terms of the bond included the following archaic but traditional language:

> "If the sub-contractor shall duly perform and observe all the terms provisions conditions and stipulations of the said sub-contract on the sub-contractor's part to be performed and observed according to the true purport, intent and meaning thereof or if on default by the sub-contractor the surety shall satisfy and discharge the damages sustained by the main contractors thereby up to the amount of the above written bond then this obligation shall be null and void . . ."

The contractor completed the work and issued proceedings against the bondsman. The Court of Appeal held the contractor entitled to summary judgment without a full hearing,[27] on the claimant's assertion that it had suffered loss exceeding the value of the bond. The Court of Appeal considered that the commercial purpose of the bond was to provide immediate funds in the event of a failure by the sub-contractor. The House of Lords, however, held that the bondsman was entitled to raise any matter of defence or cross-claim and that the contractor would have to establish liability against the sub-contractor before payment on the bond could be demanded. In effect, the House of Lords rejected the Court of Appeal's attempt to treat the guarantee bond as an on-demand bond (see below).

On-demand bonds

As an alternative to a conditional bond, the effect of which is to guarantee payment of loss once established, there has been great development in recent years in a different type of instrument, still called a bond, but having a totally different effect. The on-demand bond usually entitles the holder (the employer) to call for payment

---

[26] [1996] A.C. 199.
[27] RSC, Ord. 14 (Now CPR, Pt. 24)—see Chap. 2.

by the bondsman (usually a bank) upon giving a particular form of notice. The notice usually requires no more than an assertion of default on behalf of the contractor. The bond money will then be paid irrespective of any disputes that may exist, either in relation to the underlying contract generally, or in relation to the purported reason for calling the bond in particular. The law relating to this type of bond has developed in the English courts through a series of cases in which contractors or sellers have sought injunctions from the court to prevent the bank paying the bond or to prevent the other party to the contract calling for payment. The courts have consistently refused such injunctions, in line with decisions of courts in many other countries. The only material exception to this rule is where the contractor is able to establish (not merely allege) fraud. In *Edward Owen Engineering v. Barclays Bank*,[28] the plaintiff, English suppliers to a Libyan customer, gave an on-demand bond. The customer, when himself in default, called the bond and the plaintiff sought an injunction against the bank to restrain payment. Lord Denning explained the procedure: the customer claims from the bank (there may be an intermediary bank involved also), the bank pay "on first demand without proof or conditions", and the bank then claim against the English suppliers. Lord Denning continued:

> "It is obvious that that course of action can be followed not only where there are substantial breaches of contract, but also when the breaches are insubstantial or trivial, in which case they bear the colour of a penalty rather than liquidated damages; or even when the breaches are non-existent. The performance guarantee then bears the colour of a discount on the price of 10 per cent. or five per cent. or as the case may be. The customer can always enforce payment by making a claim on the guarantee and it will then be passed down the line to the English supplier. The possibility is so real that the English supplier, if he is wise, will take it into account when quoting his price for the contract."

The demand for this type of bond has increased as international trade and construction work in particular have grown. Considerable problems remain for a contractor seeking to recover the proceeds of a bond which has been wrongly called, or even claiming credit for the value of the bond in any subsequent dispute. It is a matter of considerable importance that these issues should be dealt with fully in the underlying construction contract, so that they may be brought before arbitrators who may become seized of a dispute under that contract.

---

[28] [1978] Q.B. 159.

CHAPTER 11

STANDARD FORMS OF CONTRACT

THERE are good reasons for the wide-spread adoption of standard forms of contract. Performance of the contract often stretches far into the future and may involve uncertainty as to the conditions which will prevail. Many possible eventualities need to be catered for so that the parties to the contract can know the consequences of various possible courses of action. Historically, parties involved in construction work have developed and adopted standard forms of contract covering many types of work. Standard forms go back to the nineteenth-century and in some respects even earlier. Many of the standard forms in use today can be traced back to public works contracts drafted by nineteenth-century lawyers, and this partly explains their anachronistic appearance.

Since their origin, a number of different trends and influences are discernible leading up to the present day. Some of the earliest draftsmen were public authority lawyers who tended to impose harsh conditions on contractors. They established the structure of contracts which is still with us. Thereafter, professional bodies such as the RIBA and latterly the ICE, and other bodies as well, drafted and promoted their own forms, often based on and containing clauses from these earlier forms. After World War II, concern was expressed about the growing number of these standard forms, and the increasing practice for main contractors to draft their own forms of sub-contract, often containing one-sided provisions. In 1964, the report of the Banwell Committee recommended that a single standard form of contract for the whole construction industry was both desirable and practicable, and that standardisation of sub-contract conditions should follow. This objection was, however, never achieved.

The decades that followed Banwell saw a great upsurge in construction disputes and a parallel increase in the numbers and variety of standard forms; and latterly there has been a great increase in the variety of procurement for which standard forms are required. More recently, with the common availability of word processors, contract forms have proliferated to the point that a substantial amount of construction work is now carried on under ad hoc forms, based on standard or traditional drafts, but tailored to the particular commercial transaction.

The latest development in the field of standard forms is the Latham Report produced in 1994.[1] This was produced following a rapid consultation process throughout the whole construction industry aimed at identifying the source of a variety of complaints. A number of conclusions pertaining to standard forms were drawn, namely that:

(1) adversarial attitudes generated by the existing forms should be reduced or controlled by the use of adjudication;

(2) contracts, and particularly sub-contracts, should contain certain "fair" provisions to ensure that payments were not held up

(3) the existing standard forms did not operate efficiently and should be superseded by one generally recognised form, for which the New Engineering Contract was suggested as a suitable model.

There were many other detailed recommendations and a substantial amount of work has been undertaken to implement the report through the newly formed Construction Industry Board and 12 Working Groups. The proposals for legislation have been partly implemented by the Housing Grants, etc., Act 1996, dealt with elsewhere in this book. These measures, however, have not been without their critics and the NEC remains to be adopted as a serious rival to either of the main forms.[2]

As a result of the Housing Grants, etc., Act 1996, all standard forms in the construction industry have had to be adapted to provide for adjudication (see Chapter 2), stage payments and the right of suspension (see Chapter 9). In addition, "pay-when-paid" clauses are no longer permitted, save on the ground of insolvency of the payer (see Chapter 10). In a few cases, the forms have been reprinted, but in most cases, at the present time, the forms are issued with supplements.

---

[1] *Constructing the Team*, HMSO.
[2] For a full discussion of the Latham report and recommendations see "Contemporary Issues in Construction", Vol. II in *Construction Contract Reform: A Plea for Sanity*, (John Uff Q.C., ed, 1997).

FORMS OF MAIN CONTRACT

The two forms most frequently encountered are the JCT Standard Form of Building Contract, and the ICE Conditions of Contract. These are each dealt with in more detail in the following two chapters. The present chapter reviews other standard forms likely to be encountered in various types of construction work.

GC/works/1

This form has a long history as the standard government document for placing building and civil engineering contracts (works). Earlier editions were produced by government and, while published for information, use was restricted to government projects. After the demise of the PSA, production of the third edition in 1990 was undertaken as a "privatised" project, although the document continued to be published by HMSO. A new edition was issued in 1998, also privately produced and published by HMSO. This exists in different versions: With Quantities, Without Quantities and Single Stage Design and Build Conditions. A model form and commentary are also published separately.

Under GC/Works/1, the work is to be carried out subject to instructions given by the PM (Project Manager) who is to be a person employed in that capacity by the employer "to act on his behalf in carrying out those duties described in the contract". The PM may issue a wide range of instructions (clause 40). They are to be valued, whether amounting to variations or not, by the QS (Quantity Surveyor), who is also appointed for the time being by the employer. The PM is empowered to grant extensions of time (clause 36). Provision is made for claiming expense by reason of delay or disruption from specified causes, and the QS is to value the claim (clause 64). Both the PM and the QS are, therefore, required to exercise their professional judgment in the same way as the engineer or architect under other forms.

The contract provides a wide range of measures which the employer may utilise at his option, including agreed measures for acceleration or cost savings (clause 38) and payment of a bonus for early completion (clause 38A). Provision is made for design work carried out by the contractor (clause 10). Where a programme is submitted, the contractor is taken to warrant that it shows the events that are critical to the satisfactory completion of the works, and that the programme is achievable and conforms with the requirements of the contract (clause 33). The contract expressly requires regular progress meetings and reports on progress and delay (clause 35). The contract allows for additional payment in respect of ground conditions or artificial obstructions which could not reasonably have been foreseen, having regard to information which ought reasonably to have been ascertained (clause 7), in terms similar to those appearing in the ICE Conditions of Contract.

The Conditions of Contract are specifically tailored to the requirements of the Housing Grants, etc., Act 1996, *e.g.* providing expressly for suspension for non-payment (clause 52) and for adjudication (clause 59). Disputes (other than relating to enforcement of any adjudicator's decision) are to be referred to arbitration, but no reference is to be made until after completion, unless the parties otherwise agree (clause 60). Consequently, the decision of an adjudicator will be binding until the award of an arbitrator is subsequently given. Upon a reference to arbitration a timetable is to be set for the hearing, which is not to exceed a period of six months from the date of the preliminary meeting; and the award is to be made within three months thereafter.

One of the recommendations of the Latham Report, which was not carried forward into the Housing Grants, etc., Act 1996, is included within the contract: clause 1A expressly requires the employer and the contractor to "deal fairly, in good faith and in mutual co-operation with one another". The contractor is required similarly to deal with all his sub-contractors and suppliers. As the law currently stands, this would be unenforceable by sub-contractors (who are not parties to the contract) but as between employer and contractor the obligation of good faith and, possibly co-operation, might well lead to substantive allegations of breach.[3] The contract makes provision for nomination (clause 63) but the main contractor is to remain fully responsible and the employer is not to be required to pay any greater sum arising from any determination or renomination except on the ground of the sub-contractor's insolvency (clause 63A).

Form GC/Works/1 is intended for large scale building and civil engineering works. Alternative versions of the form are also issued known as GC/Works/2 for medium sized projects, GC/Works/3 for M&E contracts and GC/Works/4 for small projects. Non-governmental forms are also issued known as PC/Works based on the government versions.

---

[3] See Chap.5—Good Faith, Best Endeavours and Fair Dealing.

FIDIC (International) Conditions

There has been in existence since 1956 a form of contract based on the ICE conditions, for use in engineering contracts having an international element. A fourth edition of this form was issued in 1987 by the International Federation of Consulting Engineers (FIDIC), recommended for use both in international and domestic civil engineering contracts. Up to and including the fourth edition, the clause numbering and much of the content of the FIDIC Conditions followed the ICE Form of Contract. In 1998, however, FIDIC decided upon a new course by publishing four entirely new standard forms as follows:

(1) Conditions of Contract for construction of building or engineering works designed by the employer or his representative, the engineer.

(2) Conditions of Contract for plant and design-build, for the provision of electrical or mechanical plant and for the design and execution of building or engineering works.

(3) Conditions of Contract for EPC (Estimated Prime Cost) turnkey projects, where one entity takes total responsibility for the design and execution of an engineering project.

(4) Short Form of Contract for building or engineering works of relatively small value.

These forms adopt an entirely new clause numbering pattern, even though much of the wording of earlier editions has been retained. The documents are published as a "test edition" which is to be revised after consultation in a formal first edition, proposed to be issued in 1999. The notes below relate to the first of the four standard forms.

As a departure from previous editions containing 70 clauses, the present document has 20 clauses, in which the topics are re-arranged and consolidated. Clause 1 includes extensive new definitions as well as provisions as to the contract documents, assignment, delayed receipt of information, compliance with applicable laws, etc. Clause 1.5, as with previous editions, states the order of priority of the contract documents, which include the contract agreement, particular conditions, general conditions, specification and drawings, the first having the highest priority. The engineer is empowered to give instructions to clarify an ambiguity (these latter provisions being found in clause 5 of the previous edition). Clause

2 contains consolidated provisions dealing with the rights and duties of the employer. Clause 2.4 now obliges the employer, upon request from the contractor, to provide reasonable evidence of the financial arrangements to facilitate payment.

Clause 3 sets out provisions, substantially based on the previous edition, dealing with the role of the engineer. Clause 4 sets out the general obligations of the contractor. Clause 4.4 specifically requires the contractor to take responsibility for acts or defaults of any sub-contractor. This clause is subject to particular provisions governing nominated sub-contracting (see below). Clauses 4.11 and 4.12 substantially replicate the former clauses 11 and 12 (based on the same numbered clauses of the ICE Conditions) whereby the contractor is rendered generally responsible for the sufficiency of his tender, but may claim additional payment in respect of physical conditions or obstructions which were unforeseeable. Clause 8 deals with commencement, delays and suspension, including detailed provisions as to programmes (clause 8.3) and extensions of time (clause 8.4). By clause 8.6. the engineer is entitled to require measures to expedite progress, based on clause 46 of the ICE Conditions; and by clause 8.7, the contractor is required to pay "delay damages" as specified in the appendix, in respect of any period of overrun.

Measurement, valuation and payment are dealt with compendiously in clauses 12, 13 and 14, which contain many of the provisions of the previous edition suitably re-arranged. The work is to be re-measured and valued at rates or prices specified in the contract, which may themselves be varied if circumstances require (clause 12.3). Variations are to be valued in accordance with an agreed evaluation or otherwise in accordance with clause 12 (clause 13.3). Where the contract includes a provisional sum, the engineer may instruct work to be executed by the contractor or by a nominated sub-contractor (clause 13.5). The contractor is not obliged to employ a nominated sub-contractor against whom he raises reasonable objection. In this event, the employer has the option of agreeing to indemnify the contractor against the consequences of such objection (clause 5.2). As an alternative to conventional interim payment by measurement, there are provisions for advance payment (clause 14.2) and for a schedule of payments (clause 14.4). Clause 14 includes a procedure for the contractor's final statement and for the engineer to issue the final payment certificate. The employer is not to be liable for further claims except in the case of fraud, wilful misconduct or gross negligence by the employer (clause 14.4).

The contract contains new provisions for tests on completion (clause 9) and an extended procedure for the employer's taking over and defects liability (clauses 10 and 11). The term "defects liability" is a misnomer since clause 11.10 provides that each party is to remain liable for the fulfilment of any obligation which remains unperformed after the Performance Certificate, which otherwise signifies completion. Where the works fail to pass the Tests on Completion for a second time (clause 9.4) or the contractor fails to remedy any defect or damage at the second opportunity (clause 11.4), the employer has a range of remedies. Where the defect deprives the employer substantially of the whole benefit of the works or part of the works, the employer may terminate the contract in part or in whole.

Clauses 15 and 16 make provision for termination by employer and contractor respectively. Clauses 17 and 18 make provision risks, indemnity and insurance and clause 19 for events of *force majeure*. Clause 20 contains new provisions governing claims and disputes. In place of the traditional engineer's decision on matters in dispute, the contract now requires the appointment of a Dispute Adjudication Board (DAB) which may be named in the contract or otherwise appointed. The DAB may consist of one or three persons. In the latter case, each party is to nominate one person, whom the other must approve, the third member being agreed by the parties. The DAB is then required to give a decision on any dispute which, in the absence of Notice of Dissatisfaction by either party within 28 days, will become final and binding upon both parties (clause 20.4). After Notice of Dissatisfaction, the parties are required to attempt amicable settlement, following which the dispute may be referred to arbitration in accordance with the ICC Rules.

The new form which has extensive international backing, represents a significant achievement in the modernisation of standard forms of contract, retaining the traditional formula of technical and financial control of the engineer, but without the additional role of settlement of disputes, which has, in the past, led to abuse.

Model form MF/1

A set of conditions known as the "Model form" has been issued by the Institutions of Mechanical and Electrical Engineers for many years. The current edition of 1988 is issued with the Association of Consulting Engineers and incorporates amendments up to 1998. The form is intended for contracts involving substantial elements of plant and machinery. In such forms of work, it is traditional for the supplier to perform much of the detailed design work but there is an appointed engineer with important powers. A

notable feature of the form is the extensive provisions for completion and performance tests and for release of the contractor's
liability at the end of the "defects liability period", a term which is
correctly employed under this form.

The contractor is required, with due care and diligence, to
"design manufacture deliver to site erect and test the plant, execute
the works and carry out the tests on completion within the time for
completion". The works are to be to the reasonable satisfaction of
the engineer but the contractor is responsible for the "detailed
design of the plant and of the works in accordance with the
requirements of the specification" (clause 13). The contractor is
also required to submit to the engineer drawings as called for in
the contract or as may be required by the engineer. Drawings, etc.,
are to be approved and signed by the engineer (clause 15). The
contract contains provisions for gaining access to the site, co-
operation by the employer and for the contractor to maintain
progress (clauses 11, 13). The contractor may make a claim if
obstructions are found on the site which could not reasonably have
been ascertained from an inspection (clause 5.7).

The bulk of the form is directed towards installation and testing
of plant and machinery. Thus, by clause 23, the engineer is given
extensive powers to inspect, examine and test plant at the contractor's premises where he may reject plant which he considers
defective, or not in accordance with the contract (clause 23). The
contractor must obtain permission before delivery of plant and in
default the engineer may suspend the work at the contractor's cost
(clauses 24, 25). After installation the engineer may carry out
further tests and reject any plant considered defective, or not in
accordance with the contract (clause 26). When the works are
complete, and after notice the contractor is to carry out the tests of
completion which may lead to rejection of the works (clause 28).
By clause 29.2, "when the works have passed the tests on completion and are complete (except in minor respects that do not affect
their use for the purpose for which they are intended) the engineer
shall issue a certificate to the contractor and to the purchaser".
This is the "Taking Over Certificate" which signifies effective
completion and passing of the risk to the employer. However, this
may be followed by the performance tests, if included in the
contract, which are to be carried out after the works have been
taken over. In the event of failure to pass the performance tests,
the contract may stipulate liquidated damages for failure to achieve
any guaranteed performance (note that these are not related to
delay but to failure of the works to comply with the contract
requirements). Alternatively, the purchaser may accept the works
subject to abatement of the price (clause 35.8).

The contract contains provisions of far reaching importance as to latent defects. By clause 36 the contractor is required to make good defects appearing within the defects liability period (usually 12 months), a similar liability period attaching to any repaired work. These obligations are expressed to be in lieu of any other obligation for the quality or fitness of the works and no further liability is to attach after the expiry of the defects liability period (clause 36.9). The only exception is in respect of "gross misconduct" where the contract may be liable within a period of three years after taking over (clause 36.10). The exclusion of liability is expressed to apply to sub-contractors. The equivalent clause has been held ineffective to protect sub-contractors by virtue of the doctrine of privity, but nevertheless effective to bar a claim in tort.[4] In addition, the contract provides for a final certificate which is to be exclusive evidence that the work is in accordance with the contract and that the contractor has performed all his obligations. The certificate is also final as to the value of the works, with the exception of fraud or dishonesty, or proceedings brought within three months after the certificate (clause 39.12).

The engineer is given extensive powers under the contract, including a general power to order variations. This is, however, limited to a net additional value of 15 per cent (clause 27). Decisions, instructions and orders of the engineer may be disputed within 21 days, whereupon the engineer must confirm, reverse or vary the same, with reasons. Thereafter, the contractor must challenge the decision within a further 21 days otherwise it is binding (clause 2.6). A supplement containing amendments required by the Housing Grants, etc., Act 1996 was published in 1999. This is inappropriate where the form is used for work situated abroad.

Management contracts

This refers to main contracts in which the contractor offers "management" services in lieu of full responsibility for the performance of the work. The interest in these forms of contract does not centre on the management services to be provided, but on the extent of the main contractor's responsibility for performance by sub-contractors. The usual arrangement is that the whole of the physical work is to be carried out by sub-contractors. The contractor is usually entitled to be paid the actual cost, so that he is not at

---

[4] *Southern Water v. Duvivier (No. 1)* (1985) 27 B.L.R. 111.

risk in this regard. Where default occurs in relation to quality or time, management contracts usually provide, in effect, that the management contractor is liable only if the sub-contractor can also be held liable, so that again the main contractor is not at risk.

Perhaps the best known standard form of management contract is the JCT Standard Form of Management Contract, first issued in 1987. In regard to the three elements of cost, quality and time, this contract provides as follows. The contractor is to be paid prime cost (which, by the second schedule, includes amounts due and payable under sub-contracts) plus the management fee. The major restriction on this indemnity is that the prime cost is to exclude costs incurred as a result of any negligence by the management contractor. As regards default by a sub-contractor (in the contract called a works contractor) the management contractor is to take steps in consultation with the architect to enforce the sub-contract or otherwise secure the satisfactory completion of the project. The employer is required to pay the management contractor all amounts properly incurred by him in fulfilling these obligations. The employer is then given the right to recover amounts so paid or credited, including liquidated damages, from the management contractor, but only to the extent that they have been recovered from the sub-contractor (clause 3.21). The difficulty which arises from such provisions (there are variants in other management forms) is that it can be argued that if the management contractors' own liability to the employer is contingent on recovery from the works contractor, he cannot show loss for the purpose of such recovery, with the result that the employer must always bear the loss. No definitive ruling on this point exists: the problem is that all these contracts contain arbitration clauses, and few such disputes will ever come before the courts. In *Copthorne v. Arup*[5] the Court of Appeal, considering a different aspect of clause 3.21, held that it did not exempt the management contractor from liability for his own breach of obligations under the contract.

The other point of interest under management contracts is the nature and definition of the services to be provided by the management contractor, in return for the management fee. In the JCT management contract these are defined in clause 1.5 as including the preparation of programmes, entering into works contracts, ensuring the standards of work by supervision, providing site facilities as required, keeping cost records, and ensuring that the project is carried out in an economical and expeditious manner. The third

---

[5] (1997) 85 B.L.R. 22.

schedule contains a list of specific tasks, which includes the provision of practical advice on all aspects of performance, including the drawings and specification, and specifically to advise on "buildability".

Forms of Sub-contract

Many contractors impose their own standard terms on sub-contractors. These often contain one-sided provisions which place the sub-contractors at a disadvantage in a dispute. Conversely some specialists and suppliers impose their own terms on main contractors. A fairer balance may be achieved by using one of the standard forms of sub-contract designed for use with the standard main forms.

JCT nominated sub-contract

A form of sub-contract comprising documents known as NSC/A and C is issued by the Joint Contracts Tribunal for use in nominations under the JCT forms of main contract. By the standard form of tender (NSC/T) the prospective sub-contractor agrees to use form NSC/C. The main contract nomination procedure (clause 35) also contemplates its use. The form contains articles of agreements, including an arbitration clause which is keyed into connected disputes under the main contract and under other forms. The sub-contract incorporates the tender (clause 2.1.1), which includes details of the main contract and particular conditions relating to the sub-contract work, such as programme details.

In the conditions, the sub-contractor undertakes to comply with the main contract so far as it relates to the sub-contract works (clause 5.1.1) and gives wide indemnities to the contractor in respect of liability under the main contract and third party claims (clause 5.1.2). The sub-contractor must comply with the architect's instructions which the contractor issues to the sub-contractor in writing. The contractor may also issue reasonable directions in regard to the sub-contract works (clause 4.2).

The sub-contract tender may be for a lump sum, subject to adjustment in accordance with the conditions (clause 16); or may be subject to complete re-measurement (clause 17). Detailed rules for valuation are provided. The architect is required, by clause 30 of the main contract, to include the value of sub-contract work in certificates. The sub-contractor undertakes to complete his work within the agreed programme. The contractor may grant extensions of time but only with the architect's consent (clause 11). Under clause 25 of the main contract the contractor is not liable for the sub-contractor's delay. If the sub-contractor fails to complete by the specified or extended date for completion, the contractor may claim any loss caused by the delay. But the contractor's right to claim or set-off such loss is dependent on the architect giving a certificate of delay under the main contract (clause 12). The sub-contractor is given a number or rights to obtain the benefit of, or to pursue disputes under, the main contract (clauses 11.3, 13.1, 21.7 and 22).

ICE form of sub-contract

A document, issued by the Federation of Civil Engineering Contractors (now Civil Engineering Contractors Association) for use in conjunction with the ICE Conditions of Contract, is often known (misleadingly) as the ICE form of sub-contract. It remains in current use with the ICE conditions, fifth and sixth editions. A revision was issued in September 1984. Being drawn up by one body only, the form is a model of clarity, avoiding the obscurities and length of the main forms of contract.

The form contains a short recital for the parties' names, and five schedules. These are to contain, *inter alia*, particulars of the main contract, further documents to be incorporated, a description of the sub-contract works, the contract price and the completion period. The sub-contractor undertakes (save where the sub-contract otherwise requires) to perform the obligations of the contractor under the main contract in relation to the sub-contract works (clause 3(2)), and to indemnify the contractor against liability incurred by reason of any breach of the sub-contract (clause 3(3)).

The sub-contractor's obligations are not tied directly to the operations of the main contract. Thus, extensions of time may be granted without reference to the engineer; save that where the delaying event entitles the contractor to an extension, the sub-contractor's extension is not to exceed the extension under the main contract (clause 6). Instructions and variations ordered under the main contract do not bind the sub-contractor, unless the engineer's order is confirmed in writing by the contractor (clauses 7, 8). The contractor has the same general powers to give instructions under the sub-contract as the engineer has under the main contract, and the sub-contractor has the same rights in relation to them (clause 7(2)). There is therefore wide scope, both for the exercise of such powers by the contractor and for making claims by the sub-contractors. The sub-contract also provides for vesting of the plant and materials in the contractor, so that these may vest in the employer under the main contract (clause 11).[6]

After completion, the sub-contractor is required to maintain his work until completion of the main works and further to maintain them throughout the maintenance period of the main contract (clause 13). The contractor may determine the sub-contractor's employment if the main contract is determined (clause 16) or if the sub-contractor commits specified defaults corresponding broadly to those under clause 63 of the ICE main contract (clause 17). The form is not specifically designed for use with nominated sub-contractors. It contains provisions broadly in accordance with the minimum requirements under the ICE form for a nominated sub-contract. The form is used both for nominated and direct sub-contracts.

OTHER STANDARD FORMS

Design and build contracts

As with management contracts, these come in many forms, often being ad hoc drafts. There is good reason for this, as the form of the contract is heavily dependent on the degree of design liability undertaken by the contractor. For example, if the employer wishes to specify the overall design, with the contractor being responsible for details, then the contract will need to contain performance requirements only for the elements that the contractor is to design, and these can be accommodated within a relatively conventional construction contract. Conversely, if the contractor is to undertake the conceptual design as well as the details, then there needs to be a carefully drafted list of employer's requirements, which may go beyond mere technical performance. Further, the contractor, on submitting a tender, will usually put forward detailed proposals and these will need to be incorporated. Consideration needs also to be given to the submission of details for approval as the work proceeds, and as to what is to happen if the employer is not satisfied or changes his mind.

---

[6] See also cl. 53, ICE main contract.

The latter type of contract is reflected in the JCT Standard Form of Building Contract with Contractor's Design, issued in 1981 and amended on a number of occasions. This form provides for a statement of the Employers Requirements and the Contractor's Proposals. Clause 2 of the conditions requires the contractor to carry out and complete the work to these requirements and proposals, and for this purpose to complete the design of the works and carry it out. The contractor is under no further obligation to seek the employer's consent or approval to the details of the design, provided they comply with the requirements and proposals. The employer may make a change in his requirements after the contract has been let, but the contractor has a right of reasonable objection to any such change (if permitted, the change must be treated as a variation). In regard to the contractor's design work, the contract places on the contractor the like liability as would apply in the case of a professional designer, *i.e.* a duty of reasonable professional skill and care. This is to be contrasted with the position in the absence of any such provision, where there would ordinarily be an implied term that the work would be fit for purpose (see Chapter 9).

The ICE produced their own Design and Construct Conditions in 1992 based closely on and using much of the drafting of the sixth edition of the main conditions. Some of the conditions are thus identical to, and many are simply adapted from, the main form. Conditions considered inappropriate to a design and construction contract are "not used" (clauses 13, 17, 27, 34, 57 and 59). Administration of the contract is through the employer's representative with no independent engineer. The employer's representative carries out many of the functions otherwise performed by the engineer including certification (clause 60) and granting extensions of time (clause 44).

The contract operates in the same manner as the JCT Contractor's Design Form. The contract documents incorporate the Employer's Requirements and the Contractor's Submission, the latter being defined as "the tender and all documents forming part of the contractor's offer together with such modifications and additions thereto as may be agreed between the parties prior to the award of the contract" (clause 1(1)(f)). The contractor's basic obligation is to "design construct and complete the works and provide all design services, labour materials . . . and everything whether of a temporary or permanent nature required . . . so far as the necessity for providing the same is specified in or reasonably to be inferred from the contract" (clause 8(1)). As regards the completion of the design (which often gives rise to dispute or at least dissatisfaction under design and build contracts) clause 36(2) provides that materials and workmanship, where not described in the contract, are to be "appropriate in all the circumstances". By clause 8(2), the contractor's design obligation, including checking the employer's design, is to "exercise all reasonable skill care and diligence". As with the JCT form, this effectively operates as a limitation of liability, which would ordinarily be one of fitness for purpose (see Chapter 9).

The employer's representative is given no general power to issue instructions. His power to alter the works is conditional upon first having consulted with the contractor's representative, who is required to submit a quotation for the work as varied and his estimate of any delay, these matters to be agreed before the order is issued "wherever possible" (clauses 51, 52). The variation must be to the employer's requirements, leaving the contractor to complete the design details in accordance with the contract.

New Engineering Contract

This form was first produced by the ICE in 1991, in parallel with the ICE, sixth edition. It was first issued as a consultation documents followed, in 1993, by a first edition, with a second edition in 1995 following the prominent support given in the Latham Report. The form has been renamed the New Engineering and Construction Contract. The form is intended to operate under a wide range of differing types of contract, including conventionally priced contracts with or without bills of quantities, target costs contracts, cost reimbursable contracts and management contracts. The documentation makes use of core clauses which are common to all versions and the forms are intended for use with a wide range of projects, not limited to building and civil engineering. The NECC has, since its exception, had a mixed reception, not least on account of its unconventional style of drafting. The FCEC decided not to support the new form and its promotion has been largely through government. It is reported to have been used extensively by the electricity authority of South Africa.

A notable feature of the form is that the functions otherwise fulfilled by the engineer are split into three. Inspection and checking functions are carried out by the "supervisor"; certifying and other administrative functions are carried out by the "project manager"; and decisions on disputes are given by the "adjudicator". Their functions are defined by the clauses which set out the duties which they are to perform.

The core clauses are contained in nine different sections which are:

(1) general, including definitions and requirements as to communication. The contractor and project manager are each to give the other "early warning" of matter which could increase the price or cause delay;

(2) contractor's main responsibilities, including provisions for submission of design details;

(3) time, including provisions for the programme, possession and acceleration by agreement;

(4) testing and defects, including provision for tests before delivery;

(5) payment, including provision for interim certificates;

(6) compensation events, including a long list of matters giving rise to adjustment of the costs or programme. The contractor is required to quote for such events. The project manager may accept the quotation or require a revised one or himself assess the consequences;

(7) title, covering ownership of plant and materials;

(8) risks and insurance, providing for either the employer or the contractor to carry and insure risks arising under the contract;

(9) disputes and termination, including the adjudication procedure.

Despite the unconventional drafting, in the present tense, many of the detailed contractual provisions are similar in effect to those of other standard forms, particularly the ICE conditions. For example, one of the compensation events is *"physical conditions within the site other than weather conditions which, at the contract date an experienced contractor would have judged to have such a small chance of occurring that it would have been unreasonable for him to have allowed for them"* (clause 60.1(12)). The draftsmen claim that the impact of compensation events is neutral in that the contractor is paid a reasonable price for the work required. However, in regard to unforeseen conditions, it remains the position that this substantial risk turns upon the question whether the contractor is able to establish that the particular conditions encountered (in effect) could not reasonably have been foreseen. In the event that the conditions do not fall within the event as defined, the contractor must bear the additional cost. The form is claimed to incorporate the principles of good project management and to avoid adversarial attitudes. Further experience will be required before its claims can be properly assessed.

Small works contracts

The JCT issues two forms of contract for smaller works: the minor building works form and the intermediate form. The range of work suitable for these forms is not defined.

The form of contract for minor building works, in keeping with its subject matter, is short. The form of contract was amended in 1980 in line with other JCT documents. The form deals fully with liability for, and insurance against, various risks. This is necessary since the consequential losses which may arise out of building works may bear no relation to the scale of the works. The form contains many of the provisions of the standard JCT contract in an abbreviated form. Thus, provisional sums are provided for, but not prime cost items. The architect may vary the works, including the order or period in which they are to be carried out. The contractor is entitled to interim payments at not less than four-weekly intervals.

In 1984 the JCT issued an "intermediate" form of building contract known as IFC 84. This form follows closely the format of the minor building works form, while adopting much of the wording of JCT 80. Like the minor works form, IFC 84 contains no provision for prime cost sums or nomination. There is, however, provision for sub-contract work to be placed with a "named person" to be identified in the contract or specified by the architect (clause 3.3). The effect of the contract is to place the "named person" substantially in the same position as a nominated sub-contractor under the main forms, save that the contractor is not entitled to an extension of time for his delay. Where the named person "drops out" there is machinery for recovery of the loss.

The ICE have also issued a form for minor works, which is accompanied by notes for guidance, stating that the form is intended for contracts of value not exceeding £100,000, with the completion period not exceeding six months. This form represents a concise version of the essential elements of the ICE general conditions, rearranged and expressed in much simpler language. Among features to be noted is the requirement that the engineer is to be a named individual (clause 2.1). The traditional maintenance period is more accurately renamed as the "defects correction period" (clause 5) (now adopted for the ICE, sixth edition). The contractor is made fully liable for acts or defaults of sub-contractors (clause 8.3).

Other construction industry forms

The above discussion has been limited to forms of construction contract or sub-contract. There are also a number of standard forms of warranty intended to be executed by sub-contractors and designers. The professional institutions, in addition, issue numerous forms of engagement covering architects, surveyors and consulting engineers. All of these forms are periodically revised and re-issued.

CHAPTER 12

THE STANDARD FORM OF BUILDING CONTRACT

THE Standard Form of Building Contract was formerly known as the RIBA form and continues to have an association with that institute. The form is now issued by the Joint Contracts Tribunal (JCT), whose constituent bodies include the RIBA, the RICS, the Association of Consulting Engineers, the Construction Confederation and bodies representing employers, local authorities and sub-contractors. For brevity it may be called the JCT form and is intended for use with all types of building work.

A new edition of the form was issued in 1980. It exists in versions for a private employer, or for a local authority; and with or without quantities, giving four combinations. There is also a version with approximate quantities. The different versions contain variations in detail rather than substance. The commentary which follows is based upon the local authorities' edition with quantities, but most of the provisions are common to all the forms. The issuing body makes periodic revisions of the forms which are published either as amendments or as revisions of the complete forms. There were substantial changes in 1987, 1992 and 1994, and a new "edition" was issued in 1998 incorporating amendments. Many versions of the form will therefore be encountered. The commentary which follows is based on the 1998 version. In the local authority forms "architect" becomes "architect/Contract Administrator", as this office may be filled by an employee of the authority who may not be a qualified architect.

The published form contains articles of agreement, which may be executed as a deed, an appendix, the conditions of contract, forms of bond and a supplemental VAT agreement. The conditions create a "lump sum" contract, *i.e.* the contractor undertakes to carry out the work described and measured in the bills, for a stated sum of money. This is subject, however, to many possible alterations on account, *inter alia*, of variations, price fluctuations and claims.

The conditions assume that the work will be carried out under the supervision of the architect/Contract Administrator (referred to here simply as "the architect"). Under the contract the architect is given powers and duties which are wide and important, but which are also limited in their scope. The powers of the architect are considerably less than those of the engineer under the ICE conditions (see Chapter 12), reflecting the less hazardous nature of building work compared to civil engineering construction.

This commentary is intended as an introduction to the basic working of the contract. The most important clauses or sub-clauses are printed, with notes as to their effect. Other clauses are referred to where appropriate. Some clauses are omitted as being not essential to the basic scheme of the form. These clauses may, of course, be vital to any particular issue or dispute.

THE CONTRACT

The form contemplates that a contract will be made by the parties executing the articles of agreement. This is not essential. The conditions of contract may be incorporated by reference in any other document of agreement. But the articles provide, in the recitals, for specifying the names of the parties, a brief description of the works and a list of contract drawings.

The articles themselves state:

> "Now it is hereby agreed as follows
> Article 1
> For the consideration hereinafter mentioned the Contractor will upon and subject to the Contract Documents carry out and complete the Works shown upon, described by or referred to in those Documents.
> Article 2
> The employer will pay to the Contractor the sum of .... (hereinafter referred to as 'the Contract Sum') or such other sum as shall become payable hereunder and in the manner specified in the Conditions."

In articles 3 and 4 the architect and quantity surveyor are to be named. Articles 6.1 and 6.2 provide for identification of the planning supervisor and principal contractor (in accordance with the CDM Regulations). Article 5 provides that any dispute or difference may be referred to adjudication in according with clause 41A, which contains an adjudication procedure conforming with section 108 of the Housing Grants, etc., Act 1996. Subject to adjudication, articles 7A and B provide for disputes to be resolved by arbitration where clause 41B applies; or by litigation where it is deleted.

By clause 2.1 the contractor is required to "carry out the works in compliance with the Contract Documents". Clause 1.3 contains now a list of definitions including "Contract Documents" which means "the Contract Drawings, the Contract Bills, the Articles of Agreement, the Conditions and the Appendix." The relative effect of these contract documents is provided for as follows:

> **"14.1** The quality and the quantity of the work included in the Contract Sum shall be deemed to be that which is set out in the Contract Bills.
>
> **2.2.1** Nothing contained in the Contract Bills shall override or modify the application or interpretation of that which is contained in the Articles of agreement, the Conditions or the Appendix.
>
> **2.2.2** Subject always to clause 2.2.1:
>
> **.2.1** the Contract Bills, unless otherwise specifically stated therein in respect of any specified item or items, are to have been prepared in accordance with the Standard Method of Measurement of Building Works, 7th Edition published by the Royal Institution of Chartered Surveyors and the Building Employers Confederation (now Construction Confederation);
>
> **.2.2** if in the Contract Bills . . . there is any departure from the method of preparation referred to in clause 2.2.2.1 or any error in description or in quantity or omission of items (including any error in or omission of information in any item which is the subject of a provisional sum for defined work) then such departure or error or omission shall not vitiate this Contract but the departure or error or omission shall be corrected; where the description of a provisional sum for defined work does not provide the information required by General Rule 10.3 in the Standard Method of Measurement the correction shall be made by correcting the description so that it does provide such information; and such correction under this clause 2.2.2.2 shall be treated as if it were a Variation required by an instruction of the Architect under clause 13.2." •

The effect of these clauses is that, save for questions of the quality or quantity of the work, the conditions override the contract bills. A provision intended to amend the conditions (such as one for sectional completion) may therefore be ineffective if placed in the bills. But see clauses 13.1 and 25.4.12, which contemplate that the bills or drawings may limit the contractor's access to or use of the site and the order in which the work is to be carried out. See also clause 29, below. The bills will also override the contract drawings so that, while the contractor must perform the work shown on the drawings, any part which is not included in the bills is an extra, to be paid for. In a contract of any size there is also likely to be a specification or descriptive schedule. This, however, is not of itself a contract document and must be incorporated, for example, into the bills. The effect of the articles and conditions is dealt with as follows:

> "1.2.   The Articles of Agreement, the conditions and the Appendix are to be read as a whole and the effect or operation of any article or clause in the Conditions or item in or entry in the Appendix must therefore unless otherwise specifically stated be read subject to any relevant qualification or modification in any other article or any of the clauses in the Conditions or item in or entry in the Appendix."

The articles and conditions therefore rank equally in the event of an ambiguity, and neither may be overridden by the contract bills or drawings.

The architect is given the power and duty to supply such further details as are necessary.

> "5.3.1 So soon as is possible after the execution of this Contract:
> .1.1   the Architect without charge to the Contractor shall provide him (unless he shall have been previously so provided) with two copies of any descriptive schedules or other like documents necessary for use in carrying out the Works. . . .
> 5.3.2 Nothing contained in the descriptive schedules or other like documents referred to in clause 5.3.1.1 . . . shall impose any obligation beyond those imposed by the Contract Documents."

Clause 5.4 covers the provision of necessary information and drawings. By clause 5.4.1 there is to be an Information Release Schedule which the architect is required to follow. By clause 5.4.2, however, the architect is also required to provide any other drawings or details which are reasonably necessary to explain and amplify the contract drawings. These are to be provided when reasonably necessary.

The architect is further required to provide drawings for setting out the work:

> "7.   The Architect shall determine any levels which may be required for the execution of the Works, and shall provide the Contractor by way of accurately dimensioned drawings with such information as shall enable the Contractor to set out the Works at ground

level. The Contractor shall be responsible for, and shall, at no cost to the Employer, amend any errors arising from his own inaccurate setting out. With the consent of the Employer the Architect may instruct that such errors shall not be amended and an appropriate deduction for such errors not required to be amended shall be made from the Contract Sum."

Control of the work

The conditions envisage that the work will be under the general supervision of the architect. The RIBA conditions of engagement, used by many architects, provide for periodic but not constant supervision. Day-to-day site supervision is therefore left to the contractor and to the employer.

"10. The Contractor shall constantly keep upon the Works a competent person-in-charge and any instructions given to him by the Architect or directions given to him by the clerk of works in accordance with clause 12 shall be deemed to have been issued to the Contractor.

11. The Architect and his representatives shall at all reasonable times have access to the Works and to the workshops or other places of the Contractor where work is being prepared for this other places of a Domestic Sub-Contractor or a other places of a Domestic Sub-Contractor or a Nominated Sub-Contractor the Contractor shall by a term in the sub-contract so far as possible secure a similar right of access to those workshops or places for the Architect and his representatives and shall do all things reasonably necessary to make such right effective. Access in accordance with clause 11 may be subject to such reasonable restrictions of the Contractor or any Domestic Sub-Contractor or any Nominated Sub-Contractor as are necessary to protect any proprietary right of the Contractor or of any Domestic or Nominated Sub-Contractor in the work referred to in clause 11.

12. The Employer shall be entitled to appoint a clerk of works whose duty shall be to act solely as inspector on behalf of the Employer under the directions of the Architect and the Contractor shall afford every reasonable facility for the performance of that duty. If any direction is given to the Contractor by the clerk of works the same shall be of no effect unless given in regard to a matter in respect of which the Architect is expressly empowered by the Conditions to issue instructions and unless confirmed in writing by the Architect within two working days of such direction being given. If any such direction is so given and confirmed then as from the date of issue of that confirmation it shall be deemed to be an Architect's instruction."

On larger projects the architect and employer may agree to employ a resident architect on the works, but he will have no specific power or duty such as those of the Engineers' Representative under the ICE conditions.

General Obligations of the Contractor

The contractor's basic obligations are to comply with the contract documents, which define the work and the time within which it is to be done, and to comply with proper instructions of the architect.

> "**2.1.** The Contractor shall upon and subject to the conditions carry out and complete the works in compliance with the Contract Documents, using materials and workmanship of the quality and standards therein specified, provided that where and to the extent that approval of the quality of materials or the standards of workmanship is a matter for the opinion of the architect, such quality and standards shall be the reasonable satisfaction of the architect."

The "works" are defined by clause 1.3 as "the works briefly described in the First recital and shown upon, described by or referred to in the Contract Documents and including any changes made to these works in accordance with this Contract". The recital to the articles will not add to the technical description of the work, but may be an aid to construction in case of ambiguity, for example, as to the location or purpose of the work.

Clause 2.3 requires the contractor to give notice should he find any discrepancy or divergence between the drawings, the bills, architect's instructions, and other documents, and for the architect to issue further instructions in regard thereto.

Clause 2.1 (above) emphasises that the architect has no general or absolute power of approval or control. The provisions relating to the architect's satisfaction have given rise to uncertainty as to the scope of the final certificate to be issued under Clause 30 (see below). Where there is no stipulation for the architect's approval, his function is to ensure compliance with the contract documents.

> "**4.1.1** The Contractor shall forthwith comply with all instructions issued to him by the Architect in regard to any matter in respect of which the Architect is expressly empowered by the Conditions to issue instructions; save that:
>
>   **.1.1** where such instruction is one requiring a Variation within the meaning of clause 13.1.2 the Contractor need not comply to the extent that he makes reasonable objection in writing to the Architect to such compliance . . .
>
>   **.2** If within seven days after receipt of a written notice from the Architect requiring compliance with an instruction the Contrac-

tor does not comply therewith, then the Employer may employ and pay other persons to execute any work whatsoever which may be necessary to give effect to such instruction; and all costs incurred in connection with such employment may be deducted by him from any monies due or to become due to the Contractor under this Contract or may be recoverable from the Contractor by the Employer as a debt."

This provision makes it clear that the architect's powers are limited to those expressly contained in the contract. The machinery entitling the employer to bring in another contractor in the event of non-compliance with an architect's instruction is a valuable sanction. Clause 4.2 allows the contractor to challenge an instruction by asking under what power it is given. Clause 4.3 requires instructions to be given in writing and allows either the contractor or the architect to confirm oral instructions in writing.

Clause 13.1.2 empowers the architect, by variation order, to change the specified access to or use of the site or the order in which the work is to be done. It is not clear what limits are to be placed on the grounds of a reasonable objection. Ordinarily, expense would not be a proper ground, it is thought, since the contract provides for reimbursement (clauses 13 and 26). The contractor's right to uninterrupted possession of the site may be restricted by the bills or otherwise by agreement.

> "29.1   Where the Contract Bills, in regard to any work not forming part of this Contract and which is to be carried out by the Employer himself or by persons employed or otherwise engaged by him, provide such information as is necessary to enable the Contractor to carry out and complete the Works in accordance with the Conditions, the Contractor shall permit the execution of such work.
>
> 29.2   Where the Contract Bills do not provide the information referred to in clause 29.1 and the Employer requires the execution of work not forming part of this Contract by the Employer himself or by persons employed or otherwise engaged by the Employer, then the Employer may, with the consent of the Contractor (which consent shall not be unreasonably withheld) arrange for the execution of such work.
>
> 29.3   Every person employed or otherwise engaged by the Employer as referred to in clauses 29.1 and 29.2 shall for the purpose of clause 20 be deemed to be a person for whom the Employer is responsible and not to be a sub-contractor."

Note that clause 25.4.8.1 (see below) provides for an extension of time, whether or not such work is contemplated by the bills.

Apportionment of various risks arising out of the work and requirements as to insurances are dealt with in clauses 20 to 22.

> "20.1   The Contractor shall be liable for, and shall indemnify the Employer against, any expense, liability, loss, claim or pro-

ceedings whatsoever arising under any statute or at common law in respect of personal injury to or the death of any person whomsoever arising out of or in the course of or caused by the carrying out of the Works, except to the extent that the same is due to any act or neglect of the Employer or of any person for whom the Employer is responsible including the persons employed or otherwise engaged by the Employer to whom clause 29 refers.

20.2 The Contractor shall be liable for, and shall indemnify the Employer against, any expense, liability, loss, claim or proceedings in respect of any injury or damage whatsoever to any property real or personal in so far as such injury or damage arises out of or in the course of or by reason of the carrying out of the Works, and to the extent that the same is due to any negligence, breach of statutory duty, omission or default of the Contractor, his servants or agents or of any person employed or engaged upon or in connection with the Works or any part thereof, his servants or agents or of any other person who may properly be on the site upon or in connection with the Works or any part thereof, his servants or agents, other than the Employer or any person employed, engaged or authorised by him or by any local authority or statutory undertaker executing work solely in pursuance of its statutory rights or obligations. This liability and indemnity is subject to clause 20.3 and, where clause 22C.1 is applicable, excludes loss or damage to any property required to be insured thereunder caused by a Specified Peril.

20.3.1 Subject to clause 20.3.2 the reference in clause 20.2 to 'property real or personal' does not include the Works, work executed and/or Site Materials up to and including the date of issue of the certificate of Practical Completion or up to and including the date of determination of the employment of the Contractor (whether or not the validity of that determination is disputed) under clause 27 or clause 28 or clause 28A, where clause 22C applies, under clause 27 or clause 28 or clause 28A or clause 22C.4.3, whichever is the earlier.

.2 If clause 18 has been operated then, in respect of the relevant part, and as from the relevant date such relevant part shall not be regarded as 'the Works' or 'work executed' for the purpose of clause 20.3.1."

Clause 20.1 renders the contractor liable for third party claims in respect of personal injury, unless due to any act or neglect of the employer. Clause 20.2 renders the contractor liable for third party claims in respect of damage to property only when the contractor is at fault. The words "any property" could include the works themselves, but clause 20.3.1 makes clear that this is not so. However, the effect of clause 20.3.2 is that parts of the works which have been taken over will qualify as other property for the purpose of clause 20.2. Further, the duty to carry out and complete the works, under clause 2 means that the contractor is generally liable for damage to the works, unless due to the employer's default.

Clause 21.1 requires the contractor to maintain insurance against his own liability under clause 20. Clause 21.2 requires the contractor, where a provisional sum is included in the bills, to take out insurance in the joint names of the employer and the contractor against claims for damage to property other than the works, caused (broadly) by the carrying out of the works. However, this cover is of limited use because of the wide-ranging exceptions provided, namely damage:

"**21.2.1.1** for which the Contractor is liable under clause 20.2;

.2 attributable to errors or omissions in the designing of the Works;

.3 which can reasonably be foreseen to be inevitable having regard to the nature of the work to be executed or the manner of its execution;

.4 which is the responsibility of the Employer to insure under clause 22C.1 (if applicable);

.5 arising from war risks or the Excepted Risks."

The range of cover required is, therefore, very limited, effectively covering unforeseeable accident in the absence of fault. The Excepted Risks are defined in clause 1.3 and cover, broadly, radiation, explosion and like matters.

Clause 22 sets out three alternative provisions (clauses 22A, B and C) by which the main insurance of the works is to be effected. The cover required is defined in clause 22.2 as "cover against any physical loss or damage to work executed and site materials" but excluding cost under three heads being (broadly): (1) property which is itself defective, (2) defective designs, (3) certain excluded risks, including the Excepted Risks. The three ways in which the main insurance may be placed are described in a note to the form as follows:

Clause 22A is applicable to the erection of new buildings where the Contractor is required to take out a Joint Names Policy for All Risks Insurance for the Works and clause 22B is applicable where the Employer has elected to take out such Joint Names Policy. Clause 22C is to be used for alterations of or extensions to existing structures under which the Employer is required to take out a Joint Names Policy for All Risks Insurance for the Works and also a Joint Names Policy to insure the existing structures and their contents owned by him or for which he is responsible against loss or damage thereto by the Specified Perils.

The contractor is solely responsible for his method of work and for any temporary works (such as scaffolding), plant and equipment. The architect is given no powers either of approval or control, save for the limited power under clause 13.1.2 (see below).

Workmanship and materials

"8.1.1 All materials and goods shall, so far as procurable, be of the kinds and standards described in the Contract Bills, provided that materials and goods shall be to the reasonable satisfaction of the Architect where and to the extent that this is required in accordance with clause 2.1.

  .2 All workmanship shall be of the standards described in the Contract Bills, or, to the extent that no such standards are described in the Contract Bills, shall be of a standard appropriate to the Works, provided that workmanship shall be to the reasonable satisfaction of the Architect where and to the extent that this is required in accordance with clause 2.1.

  .3 All work shall be carried out in a proper and workmanlike manner and in accordance with the Health and Safety Plan.

8.2.1 The Contractor shall upon the request of the Architect provide him with vouchers to prove that the materials and goods comply with clause 8.1.

  .2 In respect of any materials, goods or workmanship, as comprised in executed work, which are to be to the reasonable satisfaction of the Architect in accordance with clause 2.1, the Architect shall express any dissatisfaction within a reasonable time from the execution of the unsatisfactory work.

8.3 The Architect may issue instructions requiring the Contractor to open for inspection any work covered up or to arrange for or carry out any test of any materials or goods (whether or not already incorporated in the Works) or of any executed work, and the cost of such opening up or testing (together with the cost of making good in consequence thereof) shall be added to the Contract Sum unless provided for in the Contract Bills or unless the inspection or test shows that the materials, goods or work are not in accordance with this Contract.

8.4 If any work, materials or goods are not in accordance with this Contract the Architect without prejudice to the generality of his powers, may:

  .4.1 issue instructions in regard to the removal from the site of all or any of such work, materials or goods; and/or

  .2 after consultation with the Contractor (who shall immediately consult with any relevant Nominated Sub-Contractor) and with the agreement of the Employer, allow all or any of such work, materials or goods to remain and confirm this in writing to the Contractor (which shall not be construed as a Variation) and where so allowed and confirmed an appropriate deduction shall be made in the adjustment of the Contract Sum; and/or

  .3 after consultation with the Contractor (who shall immediately consult with any relevant Nominated Sub-Contractor) issue such instructions requiring a Variation as are reasonably necessary as a consequence of such an instruction under clause 8.4.1 or such confirmation under clause 8.4.2 and to the extent that such instructions are so necessary and notwithstanding clauses 13.4, 24 and 26 no addition to the Contract

Sum shall be made and no extension of time shall be given; and/or

.4 having had due regard to the Code of Practice appended to these Conditions, issue such instructions under clause 8.3 to open up for inspection or to test as are reasonable in all the circumstances to establish to the reasonable satisfaction of the Architect the likelihood or extent, as appropriate to the circumstances, of any further similar non-compliance. To the extent that such instructions are so reasonable, whatever the results of the opening up for inspection or test, and notwithstanding clauses 8.3 and 26 no addition to the Contract Sum shall be made. Clause 25.4.5.2 shall apply unless as stated therein the inspection or test showed that the work, materials or goods were not in accordance with this Contract.

8.5 Where there is any failure to comply with clause 8.1.3 in regard to the carrying out of the work in a proper and workmanlike manner the Architect/the Contract Administrator, without prejudice to the generality of his powers, may, after consultation with the Contractor (who shall immediately consult with any relevant Nominated Sub-Contractor), issue such instructions whether requiring a Variation or otherwise as are reasonably necessary as a consequence thereof. To the extent that such instructions are so necessary and notwithstanding clauses 13.4 and 25 and 26 no addition to the Contract Sum shall be made and no extension of time shall be given in respect of compliance by the Contractor with such instruction.

8.6 The Architect may (but not unreasonably or vexatiously) issue instructions requiring the exclusion from the Works of any person employed thereon."

This clause has been substantially expanded to deal with a number of potential difficulties under the original clause contained in JCT 80. The requirement for the architect to express dissatisfaction within a reasonable time (8.2.2) is new. Likewise, the provisions for allowing non-complying work materials or goods to remain with a deduction from the contract sum is new to the contract, although it probably reflects a common practice. Clause 8.4.4 provides a useful mechanism, by use of the Code of Practice, for discovering the extent of defects, without running the risk of having to pay where work is found to comply with the contract. Note that there is no power to order replacement of defective work; this obligation remains on the contractor, however, by virtue of the obligation to complete the works, under clause 2.1. If the contractor fails to comply with an instruction under clause 8.4, the architect may invoke clause 4.1.2. Alternatively, there is a final sanction of determination under clause 27.1.3. 8.1.3 and the corresponding sanction under 8.5 are also new.

Independent of the contract, the work must, by law, comply with various statutory requirements (see Chapter 14). Difficult questions can arise in the event of conflict between these requirements and the architect's design. This is dealt with as follows:

"**6.1** Subject to clause 6.1.5 the Contractor shall comply with, and give all notices required by, any Act of Parliament, any instrument, rule or order made under any Act of Parliament, or any regulation or byelaw of any local authority or of any statutory undertaker which has any jurisdiction with regard to the Works or with whose systems the same are or will be connected (all requirements to be so complied with being referred to in the Conditions as 'the Statutory Requirements').

.2 If the Contractor shall find any divergence between the Statutory Requirements and all or any of the documents referred to in clause 2.3 or between the Statutory Requirements and any instruction of the Architect requiring a Variation issued in accordance with clause 13.2, he shall immediately give to the Architect a written notice specifying the divergence.

.3 If the Contractor gives notice under clause 6.1.2 or if the Architect shall otherwise discover or receive notice of a divergence between the Statutory Requirements and all or any of the documents referred to in clause 2.3 or between the Statutory Requirements and any instructions requiring a Variation issued in accordance with clause 13.2, the Architect shall within 7 days of the discovery or receipt of a notice issue instructions in relation to the divergence. If and insofar as the instructions require the Works to be varied, they shall be treated as if they were Architect's instructions requiring a Variation issued in accordance with clause 13.2.

.4 . . .

.5 Provided that the Contractor complies with clause 6.1.2 the Contractor shall not be liable to the Employer under this Contract if the Works do not comply with the Statutory Requirements where and to the extent that such non-compliance of the Works results from the Contractor having carried out work in accordance with the documents referred to in clause 2.3 or with any instruction requiring a Variation issued by the Architect in accordance with clause 13.2."

Provided the contractor adheres to the contract documents, his obligation is limited to giving notice to the architect of any breach of statutory requirements which he in fact discovers. The additional cost of taking down and rebuilding work so as to comply with byelaws is then likely to fall on the architect if it is shown that he was at fault. Clause 6.1.4 deals with emergency works, carried out before receiving instructions.

The contractor's obligations are generally limited to physical performance of the work described in the Contract Documents. Any provision requiring design work is likely to be held ineffective and not binding upon the contractor. Where the contractor is required to carry out design work, this may be achieved to a limited degree by the use of the new Part 5: Performance specified work. This is issued as Amendment 12 and comprises a new clause 42. The amendment involves substantial alterations to other clauses within the conditions.

Completion and maintenance

The contractor's obligations fall into two separate periods: the period up to the certificate of practical completion, when the work is carried out; and the defects liability period, during which the contractor must make good any defects. These are dealt with as follows:

"17.1 When in the opinion of the Architect Practical Completion of the Works is achieved . . . he shall forthwith issue a certificate to that effect and Practical Completion of the Works shall be deemed for all the purposes of this Contract to have taken place on the day named in such certificate.

17.2 Any defects, shrinkages or other faults which shall appear within the Defects Liability Period and which are due to materials or workmanship not in accordance with this Contract or to frost occurring before Practical Completion of the Works, shall be specified by the Architect in a schedule of defects which he shall deliver to the Contractor as an instruction of the Architect not later than 14 days after the expiration of the said Defects Liability Period, and within a reasonable time after receipt of such schedule the defects, shrinkages and other faults therein specified shall be made good by the Contractor at no cost to the Employer unless the Architect with the consent of the Employer shall otherwise instruct; and if the Architect does so otherwise instruct then an appropriate deduction in respect of any such defects, shrinkages or other faults not made good shall be made from the Contract Sum.

17.3 Notwithstanding clause 17.2 the Architect may whenever he considers it necessary so to do, issue instructions requiring any defect, shrinkage or other fault which shall appear within the Defects Liability Period and which is due to materials or workmanship not in accordance with this Contract or to frost occurring before Practical Completion of the Works, to be made good, and the Contractor shall within a reasonable time after receipt of such instructions comply with the same and at no cost to the Employer unless the Architect with the consent of the Employer shall otherwise instruct; and if the Architect does so otherwise instruct then an appropriate deduction in respect of any such defects, shrinkages or other faults no made good shall be made from the Contract Sum. Provided that no such instructions shall be issued after delivery of a schedule of defects or after 14 days from the expiration of the Defects Liability Period.

17.4 When in the opinion of the Architect any defects, shrinkages or other faults which he may have required to be made good under clauses 17.2 and 17.3 shall have been made good he shall issue a certificate to that effect, and completion of making good defects shall be deemed for all the purposes of this Contract to have taken place on the day named in such certificate (the 'Certificate of Completion of Making Good Defects').

17.5 In no case shall the Contractor be required to make good at his own cost any damage by frost which may appear after Practical Completion, unless the Architect shall certify that such damage is due to injury which took place before Practical Completion."

Clauses 17.2 and 17.3 have been amended so as to provide expressly that, with the employer's consent, there may be a deduction from the contract sum in lieu of making good.

The contractor's liability for defects is limited to those which are "due to materials or workmanship not in accordance with the contract or to frost." If defects become manifest due to some other cause, such as unsuitability of the design, the contractor is not obliged to carry out rectification; nor can the architect exercise his power to order a variation after practical completion, it is thought. The contractor becomes entitled to payment of the retention money, half on the certificate of practical completion and the remainder on the certificate of completion of making good defects (clause 30.4.1).

Clause 18 provides for the consequences of the employer (with the contractor's consent) taking possession of any completed part of the works before completion of the whole. This stipulates for a separate defects liability period in respect of the part, starting when possession is taken; and for proportional reductions in the value of the works to be insured and in liquidated damages. The retention money will be reduced by virtue of clause 30.4.1.2.

Time

The speed with which the contractor carries out the work is likely to be an important element in the performance of the contract. The contract stipulates, in the Appendix, for fixed dates of possession and completion, but the period of the work may become extended under the contract.

"**23.1.1** On the Date of Possession possession of the site shall be given to the Contractor who shall thereupon begin the Works, regularly and diligently proceed with the same and shall complete the same on or before the Completion Date.

   .2 Where clause 23.1.2 is stated in the Appendix to apply the Employer may defer the giving of possession for a period not exceeding six weeks or such lesser period stated in the Appendix calculated from the Date of Possession.

23.2. The Architect may issue instructions in regard to the postponement of any work to be executed under the provisions of this Contract."

Failure to proceed regularly and diligently may lead to determination under clause 27.1.2. Apart from this major sanction and any stipulation in the bills (see note to clause 2.2.1 above), the contractor is entitled to proceed at his own pace and pay damages for any delay at completion. The architect is entitled to be provided with a programme, but this does not bind the contractor to perform in accordance with it. The option to defer giving possession under clause 23.1.2 is new; if exercised it qualifies for an extension of time, by clause 25.4.13.

> "5.3.1.2   the Contractor without charge to the Employer shall provide the Architect (unless he shall have been previously so provided) with 2 copies of his master programme for the execution of the Works and within 14 days of any decision by the Architect under clause 25.3.1 or of the date of issue of a confirmed acceptance of a 13A Quotation with 2 copies of any amendments and revisions to take account of that decision or of that confirmed acceptance.
>
> "5.3.2.   Nothing contained in the . . . (. . . master programme for the execution of the Works or any amendment to that programme or revision therein referred to in clause 5.3.1.2) shall impose any obligation beyond those imposed by the Contract Documents."

The contractor's programme is of use, however, in giving notice to the architect as to when instructions may be needed (note that clause 26.2.1 still requires written notice); the programme is also some evidence of what constitutes diligent progress.

> "24.1   If the Contractor fails to complete the Works by the Completion Date then the Architect shall issue a certificate to that effect. In the event of a new Completion Date being fixed after the issue of such a certificate such fixing shall cancel that certificate and the Architect shall issue such further certificate under clause 24.1 as may be necessary."

Clause 24.2 provides that, where the architect has issued a certificate under clause 24.1 and the employer has informed the contractor of his intention to deduct liquidated damages, the employer may then either require the contractor to pay or deduct liquidated damages at the rate stated in the appendix (or a lesser rate if so notified). Provision is made for liquidated damages to be repaid in the event of a further extension of time.

The completion date is defined by clause 1.3 as:

> "The Date for Completion as fixed and stated in the Appendix or any later date fixed under either clause 25 or 33.1.3."

Completion of the works refers to "practical completion" under clause 17.1. The works may be deemed completed although they contain defects which come to light later (see Chapter 9). In this event the employer, if deprived of use of the works, is not limited to the liquidated damages for delay, but may prove his actual loss. When liquidated damages are recoverable the employer is "empowered" to recover part only of the sum due. The reason for introducing this provision is obscure. It may be intended to avoid repayment where a further extension is likely. It is not clear whether election to recover part only operates as waiver of the balance.

The list of grounds entitling the contractor to an extension of time are set out in clause 25. If the work is delayed by a default on the part of the employer which does not permit an extension of time, the liquidated damages clause becomes unenforceable (see Chapter 9).

> "**25.1** In clause 25 any reference to delay, notice or extension of time includes further delay, further notice or further extension of time.
>
> **25.2.1.1** If and whenever it becomes reasonably apparent that the progress of the Works is being or is likely to be delayed the Contractor shall forthwith give written notice to the Architect of the material circumstances including the cause or causes of the delay and identify in such notice any event which in his opinion is a Relevant Event.
>
> **.1.2** Where the material circumstances of which written notice has been given under clause 25.2.1.1 include reference to a Nominated Sub-Contractor, the Contractor shall forthwith send a copy of such written notice to the Nominated Sub-Contractor concerned.
>
> **.2** In respect of each and every Relevant Event identified in the notice given in accordance with clause 25.2.1.1 the Contractor shall, if practicable in such notice, or otherwise in writing as soon as possible after such notice:
>
> **.2.1** give particulars of the expected effects thereof; and
>
> **.2.2** estimate the extent, if any, of the expected delay in the completion of the Works beyond the Completion Date resulting therefrom whether or not concurrently with delay resulting from any other Relevant Event
> and shall give such particulars and estimate to any Nominated Sub-Contractor to whom a copy of any written notice has been given under clause 25.2.1.2.
>
> **.3** The Contractor shall give such further written notice to the architect and send a copy to any Nominated Sub-Contractor to whom a copy of any written notice has been given under clause 25.2.1.2, as may be reasonably necessary or as the Architect may reasonably require for keeping up-to-date the particulars and estimate referred to in clauses 25.2.2.1 and 25.2.2.2 including any material change in such particulars or estimate.

25.3.1 If, in the opinion of the Architect, upon receipt of any notice, particulars and estimate under clauses 25.2.1.1 and 25.2.2,

.1.1 any of the events which are stated by the Contractor to be the cause of the delay is a Relevant Event, and

.1.2 the completion of the Works is likely to be delayed thereby beyond the Completion Date

the Architect shall in writing to the Contractor give an extension of time by fixing such later date as the Completion Date as he then estimates to be fair and reasonable. The Architect shall, in fixing such new Completion Date, state:

.1.3 which of the Relevant Events he has taken into account, and

.1.4 the extent, if any, to which he has had regard to any instruction under clause 13.2 requiring as a Variation the omission of any work issued since the fixing of the previous Completion Date,

and shall, if reasonably practicable having regard to the sufficiency of the aforesaid notice, particulars and estimates, fix such new Completion Date not later than 12 weeks from receipt of the notice and of reasonably sufficient particulars and estimate, or, where the period between receipt thereof and the Completion Date is less that 12 weeks, not later than the Completion Date.

If in the opinion of the Architect, upon receipt of any such notice, particulars and estimate it is not fair and reasonable to fix a later date as a new Completion Date, the Architect shall if reasonably practicable having regard to the sufficiency of the aforesaid notice, particulars and estimate so notify the Contractor in writing not later than 12 weeks from receipt of the notice, particulars and estimate, or, where the period between receipt thereof and the Completion Date is less than 12 weeks, not later than the Completion Date."

Clause 25.3.2 provides for the architect to review extensions of time granted. They may be reviewed by fixing an earlier completion date where contract work is ommitted. After completion the architect may, within 12 weeks of practical completion, review extensions granted taking into account any relevant event (whether or not notified). The contractor is required to use best endeavours to prevent delay. The extensions of time are to be notified to nominated sub-contractors. The relevant events which may give rise to extensions of time are set out in the following provisions.

"**25.4** The following are the Relevant Events referred to in clause 25:

.4.1 force majeure;

.4.2 exceptionally adverse weather conditions;

    .4.3 loss or damage occasioned by any one or more of the Specified Perils;

    .4.4 civil commotion, local combination of workmen, strike or lock-out affecting any of the trades employed upon the Works or any of the trades engaged in the preparation, manufacture or transportation of any of the goods or materials required for the Works;

    .4.5 compliance with the Architect's instructions

        .5.1 under clause 2.3, 2.4.1, 13.2, (except for a confirmed acceptance of a 13A Quotation), 13.3 (except compliance with an Architect's instruction for the expenditure of a provisional sum for defined work or of a provisional sum for Performance Specified Work), 13A.4.1, 23.2, 34, 35 or 36; or

        .5.2 in regard to the opening up for inspection of any work covered up or the testing of any of the work, materials or goods in accordance with clause 8.3 (including making good in consequence of such opening up or testing) unless the inspection or test showed that the work, materials or goods were not in accordance with this Contract;

    4.6.1 where an Information Release Schedule has been provided, failure of the Architect to comply with clause 5.4.1;

    .6.2 failure of the Architect to comply with clause 5.4.2;

25.4.7 delay on the part of Nominated Sub-Contractors or Nominated Suppliers which the Contractor has taken all practicable steps to avoid or reduce;

    4.8.1 the execution of work not forming part of this Contract by the Employer himself or by persons employed or otherwise engaged by the Employer as referred to in clause 29 or the failure to execute such work;

    .8.2 the supply by the Employer of materials and goods which the Employer has agreed to provide for the Works or the failure so to supply;

    .4.9 the exercise after the Base Date by the United Kingdom Government of any statutory power which directly affects the execution of the Works by restricting the availability or use of labour which is essential to the proper carrying out of the Works or preventing the Contractor from, or delaying the Contractor in, securing such goods or materials or such fuel or energy as are essential to the proper carrying out of the Works;

    .4.10.1 the Contractor's inability for reasons beyond his control and which he could not reasonably have foreseen at the Base Date to secure such labour as is essential to the proper carrying out of the Works; or

    .10.2 the Contractor's inability for reasons beyond his control and which he could not reasonably have foreseen at the Base Date to secure such goods or materials as are essential to the proper carrying out of the Works;

.11 the carrying out by a local authority or statutory undertaker of work in pursuance of its statutory obligations in relation to the Works, or the failure to carry out such work;

.12 failure of the Employer to give in due time ingress to or egress from the site of the Works or any part thereof through or over any land, buildings, way or passage adjoining or connected with the site and in the possession and control of the Employer, in accordance with the Contract Bills and/or the Contract Drawings, after receipt by the Architect of such notice, if any, as the Contractor is required to give, or failure of the Employer to give such ingress or egress as otherwise agreed between the Architect and the Contractor.

.13 where clause 23.1.2 is stated in the Appendix to apply, the deferment by the Employer of giving possession of the site under clause 23.1.2.

.14 by reason of the execution of work for which an Approximate Quantity is included in the Contract Bills which is not a reasonably accurate forecast of the quantity of work required.

.15 delay which the Contractor has taken all practicable steps to avoid or reduce consequent upon a change in the Statutory Requirements after the Base Date which necessitates some alteration or modification to any Performance Specified Work;

.16 the use or threat of terrorism and/or the activity of the relevant authorities in dealing with such use or threat;

.17 compliance or non-compliance by the Employer with clause 6A.1;

.18 delay arising from a suspension by the Contractor of the performance of his obligations under the Contract to the Employer pursuant to clause 30.1.4."

This clause has been substantially enlarged from the earlier editions, as part of the policy of attempting to legislate for every eventuality. Points to be noted include the requirement for the contractor's notice to specify the relevant event, the expected effects and the delay, and for the notice to be up-dated as necessary. The architect must make his decision "if reasonably practicable" within 12 weeks or before completion. All extensions are subject to review within 12 weeks of completion. There may be a reduction where omissions are ordered, but only in cases where an extension has initially been allowed, and the date for completion stated in the appendix may not be advanced.

Failure by the contractor to give notice of delay will deprive him of extensions during the currency of the work. But the architect is bound to consider any relevant events, whether notified or not, in his review after completion. Under clause 26.3 the architect is required to specify those extensions relevant to time-dependent loss and expense claims. Note that under clauses 38 to 40 (see below) the employer's right to "freeze" fluctuation payments at the contractual completion date is dependent on timely exercise of the architect's powers to extend time.

POWERS AND REMEDIES

The limitations of the architect's powers are commented on above. Specific powers already referred to are those under clause 8.3 to order opening up of work; under clause 8.4 to order removal of improper work; under clause 8.5 to order removal of workmen; under clauses 17.2 and 17.3 to require defects to be made good; and under clause 23.2 to order postponement of work.

Instructions must normally be in writing. Clause 4.3 lays down an elaborate provision for dealing with oral instructions:

> "**4.3.1** All instructions issued by the Architect shall be issued in writing.
>
> .2 If the Architect purports to issue an instruction otherwise than in writing it shall be of no immediate effect, but shall be confirmed in writing by the Contractor to the Architect within seven days, and if not dissented from in writing by the Architect to the Contractor within seven days from receipt of the Contractor's confirmation shall take effect as from the expiration of the latter said seven days. Provided always:
>
> .2.1 that if the Architect within seven days of giving such an instruction otherwise than in writing shall himself confirm the same in writing, then the Contractor shall not be obliged to confirm as aforesaid, and the said instruction shall take effect as from the date of the Architect's confirmation; and
>
> .2.2 that if neither the Contractor nor the Architect shall confirm such an instruction in the manner and at the time aforesaid but the Contractor shall nevertheless comply with the same, then the Architect may confirm the same in writing at any time prior to the issue of the Final Certificate, and the said instruction shall thereupon be deemed to have taken effect on the date on which it was issued otherwise than in writing by the Architect.

The widest power given to the architect is to order variations to the works:

> **13.1** The term "Variation" as used in the Conditions means:
>
> **13.1.1** the alteration or modification of the design, quality or quantity of the Works as shown upon the Contract Drawings and described by or referred to in the Contract Bills; including

.1.1 the addition, omission or substitution of any work,

.1.2 the alteration of the kind or standard of any of the materials or goods to be used in the Works,

.1.3 the removal from the site of any work executed or materials or goods brought thereon by the Contractor for the purposes of the Works other than work materials or goods which are not in accordance with this Contract;

13.1.2 the imposition by the Employer of any obligations or restrictions in regard to the matters set out in clauses 13.1.2.1 to 13.1.2.4 or the addition to or alteration or omission of any such obligation or restrictions so imposed or imposed by the Employer in the Contract Bills in regard to:

.2.1 access to the site or use of any specific parts of the site;

.2.2 limitations of working space;

.2.3 limitations of working hours;

.2.4 the execution or completion of the work in any specific order; but excludes

.3 nomination of a Sub-Contractor to supply and fix materials or goods or to execute work of which the measured quantities have been set out and priced by the Contractor in the Contract Bills for supply and fixing or execution by the Contractor.

13.2.1 The Architect may issue instructions requiring a Variation.

.2 Any instruction under clause 13.2.1 shall be subject to the Contractor's right of reasonable objection set out in clause 4.1.1.

. . .

.4 The Architect may sanction in writing any Variation made by the Contractor otherwise than pursuant to an instruction of the Architect.

.5 No Variation required by the Architect or subsequently sanctioned by him shall vitiate this Contract.

13.3 The Architect shall issue instructions in regard to:

3.1 the expenditure of provisional sums included in the Contract

3.2 the expenditure of provisional sums included in a Sub-Contract."

Clauses 13.2.3 and 13.4 deal with the valuation of variations (see below).

The employer will be bound by any variation order given by the architect. However, the architect must ensure that he has authority to order the variation, otherwise he may be liable to the employer for its cost. The effect of clause 4.3 is that a variation order must normally be in writing. Although clause 13.1 and 13.2 contemplates that the power to give variations is unlimited, there will be some implied limit, beyond which the contractor may say that the contract has ceased to apply, and that he is entitled to re-price the work (see Chapter 9).

By clause 4.1.1 the contractor is bound to comply with instructions forthwith. But the contractor may make "reasonable objection" to compliance with an instruction given under clause 13.1.2 (see notes to clause 4 above).

Sub-contractors

It is very rare in practice for the contractor to wish to perform the whole of the work himself. The right to have the work performed by others is restricted as follows:

> "**19.1.1** Neither the Employer nor the Contractor shall, without the written consent of the other, assign this Contract.
>
> **.2** Where clause 19.1.2 is stated in the Appendix to apply then, in the event of transfer by the Employer of his freehold or leasehold interest in, or of a grant by the Employer of a leasehold interest in, the whole of the premises comprising the Works, the Employer may at any time after Practical Completion of the Works assign to any such transferee or lessee the right to bring proceedings in the name of the Employer (whether by arbitration or litigation) to enforce any of the terms of this contract made for the benefit of the Employer hereunder. The assignee shall be estopped from disputing any enforceable agreements reached between the Employer and the Contractor and which arise out of and relate to this Contract (whether or not they are or appear to be a derogation from the right assigned) and made prior to the date of any assignment.
>
> **19.2** The Contractor shall not without the written consent of the Architect (which consent shall not be unreasonably withheld) sub-let any portion of the Works. A person to whom the Contractor sub-lets any portion of the Works other than a Nominated Sub-Contractor is in this Contract referred to as a 'Domestic Sub-Contractor.'
>
> **19.3** . . ."

Clause 19.1.2 applies only to a transfer of the whole of the premises. The clause does not deal with the question of what damages might be recoverable by the assignee (see further Chapter 10). Substantial sections of work are frequently intended to be sub-let to particular sub-contractors. Where such work is described in the bills as prime cost or provisional sums, the contract allows the architect to nominate the sub-contractor to perform the work. Nomination may also arise in other ways, including naming the sub-contractor in the bills:

"**35.1** Where

.1 in the Contract Bills; or

.2 in any instruction of the Architect under clause 13.3 on the expenditure of a provisional sum included in the Contract Bills; or

.3 in any instruction of the Architect under clause 13.2 requiring a Variation to the extent, but not further or otherwise,

.3.1 that it consists of work additional to that shown upon the Contract Drawings and described by or referred to in the Contract Bills and

.3.2 that any supply and fixing of materials or goods or any execution of work by a Nominated Sub-Contractor in connection with such additional work is of a similar kind to any supply and fixing of materials or the execution of work for which the Contract Bills provided that the Architect would nominate a sub-contractor; or

.4 by agreement (which agreement shall not be unreasonably withheld) between the Contractor and the Architect on behalf of the Employer

the Architect has, whether by the use of a prime cost sum or by naming a sub-contractor, reserved to himself the final selection and approval of the sub-contractor to the Contractor who shall supply and fix any materials or goods or execute work, the sub-contractor so named or to be selected and approved shall be nominated in accordance with the provisions of clause 35 and a sub-contractor so nominated shall be a Nominated Sub-Contractor for all the purposes of this Contract. The provisions of clause 35.1 shall apply notwithstanding the requirement in rule A51 of the Standard Method of Measurement, 7th Edition for a PC sum to be included in the Bills of Quantities in respect of Nominated Sub-contractors; where however such sum in included in the Contract Bills the provisions of the aforesaid rule A51 shall apply in respect thereof."

Clause 35.4 to 35.9 sets out the detailed procedure for nomination. This requires the architect to obtain the sub-contractor's tender and performance warranty on the standard forms provided for the purpose. The contractor is then given a preliminary notice of nomination with instructions to settle with the sub-contractor important terms, including the timing and order of the work. These are matters which often produced dispute under the former procedure. Only when these terms are agreed may the architect make his nomination. Alternatively, by a provision in the bills or an instruction, the architect may opt out of the preliminary procedure.

Provisions dealing with payment to nominated sub-contractors, extensions of time, delay in completion, practical completion and final payment are set out in clause 35.13 to 35.19. The employer has the benefit of a Standard Form of Warranty to be given by the nominated sub-contractor, known as NSC/W. In addition, clause 35 contains or incorporates the following package of benefits for the main contractor:

(1) the benefit of the standard form of nominated sub-contract NSC/A and /C including indemnities for non-performance by the sub-contractor;

(2) a discount of two-and-a-half per cent on the sub-contract price for payment within the period stipulated in the sub-contract;

(3) a right of reasonable objection to any proposed sub-contractor;

(4) limitation of liability where the sub-contractor's liability is limited;

(5) an express right of re-nomination where the sub-contractor defaults.

By this latter provision the JCT has adopted the construction placed on the 1963 edition by the House of Lords in N.W. *Metropolitan Hospital Board v. Bickerton.*[1] The sub-contractor has the benefit of the right of direct payment, reserved to the employer by clause 35.13.5. Proof of payment is made mandatory on the contractor by clause 35.13.3. The employer is, by clause 35.13.5.1, bound to operate the direct payment procedure subject, *inter alia,* to claims by more than one sub-contractor. Parts of this long clause have been re-drafted in the 1998 edition including the provisions as to re-nomination (now clause 35.24 to 35.26). Clause 35.20 confirms that the employer's liability to a sub-contractor is limited to the terms of NSC/W; and clause 35.21 confirms that the contractor is not to be responsible for matters of design, etc., taken on by the sub-contractor.

Clause 36 covers nominated suppliers, which have traditionally been dealt with separately in the RIBA forms (there is no distinction between sub-contractor and supplier in the ICE form). Nominated suppliers are defined in the following terms:

"**36.1.1** In the Conditions 'Nominated Supplier' means a supplier to the Contractor who is nominated by the Architect in one of the following ways to supply materials or goods which are to be fixed by the Contractor:

   .1.1 where a prime cost sum is included in the Contract Bill in respect of those materials or goods and the supplier is either named in the Contract Bills or subsequently named by the Architect in an instruction issued under clause 36.2;

---

[1] [1970] 1 W.L.R. 607.

.1.2 where a provisional sum is included in the Contract Bills and in any instruction by the Architect in regard to the expenditure of such sum the supply of materials or goods is made the subject of a prime cost sum and the supplier is named by the Architect in that instruction or in an instruction issued under clause 36.2;

.1.3 where a provisional sum is included in the Contract Bills and in any instruction by the Architect in regard to the expenditure of such a sum materials or goods are specified for which there is a sole source of supply in that there is only one supplier from whom the Contractor can obtain them, in which case the supply of materials or goods shall be made the subject of a prime cost sum in the instructions issued by the Architect in regard to the expenditure of the provisional sum and the sole supplier shall be deemed to have been nominated by the Architect;

.1.4 where the Architect requires under clause 13.2, or subsequently sanctions, a Variation and specifies materials or goods for which there is a sole supplier as referred to in clause 36.1.1.3, in which case the supply of the materials or goods shall be made the subject of a prime cost sum in the instruction or written sanction issued by the Architect under clause 13.2 and the sole supplier shall be deemed to have been nominated by the Architect.

.2 In the Conditions the expression 'Nominated Supplier' shall not apply to a supplier of materials or goods which are specified in the Contract Bills to be fixed by the Contractor unless such materials or goods are the subject of a prime cost sum in the Contract Bills, notwithstanding that the supplier has been named in the Contract Bills or that there is a sole supplier of such materials or goods as defined in clause 36.1.1.3."

The contractor is entitled to terms in the contract of sale covering the quality of the goods, rectification of defects and the delivery programme. The supplier must offer five per cent discount for prompt payment; and ownership of the goods is to pass on delivery to the contractor, whether or not payment has been made (clause 36.4). The latter provision seeks to overcome the effect of so-called "Romalpa clauses", often inserted by suppliers of goods (see Chapter 7). Where the supplier insists on restricting his liability, the architect must approve the restrictions, whereupon the contractor's liability is to be limited to the same extent (clause 36.5).

An important matter not dealt with is the effect of repudiation by the supplier. Under the 1963 forms it was thought that the *Bickerton* principle applied (see Chapter 10). However, the express provision for re-nomination in clause 35 is a material indication that the procedure may not be intended to apply to suppliers. In the *Bickerton* case the House of Lords placed weight on the words "such sums [Prime Cost] shall be expended in favour of such persons as the architect shall instruct." This is now replaced by a less specific formula:

> **"36.2** The Architect shall issue instructions for the purpose of nominating a supplier for any materials or goods in respect of which a prime cost sum is included in the Contract Bills or arises under clause 36.1."

The result may be that the contractor must bear the loss caused by the failure of a nominated supplier.

In addition to nominated and domestic sub-contractors, the bills may place restriction on who may be employed for specific work, thereby creating a new category which may be called a "listed sub-contractor."

> **"19.3.1** Where the Contract Bills provide that certain work measured or otherwise described in those Bills and priced by the Contractor must be carried out by persons named in a list in or annexed to the Contract Bills, and selected therefrom by and at the sole discretion of the Contractor the provisions of clause 19.3 shall apply in respect of that list.
>
> **.2.1** The list referred to in clause 19.3.1 must comprise not less than three persons. Either the Employer (or the Architect on his behalf) or the Contractor shall be entitled with the consent of the other, which consent shall not be unreasonably withheld, to add additional persons to the list at any time prior to the execution of a binding sub-contract agreement.
>
> **.2.2** If at any time prior to the execution of a binding sub-contract agreement and for whatever reason less than three persons named in the list are able and willing to carry out the relevant work then
> either the Employer and the Contractor shall by agreement (which agreement shall not be unreasonably withheld) add the names of other persons so that the list comprises not less than three such persons
> or the work shall be carried out by the Contractor who may sub-let to a Domestic Sub-Contractor in accordance with clause 19.2.
>
> **.3** A person selected by the Contractor under clause 19.3 from the aforesaid list shall be a Domestic Sub-Contractor."

In respect both of nominated suppliers and sub-contractors the contractor's liability for the suitability of the work is likely to be excluded by the architect's selection of materials.[2] Further, by virtue of clause 25.4.7, the contractor is not liable for delay by nominated sub-contractors or suppliers. Clause 19.3.3, however, ensures that, in regard to listed sub-contractors, the contractor remains liable to the employer for their default, as though the work had not been sub-let.

---

[2] *Young & Marten v. McManus Childs* [1969] 1 A.C. 454.

Default and determination

In the event of the contractor failing to comply with the contract, the employer has a number of remedies. In respect of defective work or work suspected to be defective, the architect may order the removal of the work or opening up or testing under clause 8 (see above). If the contractor fails to complete by the date for completion the employer is entitled to deduct liquidated damages (clause 24). Note that the work cannot be complete if it contains material patent defects.

If the contractor's default is more serious or if he becomes unable to perform the contract, the employer may become entitled to determine (terminate) the contractor's employment:

"27.2 .1 If, before the date of Practical Completion, the Contractor shall make a default in any one or more of the following respects:

　.1.1 without reasonable cause he wholly or substantially suspends the carrying out of the Works; or

　.1.2 he fails to proceed regularly and diligently with the Works; or

　.1.3 he refuses or neglects to comply with a written notice or instruction from the Architect requiring him to remove any work, materials or goods not in accordance with this Contract and by such refusal or neglect the Works are materially affected; or

　.1.4 he fails to comply with the provisions of clause 19.1.1 or clause 19.2.2; or

　.1.5 he fails pursuant to the Conditions to comply with the requirements of the CDM Regulations,

the Architect may give to the Contractor a notice specifying the default or defaults (the 'specified default or defaults').

27.2.2 If the Contractor continues a specified default for 14 days from receipt of the notice under clause 27.2.1 then the Employer may on, or within 10 days from, the expiry of that 14 days by a further notice to the Contractor determine the employment of the Contractor under this Contract. Such determination shall take effect on the date of receipt of such further notice.

27.2.3 If the Contractor ends the specified default or defaults, or the Employer does not give the further notice referred to in clause 27.2.2

and the Contractor repeats a specified default (whether previously repeated or not) then, upon or within a reasonable time after such repetition, the Employer may by notice to the Contractor determine the employment of the Contractor under this Contract. Such determination shall take effect on the date of receipt of such notice.

27.2.4 A notice of determination under clause 27.2.2 or clause 27.2.3 shall not be given unreasonably or vexatiously."

Clause 27.3 contains provisions relating to the insolvency of the contractor. The contract formerly provided for automatic determination subject to reinstatement. The present clause provides for automatic determination in the case of bankruptcy or winding up, subject to reinstatement. In other cases, determination may be effected by notice from the employer.

The operation of this clause does not terminate the contract; the parties remain bound by its terms. After determination, clause 27.6 entitles the employer to complete the work by others, and to claim from the contractor any additional cost so incurred. Clause 27.4 gives the employer a further right of determination if the contractor or any employee of his gives or offers any gift or consideration as an inducement. Note that two notices are required to effect a determination under clause 27.2. Clause 27.1 now deals expressly with the means of delivery of any notice under the clause, providing that it is to be *"given by actual delivery or by special delivery or by recorded delivery"*. This provision is intended to avoid argument as to the effectiveness of notice. It does not, however, appear to answer the question whether delivery by fax or E-mail is to be regarded as "actual delivery".[3] If determination is to be effected under the clause, it is vital that the machinery is correctly operated since the grounds may not amount to repudiation by the contractor. An incorrect determination may render the employer liable to the contractor in damages (see Chapter 6).

Clause 28 gives the contractor rights to determine in the following circumstances:

(1) Non-payment of sums due

(2) Interference with the issue of a certificate

(3) Assignment without consent

(4) Failure to comply with CEM regulations

(5) Suspension of the works for reasons other than the contractor's default.

The clause requires service of notice, continuance of the ground relied on for 14 days (or subsequent repeat of the ground in question) after which notice of determination may be given. The clause contains similar provisions to those in clause 27 in relation to the employer's insolvency. Clause 28.1 contains an identical

---

[3] *Hastie & Jenkerson v. McMahon* [1990] 1 W.L.R. 1575

Notice Provision to that applying in the case of determination by the employer by clause 27.1.

Upon determination, clause 28.4 entitles the contractor to be paid the value of all work done and materials supplied, the cost of removal and any loss caused to him. Note that the period of suspension after which the contractor becomes entitled to serve notice of determination is to be specified in the Appendix, and if none is stated the applicable period is one month. This places the employer in a vulnerable position where suspension is unavoidable, and it is clearly in his interest to stipulate for a substantially longer period.

Alternatively, in certain circumstances either party may acquire the right to terminate the work:

> "**28A.1.1** If, before the date of Practical Completion, the carrying out of the whole or substantially the whole of the uncompleted Works is suspended for the relevant continuous period of the length stated in the Appendix by reason of one or more of the following events:
>
> **.1.1** force majeure; or
> **.1.2** loss or damage to the Works occasioned by any one or more of the Specified Perils; or
> **.1.3** civil commotion; or
> **.1.4** Architect's instructions issued under clause 2.3, 13.2 or 23.2 which have been issued as a result of the negligence or default of any local authority or statutory undertaker executing work solely in pursuance of its statutory obligations; or
> **.1.5** hostilities involving the United Kingdom (whether war be declared or not); or
> **.1.6** terrorist activity
>
> then the Employer or the Contractor may upon the expiry of the aforesaid relevant period of suspension give notice in writing to the other by actual delivery or by special delivery or recorded delivery that unless the suspension is terminated within 7 days after the date of receipt of that notice the employment of the Contractor under this Contract will determine 7 days after the date of receipt of the aforesaid notice; and the employment of the Contractor shall so determine 7 days after receipt of such notice. If sent by special delivery or recorded delivery the notice shall, subject to proof to the contrary, be deemed to have been received 48 hours after the date of posting (excluding Saturday and Sunday and Public Holidays)."

The contractor is not entitled to terminate under clause 28A.1.1.2 where the loss was caused by his negligence of default. Notice from either side is not to be given unreasonably or vexatiously.[4] Upon

---

[4] *John Jarvis v. Rockdale Housing Association* (1986) 36 B.L.R. 48.

determination by either party, the contractor is entitled to be paid as though the determination were effected on grounds of the employer's default, including payment of any direct loss and/or damage caused by the determination.

CERTIFICATION AND PAYMENT

The sums payable to the contractor may be subject to many alterations, which are discussed below. The contract sum is not, however, to be altered on account of any error:

> "**14.2** The Contract Sum shall not be adjusted or altered in any way whatsoever otherwise than in accordance with the express provisions of the Conditions, and subject to clause 2.2.2.2 any error whether of arithmetic or not in the computation of the Contract Sum shall be deemed to have been accepted by the parties hereto."

The contract contains extensive provisions for payment as the work proceeds, which are vital to the contractor. He will often, during the course of a contract, have a cash turnover exceeding the value of his assets, so that lack of interim payments will create great difficulties. Cash flow has been referred to by the courts as the life-blood of the industry. Clause 30, which deals with certificates and payment, has been extensively amended, *inter alia,* to comply with the Housing Grants, etc., Act 1996. The essential payment mechanism is as follows:

> "**30.1.1.1** The Architect shall from time to time as provided in clause 30 issue Interim Certificates stating the amount due to the Contractor from the Employer specifying to what the amount relates and the basis on which that amount was calculated; and the final date for payment pursuant to an Interim Certificate shall be 14 days from the date of issue of each Interim Certificate.
>
> . . .
>
> **30.1.2.1** Interim valuations shall be made by the Quantity Surveyor whenever the Architect considers them to be necessary for the purpose of ascertaining the amount to be stated as due in an Interim Certificate.
>
> . . .
>
> **30.1.3** Interim Certificates shall be issued at the Period of Interim Certificates specified in the Appendix up to and including the end of the period during which the certificate of Practical Completion is issued. Thereafter Interim Certificates shall be issued as and when further amounts are

ascertained as payable to the Contractor from the Employer and after the expiration of the Defects Liability Period named in the Appendix or upon the issue of the Certificate of Completion of Making Good Defects (whichever is the later) provided always that the Architect shall not be required to issue an Interim Certificate within one calendar month of having issued a previous Interim Certificate.

. . .

30.2 The amount stated as due in the Interim Certificate, subject to any agreement between the parties as to stage payments, shall be the gross valuation as referred to in clause 30.2 less any amount which may be deducted and retained by the Employer as provided in clause 30.4 (in the Conditions called the 'Retention') and

the amount of any advance payment or part thereof due for reimbursement to the Employer in accordance with the terms for such reimbursement stated in the Appendix pursuant to clause 30.1.1.6 and

the total amount stated as due in Interim Certificates previously issued under the Conditions.

The gross valuation shall be the total of the amounts referred to in clauses 30.2.1 and 30.2.2. less the total of the amounts referred to than 7 days before the date of the Interim Certificate."

Clause 30.2.1 lists the matters which are subject to retention, including the value of work properly executed and claims under clause 26; clause 30.2.2 lists matters to be included which are not subject to retention; and clause 30.2.3 lists permitted deductions from the gross valuation.

Clause 16.1 provides that where materials and goods, in accordance with clause 30.2, are included in an interim certificate under which the contractor has received payment, the materials or goods become the property of the employer. Clause 30.3 gives the architect a discretion to include payment for materials or goods before delivery to the site, provided specified conditions are met. Clause 16.2 provides similarly for such materials or goods to become the employer's property when paid for.

Clause 30.4 provides for the employer to hold a retention of five per cent or any lower agreed rate. The full retention is to be levied on work which has not reached completion, and on materials. Work which is certified as practically complete carries half the retention percentage, which is released on the certificate of making good defects. Items which are not subject to retention are set out in clause 30.2.2 above. Interim certificates are required to set out, separately, retentions in respect of nominated subcontractors.

The procedure for valuing variations is set out in clause 13.5. Where the extra work is of similar character to billed work, such rates are to be used, with allowance for different conditions or significant changes in quantity. Where no similar work is billed, fair rates are to be used. Where extra work cannot be valued by measurement, the contractor is to be paid at daywork rates. By a new clause 13A, extensive provisions are made for architect's instructions to be valued in accordance with a quotation submitted by the contractor, including any adjustment to the time required and any payment claimed in respect of direct loss and/or expense under clause 26. The employer may accept the contractor's clause 13A quotation, but if not accepted, the contractor is entitled to be paid for the cost of preparing it.

The provisions for loss and/or expense claims are now consolidated into a single clause;

"**26.1** If the Contractor makes written application to the Architect stating that he has incurred or is likely to incur direct loss and/ or expense in the execution of this Contract for which he would not be reimbursed by a payment under any other provision in this Contract due to deferment of giving possession of the site under clause 23.1.2 where clause 23.1.2 is stated in the Appendix to be applicable or because the regular progress of the Works or of any part thereof has been or is likely to be materially affected by any one or more of the matters referred to in clause 26.2; and if and as soon as the Architect is of the opinion that the direct loss and/or expense has been incurred or is likely to be incurred due to any such deferment of giving possession or that the regular progress of the Works or of any part thereof has been or is likely to be so materially affected as set out in the application of the Contractor then the Architect from time to time thereafter shall ascertain, or shall instruct the Quantity Surveyor to ascertain, the amount of such loss and/or expense which has been or is being incurred by the Contractor; provided always that:

.1 the Contractor's application shall be made as soon as it has become, or should reasonably have become, apparent to him that the regular progress of the Works or any part thereof has been or was likely to be affected as aforesaid, and

.2 the Contractor shall in support of his application submit to the Architect upon request such information as should reasonably enable the Architect to form an opinion as aforesaid, and

.3 the Contractor shall submit to the Architect or to the Quantity Surveyor upon request such details of such loss and/or expense as are reasonably necessary for such ascertainment as aforesaid.

26.2 The following are the matters referred to in clause 26.1:

.1.1 where an Information Release Schedule has been provided, failure of the Architect to comply with clause 5.4.1;

.1.2 failure of the Architect to comply with clause 5.4.2;

.2 the opening up for inspection of any work covered up or the testing of any of the work, materials or goods in accordance with clause 8.3 (including making good in consequence of such opening up or testing), unless the inspection or test showed that the work, materials or goods were not in accordance with this Contract;

.3 any discrepancy in or divergence between the Contract Drawings and/or the Contract Bills and/or the Numbered Documents;

.4.1 the execution of work not forming part of this Contract by the Employer himself or by persons employed or otherwise engaged by the Employer as referred to in clause 29 or the failure to execute such work;

.4.2 the supply by the Employer of materials and goods which the Employer has agreed to provide for the Works or the failure so to supply;

.5 Architect's instructions under clause 23.2 issued in regard to the postponement of any work to be executed under the provisions of this Contract;

.6 failure of the Employer to give in due time ingress to or egress from the site of the Works, or any part thereof through or over any land, buildings, way or passage adjoining or connected with the site and in the possession and control of the Employer, in accordance with the Contract Bills and/or the Contract Drawings, after receipt by the Architect of such notice, if any, as the Contractor is required to give, or failure of the Employer to give such ingress or egress as otherwise agreed between the Architect and the Contractor;

.7 Architect's instructions issued under clause 13.2 or clause 13.A.4.1 requiring a Variation (except for a Variation for which the Architect has given a confirmed acceptance of a 13A Quotation or for a Variation thereto) or
under clause 13.3 in regard to the expenditure of provisional sums (other than instructions to which clause 13.4.2 refers or an instruction for the expenditure of a provisional sum for defined work or of a provisional sum for Performance Specified Work);

.8 the execution of work for which an Approximate Quantity is included in the Contract Bills which is not a reasonably accurate forecast of the quantity of work required.

.9 compliance or non-compliance by the Employer with clause 6A.1;

.10 suspension by the Contractor of the performance of his obligations under the Contract to the Employer pursuant to clause 30.1.4 provided the suspension was not frivolous or vexatious.

26.3 If and to the extent that it is necessary for ascertainment under clause 26.1 of loss and/or expense the Architect shall state in writing to the Contractor what extension of time, if any, has been made under clause 25 in respect of the Relevant Event or Events referred to in clause 25.4.5.1 (so far as that clause refers to clauses 2.3, 13.2, 13.3 and 23.2) and in clauses 25.4.5.2, 25.4.6, 25.4.8 and 25.4.12.

. . .

26.5 Any amount from time to time ascertained under clause 26 shall be added to the Contract Sum.

26.6 The provisions of clause 26 are without prejudice to any other rights and remedies which the Contractor may possess."

Timely notice under clause 26.1 is a condition precedent to the contractor's right to payment. The latest time for such notice is as soon as it should reasonably be apparent that progress has been delayed. However, by clause 26.6, these provisions are without prejudice to other rights, which may include a right to sue for damages, for example, for breach of clause 5.4 as an alternative to claiming under clause 26.2.1. In practice, contractual claims are often pleaded with an alternative breach claim to cover any deficiency in notices. An advantage of the claims procedure is that sums ascertained should be paid as the work proceeds. Numbered Documents referred to in clause 26.2.3, are those incorporated into a nominated sub-contract. The addition of amounts to the contract sum is dealt with under clause 3:

"**3.** Where in the Conditions it is provided that an amount is to be added to or deducted from the Contract Sum or dealt with by adjustment of the Contract Sum then as soon as such amount is ascertained in whole or in part such amount shall be included in the computation of the next Interim Certificate following such whole or partial ascertainment."

The architect has no power under the contract to deal with claims for breach, which will usually be dealt with after completion. Clause 26.4 contains provisions similar to the above, in regard to claims by nominated sub-contractors. For the right to add interest to claims see Chapter 2.

Clauses 37 to 40 (now appearing as a separately published Part 3 of the conditions) deal with fluctuation payments. Clause 37 requires the parties to stipulate in the Appendix which clause is to apply: in default of choice, clause 38 is to apply. This allows tax fluctuations only. Clause 39 allows fluctuations on labour and materials as well. Clause 40 provides for full fluctuations, based on formula adjustments, subject to any non-adjustable elements.

The operation of clauses 38 and 39 depends on the occurrence of increases (or decreases) in rates or prices. It is a condition precedent to the contractor's entitlement that written notice of any such "event" is given within a reasonable time (clauses 38.4 and 39.5). Conversely formula adjustments are independent of the actual constituents of the work, and do not require notice.

In the event of delayed completion of the works, the employer's liability is limited to rates and prices applying up to the contractual completion date, extended as necessary:

> "**39.5.7** Subject to the provisions of clause 39.5.8 no amount shall be added or deducted in the computation of the amount stated as due in an Interim Certificate or in the Final Certificate in respect of amounts otherwise payable to or allowable by the Contractor by virtue of clause 39.1 to .3 or clause 39.4 if the event (as referred to in the provisions listed in clause 39.5.1) in respect of which the payment or allowance would be made occurs after the Completion date."

The rates and prices are therefore frozen at the date for completion required under the contract. Similar provisions appear in clauses 38.4.7 and 40.7.1. This benefit to the employer is conditional, however, on proper exercise of the extension of time clause by the architect:

> "**39.5.8** Clause 39.5.7 shall not be applied unless:
> .1 the printed text of clause 25 is unamended and forms part of the Conditions; and
> .2 the Architect has, in respect of every written notification by the Contractor under clause 25, fixed or confirmed in writing such Completion Date as he considers to be in accordance with clause 25."

Note that the architect is required only to give a timely decision on the contractor's applications for extension. If his decision is later revised by an arbitrator the contractor will be entitled to recover further fluctuations *pro tanto*. Similar provisions appear in clause 38.4.8 and 40.7.2.

Final accounting

During the period following practical completion, the architect is required to undertake the final measurement and valuation of the work (clause 30.6). By clause 14.1 (see above) the quantities of work are deemed to be those set out in the contract bills. If there is any error in the bill, either party may ask for re-measurement. The difference in quantity is to be valued as a variation (clause 2.2.2.2). Clause 30.6.2 sets out a list of adjustments required by the conditions to be made to the contract sum. By clause 30.7 the architect is to issue a special interim certificate setting out final adjustments to nominated sub-contractors' accounts. The last certificate, which must be given within specified time limits, is the final certificate.

> "**30.8** The Architect shall issue the Final Certificate (and inform each Nominated Sub-Contractor of the date of its issue) not later than two months after whichever of the following occurs last:
>
> > the end of the Defects Liability Period;
> > the date of issue of the Certificate of Completion of Making Good Defects under clause 17.4;
> > the date on which the Architect sent a copy to the Contractor of any ascertainment to which clause 30.6.1.2.1 refers and of the statement prepared in compliance with clause 30.6.1.2.2.
>
> The Final Certificate shall state:
>
> **.1.1** the sum of the amounts already stated as due in Interim Certificates plus the amount of any advance payment paid pursuant to clause 30.1.6, and
>
> **.1.2** the Contract Sum adjusted as necessary in accordance with clause 30.6.2, and
>
> **.1.3** to what the amount relates and the basis on which the statement in the Final Certificate has been calculated
>
> and the difference (if any) between the two sums shall (without prejudice to the rights of the Contractor in respect of any Interim Certificates which have subject to any notice issued pursuant to clause 30.1.4 not been paid in full by the Employer by the final date for payment of such Certificate) be expressed in the said Certificate as a balance due to the Contractor from the Employer or to the Employer from the Contractor as the case may be.
>
> **30.9.1** Except as provided in clauses 30.9.2 and 30.9.3 (and save in respect of fraud), the Final Certificate shall have effect in any proceedings under or arising out of or in connection with this Contract (whether by adjudication under article 5 or by arbitration under article 7A or by legal proceedings under article 7B) as
>
> **.1.1** conclusive evidence that where and to the extent that any of the particular qualities of any materials or goods or any particular standard of an item or workmanship was described expressly by the Contract Drawings or the Contract Bills, or in any of the Numbered Documents, or in any instruction issued by the Architect under the Conditions, or in any drawings or documents issued by the Architect under clause 5.3.1.1 or 5.4 or 7, to be for the approval of the Architect, the particular quality or standard was to the reasonable satisfaction of the Architect, but such Certificate shall not be conclusive evidence that such or any other

materials or goods or workmanship comply or complies with any other requirement or term of this Contract, and

.1.2 conclusive evidence that any necessary effect has been given to all the terms of this Contract which require that an amount is to be added to or deducted from the Contract Sum or an adjustment is to be made of the Contract Sum save where there has been any accidental inclusion or exclusion of any work, materials, goods or figure in any computation or any arithmetical error in any computation in which event the Final Certificate shall have effect as conclusive evidence as to all other computations, and

.1.3 conclusive evidence that all and only such extensions of time, if any, as are due under clause 25 have been given, and

.1.4 conclusive evidence that the reimbursement of direct loss and/or expense, if any, to the Contractor pursuant to clause 26.1 is in final settlement of all and any claims which the Contractor has or may have arising out of the occurrence of any of the matters referred to in clause 26.2 whether such claim be for breach of contract, duty of care, statutory duty or otherwise.

30.9.2 If any adjudication, arbitration or other proceedings have been commenced by either party before the Final Certificate has been issued the Final Certificate shall have effect as conclusive evidence as provided in clause 30.9.1 after either:

.2.1 such proceedings have been concluded, whereupon the Final Certificate shall be subject to the terms of any decision, award or judgment in or settlement of such proceedings or

.2.2 a period of 12 months after the issue of the Final Certificate during which neither party has taken any further step in such proceedings, whereupon the Final Certificate shall be subject to any terms agreed in partial settlement,

whichever shall be the earlier.

30.9.3 If any adjudication, arbitration or other proceedings have been commenced by either party within 28 days after the Final Certificate has been issued, the Final Certificate shall have effect as conclusive evidence as provided in clause 30.9.1 save only in respect of all matters to which those proceedings relate.

30.10 Save as aforesaid no certificate of the Architect shall of itself be conclusive evidence that

.1 any works, materials or goods

or

.2 any Performance Specified Work

to which it relates are in accordance with this Contract."

In versions of the JCT form before 1977 the final certificate was expressed to be conclusive evidence that the work had been properly carried out and completed in accordance with the terms of the contract. This provision was held binding on the courts when the certificate was given in the course of litigation about defective works.[5] The final certificate clause was capable of creating serious injustice. In the present form, these provisions are largely emasculated. The certificate is not evidence of compliance with the terms of the contract save in regard to the architect's reasonable satisfaction. In *Colbart v. Kumar*,[6] it was held that clause 30.9 covered matters which were subject to the implied satisfaction of the architect, giving the certificate a much wider effect and this decision was approved by the Court of Appeal in *Crown Estate Commissioners v. Mowlem*.[7] The present clause, however, includes the words "described expressly", thereby giving the clause a much narrower effect.

Clause 30.10 confirms that no other certificate is to be evidence of compliance with the contract. The final certificate is also expressed to be conclusive as to the contractor's claims. This is a fairer notion since the contractor ought to be aware of his need to make a claim well before the final certificate (there may be many reasons why the employer is unaware of defects). Note that the conclusive effect covers extensions of time; and that the certificate cannot be avoided by pleading a claim for damages instead of loss or expense (clause 30.9.1.4).The conclusive effect of the Final Certificate is subject to adjudication, arbitration or other proceedings being commenced before or within 28 days after the Final Certificate.

DISPUTES

The dispute resolution machinery under JCT 80 has been substantially amended in the light of the Housing Grants, etc., Act 1996. It was formerly provided by clause 41 that a reference to arbitration might not be opened until after practical completion with a number of important exceptions. The effect of the Housing Grants, etc., Act has been to make available immediate reference to adjudication in respect of any dispute. The contract has, therefore,

---

[5] *Kaye v. Hosier & Dickinson* [1972] 1 W.L.R. 146.
[6] (1992) 59 B.L.R 89.
[7] (1994) 40 Con. L.R. 36.

logically been amended to permit arbitration or litigation on any
dispute to follow without time limit (it would have been illogical to
allow any dispute to be adjudicated but only some to be arbitrated
before completion). The right to adjudication is provided by article
by article 5. Arbitration is now officially optional by providing for
the deletion of clause 41B (see below). Where deleted, litigation is
available as of right, as confirmed by article 7B and clause 41C.
Where arbitration is provided for, article 7A excludes any question
of enforcement of the decision of an adjudicator (which may
therefore proceed through the courts). Arbitration is further
excluded in respect of a dispute under clause 31(statutory tax
deduction) or clause 3 (VAT).

Clause 41A provides detailed machinery for adjudication.
Clauses 41A.1 to .3 make provision for the appointment of an
adjudicator, who is required to execute a standard form of JCT
Adjudicator Agreement. The procedure for adjudication is provided by clause 41A.4 to .5 as follows:

"**41A.4 .1** When pursuant to Article 5 a Party requires a dispute or
difference to be referred to adjudication then that Party
shall give notice to the other Party of his intention to refer
the dispute or difference, briefly identified in the notice, to
adjudication. Within 7 days from the date of such notice or
the execution of the JCT Adjudication Agreement by the
Adjudicator if later, the Party giving the notice of intention
shall refer the dispute or difference to the Adjudicator for
his decision ("the referral"); and shall include with that
referral particulars of the dispute or difference together
with a summary of the contentions on which he relies, a
statement of the relief or remedy which is sought and any
material he wishes the Adjudicator to consider. The referral
and its accompanying documentation shall be copied simultaneously to the other Party.

41A.4.2 The referral by a Party with its accompanying documentation to the Adjudicator and the copies thereof to be
provided to the other Party shall be given by actual delivery
or by FAX or by registered post or recorded delivery. If
given by FAX then, for record purposes, the referral and its
accompanying documentation must forthwith be sent by
first class post or given by actual delivery. If sent by
registered post or recorded delivery the referral and its
accompanying documentation shall, subject to proof to the
contrary, be deemed to have been received 48 hours after
the date of posting subject to the exclusion of Sundays and
any Public Holiday.

41A.5.1 The Adjudicator shall immediately upon receipt of the
referral and its accompanying documentation confirm the
date of that receipt to the Parties.

41A.5.2 The Party not making the referral may, by the same means
stated in clause 41A.4.2, send to the Adjudicator within 7

days of the date of the referral with a copy to the other Party, a written statement of the contentions on which he relies and any material he wishes the Adjudicator to consider.

41A.5.3   The Adjudicator shall within 28 days of his receipt of the referral and its accompanying documentation under clause 41A.4.1 and acting as an Adjudicator for the purposes of S.108 of the Housing Grants, Construction and Regeneration Act 1996 and not as an expert or an arbitrator reach his decision and forthwith send that decision in writing to the Parties. Provided that the Party who has made the referral may consent to allowing the Adjudicator to extend the period of 28 days by up to 14 days; and that by agreement between the Parties after the referral has been made a longer period than 28 days may be notified jointly by the Parties to the Adjudicator within which to reach his decision.

41A.5.4   The Adjudicator shall not be obliged to give reasons for his decision.

41A.5.5   In reaching his decision the Adjudicator shall act impartially, set his own procedure and at his absolute discretion may take the initiative in ascertaining the facts and the law as he considers necessary in respect of the referral which may include the following:

.5.1   using his own knowledge and/or experience;

.5.2   opening up, reviewing and revising any certificate, opinion, decision, requirement or notice issued given or made under the Contract as if no such certificate, opinion, decision, requirement or notice had been issued given or made;

.5.3   requiring from the Parties further information than that contained in the notice of referral and its accompanying documentation or in any written statement provided by the Parties including the results of any tests that have been made or of any opening up;

.5.4   requiring the Parties to carry out tests or additional tests or to open up work of further open up work;

.5.5   visiting the site of the Works or any workshop where work is being or had been prepared for the Contract;

.5.6   obtaining such information as he considers necessary from any employee or representative of the Parties provided that before obtaining information from an employee of a Party he has given prior notice to that Party;

.5.7   obtaining from others such information and advice as he considers necessary on technical and on legal matters subject to giving prior notice to the Parties together with a statement or estimate of the cost involved;

.5.8   having regard to any term of the contract relating to the payment of interest deciding the circumstances in which or the period for which a simple rate of interest shall be paid."

Provision is then made for the adjudicators costs, for which the parties are jointly and severally reliable. Clause 41A.8 provides for immunity of the adjudicator in accordance with section 108(4) of the Act. The effect of the adjudicator's decision is set out in clause 41A.7 as follows:

> **"41A.7.1** The decision of the Adjudicator shall be binding on the Parties until the dispute or difference is finally determined by arbitration or by legal proceedings or by an agreement in writing between the Parties made after the decision of the Adjudicator has been given.
>
> **41A.7.2** The Parties shall, without prejudice to their other rights under the Contract, comply with the decisions of the Adjudicator, and the Employer and the Contractor shall ensure that the decisions of the Adjudicator are given effect.
>
> **41A.7.3** If either Party does not comply with the decision of the Adjudicator the other Party shall be entitled to take legal proceedings to secure such compliance pending any final determination of the referred dispute or difference pursuant to clause 41A.7.1."

Provisions for arbitration are set out in clause 41B. The giving of notice, joinder and service of further notices of arbitration are dealt with by Rules 2 and 3 of the CIMA Rules which are incorporated into the JCT form (see chapter 3). These are referred to in clause 41B.1. Provision is made for the arbitration proceedings in clause 41B.2 to .6 as follows:

> **"41B.2** Subject to the provisions of Article 7A and clause 30–9 the Arbitrator shall, without prejudice to the generality of his powers, have power to rectify this Contract so that it accurately reflects the true agreement made by the Parties, to direct such measurements and/or valuations as may in his opinion be desirable in order to determine the rights of the Parties and to ascertain and award any sum which ought to have been the subject of or included in any certificate and to open up, review and revise any certificate, opinion, decision, requirement or notice and to determine all matters in dispute which shall be submitted to him in the same manner as if no such certificate, opinion, decision, requirement or notice had been given.
>
> **41B.3** Subject to clause 41B.4 the award of such Arbitrator shall be final and binding on the Parties.
>
> **41B.4** The Parties hereby agree pursuant to Section 45(2)(a) and Section 69(2)(a) of the Arbitration Act, 1996, that either Party may (upon notice to the other party and to the Arbitrator):
>
> **41B.4.1** apply to the courts to determine any question or law arising in the course of the reference; and
>
> **41B.4.2** appeal to the courts on any question of law arising out of an award made in an arbitration under this Arbitration Agreement.

41B.5 The provisions of the Arbitration Act 1996 or any amendment thereof shall apply to any arbitration under this Contract wherever the same, or any part of it, shall be conducted.

41B.6 The arbitration shall be conducted in accordance with the JCT 1988 edition of the Construction Industry Model Arbitration Rules (CIMAR) current at the Base Date. Provided that if any amendments to the Rules so current have been issued by the Joint Contracts Tribunal after the Base Date the Parties may, by a joint notice in writing to the Arbitrator, state that they wish the arbitration to be conducted in accordance with the Rules as so amended."

Clause 41C seeks to give the court full jurisdiction over any dispute that may be referred to it. This provision has been overtaken by the decision in *Beaufort Developments v. Gilbert-Ash N.I.*[8] which confirms that the court will ordinarily possess the same powers as an arbitrator.

---

[8] [1999] 1 A.C. 266; 88 B.L.R. 1.

CHAPTER 13

THE ICE CONDITIONS OF CONTRACT

THE Institution of Civil Engineers' (ICE) form of contract is intended for use in works of civil engineering construction. The form is issued by the ICE, the Association of Consulting Engineers and the Federation of Civil Engineering Contractors. A sixth edition of the form was issued in 1991. The new edition follows the clause numbering and much of the style of previous editions. This accounts for the rather archaic tone of much of the wording. The conditions are published together with forms of tender, agreement and bond. Unlike the JCT forms, the ICE form exists in one version only, for use by private or other employers. The sponsoring bodies have set up a permanent joint committee (CCSJC) to keep the conditions under review. The form has been revised and reprinted a number of times, most recently in November 1997. Further amendments were issued in February 1998 to take account of the Housing Grants, etc., Act 1996.

The conditions create a "re-measurement" or "measure and value" contract, *i.e.* the contractor is to be paid at the contract rates (which are themselves subject to variation) for the actual quantities of work executed. This is recognised by omission of any reference to a "contract sum." Instead the conditions refer to the "tender total."

Under the conditions, the work is required to be carried out to the satisfaction of the engineer. He is given powers of control and direction which are both extensive and apparently arbitrary; although as agent of the employer, the engineer must act in the best interest of his principal. The conditions also contain wide-ranging provisions under which the sums payable to the contractor are subject to alteration, usually in favour of the contractor.

This chapter gives an introduction to the basic working provisions of the form. The most important clauses or sub-clauses are printed with notes as to their effect with reference to other clauses where appropriate.

The Contract

The conditions contemplate that the form of contract will be accompanied by drawings and a specification in which the work is described, and by bills of quantities in which the work is measured and priced. The way in which these documents operate is as follows:

> "1(1)(e) 'Contract' means the Conditions of Contract Specification Drawings Priced Bill of Quantities the Tender the written acceptance thereof and the Contract Agreement (if completed).
>
> (f) 'Specification' means the specification referred to in the Tender and any modification thereof or addition thereto as may from time to time be furnished or approved in writing by the Engineer.
>
> (g) 'Drawings' means the drawings referred to in the Specification and any modification of such drawings approved in writing by the Engineer and such other drawings as may from time to time be furnished or approved in writing by the Engineer.

> The contract itself is thus to be found in the tender (when accepted), which provides as follows:

GENTLEMEN:

Having examined the Drawings, Conditions of Contract, Specification and Bill of Quantities for the construction of the above-mentioned Works (and the matters set out in the Appendix hereto) we offer to construct and complete the whole of the said Works in conformity with the said Drawings, Conditions of Contract, Specification and Bill of Quantities for such sum as may be ascertained in accordance with the said Conditions of Contract.

We undertake to complete and deliver the whole of the Permanent Works comprised in the Contract within the time stated in the Appendix hereto.

If our tender is accepted we will, if required, provide security for the due performance of the Contract as stipulated in the Conditions of Contract and the Appendix hereto.

Unless and until a formal Agreement is prepared and executed this Tender together with your written acceptance thereof, shall constitute a binding Contract between us.

We understand that you are not bound to accept the lowest or any tender you may receive.

We are, Gentlemen,

Yours faithfully."

By clause 9 of the conditions, the contractor undertakes to execute the contract agreement, which may be under seal. If executed, the agreement becomes the primary contract document. The agreement is in the following form:

"WHEREAS the Employer is desirous that certain Works should be constructed, namely the Permanent and Temporary Works in connection with .... and has accepted a Tender by the Contractor for the construction and completion of such Works

NOW THIS AGREEMENT WITNESSES as follows:

1. In this Agreement words and expressions shall have the same meanings as are respectively assigned to them in the Conditions of Contract hereinafter referred to.
2. The following documents shall be deemed to form and be read and construed as part of this Agreement, namely:-

   (a) The said Tender and the written acceptance thereof.
   (b) The Drawings.
   (c) The Conditions of Contract.
   (d) The Specification.
   (e) The Priced Bill of Quantities.

3. In consideration of the payments to be made by the Employer to the Contractor as hereinafter mentioned the Contractor hereby covenants with the Employer to construct and complete the Works in conformity in all respects with the provisions of the Contract.
4. The Employer hereby covenants to pay to the Contractor in consideration of the construction and completion of the Works the Contract Price at the times and in the manner prescribed by the Contract.

IN WITNESS whereof the parties hereto have caused this Agreement to be executed the day and year first above written.

SIGNED"

The sums payable to the contractor by the conditions are defined as follows:

"1(1)(i) 'Tender Total' means the total of the Bill of Quantities at the date of award of the Contract or in the absence of a Bill of Quantities the agreed estimated total value of the Works at that date.

(j) 'Contract Price' means the sum to be ascertained and paid in accordance with the provisions hereinafter contained for the construction and completion of the Works in accordance with the Contract."

Clause 5 of the conditions defines the effect of the contract documents:

"5 The several documents forming the Contract are to be taken as mutually explanatory of one another and in case of ambiguities or discrepancies the same shall be explained and adjusted by the Engineer who shall thereupon issue to the Contractor appropriate issued in accordance with Clause 13."

The engineer's power under this clause is limited to the technical descriptions of the work, it is thought. He is not empowered to rewrite the conditions. The contract documents usually do not contain full working details. The engineer is therefore given both the power and duty to issue such further details as necessary:

"7(1) The Engineer shall from time to time during the progress of the Works supply to the Contractor such modified or further Drawings Specifications and instructions as shall in the Engineer's opinion be necessary for the purpose of the proper and adequate construction and completion of the Works and the Contractor shall carry out and be bound by the same.

  If such Drawings Specifications or instructions require any variation to any part of the Works the same shall be deemed to have been issued pursuant to Clause 51.

(3) The Contractor shall give adequate notice in writing to the Engineer of any further Drawing or Specification that the Contractor may require for the construction and completion of the Works or otherwise under the Contract."

Sub-clause (2) provides for the reverse situation where the contractor is to supply the design details. Where instructions are issued late clause 7(4) allows the contractor to claim additional payment and an extension of time (see below for these sub-clauses).

Control of the work

The conditions deal expressly with the authority and independence of the engineer, as follows:

"2(1)(a) The Engineer shall carry out the duties specified in or necessarily to be implied from the Contract.

(b) The Engineer may exercise the authority specified in or necessarily to be implied from the Contract. If the Engineer is required under the terms of his appointment by the Employer to obtain the specific approval of the Employer before exercising any such authority particulars of such requirements shall be those set out in the Appendix to the Form of Tender. Any requisite approval shall be deemed to have been given by the Employer for any such authority exercised by the Engineer.

(c) Except as expressly stated in the Contract the Engineer shall have no authority to amend the Terms and Conditions of the Contract nor to relieve the Contractor of any of his obligations under the Contract.

(2)(a) Where the Engineer as defined in Clause 1(1)(c) is not a single named Chartered Engineer the Engineer shall within seven days of the award of the Contract and in any event before the Works Commencement Date notify to the Contractor in writing the name of the Chartered Engineer who will act on his behalf and assume the full responsibilities of the Engineer under the Contract.

(b) The Engineer shall thereafter in like manner notify the Contractor of any replacement of the named Chartered Engineer."

The conditions envisage that work on site will be given full-time supervision on behalf of the engineer and the contractor.

"2(3)(a) The Engineer's Representative shall be responsible to the Engineer who shall notify his appointment to the Contractor in writing.

(b) The Engineer's Representative shall watch and supervise the construction and completion of the Works. He shall have no authority

(i) to relieve the Contractor of any of his duties or obligations under the Contract nor except as expressly provided hereunder

(ii) to order any work involving delay or any extra payment by the Employer or

(iii) to make any variation of or in the Works.

(4) The Engineer may from time to time delegate to the Engineer's Representative or any other person responsible to the Engineer any of the duties and authorities vested in the Engineer and he may at any time revoke such delegation. Any such delegation

(a) shall be in writing and shall not take effect until such time as a copy thereof has been delivered to the Contractor or his agent appointed under Clause 15(2)

(b) shall continue in force until such time as the Engineer shall notify the Contractor in writing that the same has been revoked

(c) shall not be given in respect of any decision to be taken or certificate to be issued under Clauses 12(6), 44, 46(3), 48, 60(4), 61, 63 or 66.

(5)(a) The Engineer or the Engineer's Representative may appoint any number of persons to assist the Engineer's

Representative in the carrying out of his duties under sub-clause (3)(b) or (4) of this Clause. He shall notify to the Contractor the names duties and scope of authority of such persons.

(b) Such assistants shall have no authority to issue any instructions to the Contractor save in so far as such instructions may be necessary to enable them to carry out their duties and to secure their acceptance of materials and workmanship as being in accordance with the Contract. Any instructions given by an assistant for these purposes shall where appropriate be in writing and be deemed to have been given by the Engineer's Representative.

(c) If the Contractor is dissatisfied by reason of any instruction of any assistant of the Engineer's Representative appointed under sub-clause (5)(a) of this Clause he shall be entitled to refer the matter to the Engineer's Representative who shall thereupon confirm reverse or vary such instruction.

(6)(a) Instructions given by the Engineer or by the Engineer's Representative exercising delegated duties and authorities under sub-clause (4) of this Clause shall be in writing. Provided that if for any reason it is considered necessary to give any such instruction orally the Contractor shall comply with such instruction.

(b) Any such oral instruction shall be confirmed in writing by the Engineer or the Engineer's Representative as soon as is possible under the circumstances. Provided that if the Contractor shall confirm in writing any such oral instruction and such confirmation is not contradicted in writing by the Engineer or the Engineer's Representative forthwith it shall be deemed to be an instruction in writing by the Engineer.

(c) Upon the written request of the Contractor the Engineer or the Engineer's Representative exercising delegated duties or authorities under sub-clause (4) of this Clause shall specify in writing under which of his duties and authorities any instruction is given.

(7) The Engineer shall, except in connection with matters requiring the specific approval of the Employer under sub-clause (1)(b) of this Clause, act impartially within the terms of the Contract having regard to all the circumstances."

This clause is over-complicated for the matters dealt with. In substance it requires delegation of powers to the engineer's representative (the E.R.) to be in writing, with notice to the contractor. Such powers may not include giving extensions of time, requesting acceleration, certifying completion, giving the final certificate or the Defects Correction certificate, nor a certificate of default (clauses 44, 46(3), 48, 60(4) 61, 63), nor giving decisions under clause 12 or 66.

The contractor is also required to supervise the work:

"15(1) The Contractor shall give or provide all necessary superintendence during the construction and completion of the Works and as long thereafter as the Engineer may consider necessary. Such superintendence shall be given by sufficient persons having adequate knowledge of the operations to be carried out (including the methods and techniques required, the hazards likely to be encountered and methods of preventing accidents) as may be requisite for the satisfactory and safe construction of the Works.

(2) The Contractor or a competent and authorised agent or representative approved of in writing by the Engineer (which approval may at any time be withdrawn) is to be constantly on the Works and shall give his whole time to the superintendence of the same. Such authorised agent or representative shall be in full charge of the Works and shall receive on behalf of the Contractor directions and instructions from the Engineer or (subject to the limitations of Clause 2) the Engineer's Representative. The Contractor or such authorised agent or representative shall be responsible for the safety of all operations."

The contractor is required to give the engineer opportunity to inspect the work:

"37 The Engineer and any person authorised by him shall at all times have access to the Works and to the Site and to all workshops and places where work is being prepared or whence materials manufactured articles and machinery are being obtained for the Works, and the Contractor shall afford every facility for and every assistance in obtaining such access or the right to such access."

GENERAL OBLIGATIONS OF THE CONTRACTOR

These are contained in a number of clauses, which must be read together.

"8(1) The Contractor shall subject to the provisions of the Contract

(a) construct and complete the Works; and
(b) provide all labour materials Contractor's Equipment Temporary Works transport to and from and in or about the Site and everything whether of a temporary or permanent nature required in and for such construction and completion so far as the necessity for providing the same is specified in or reasonably to be inferred from the Contract.

(2) The Contractor shall not be responsible for the design or specification of the Permanent Works or any part thereof (except as may be expressly provided in the Contract) or of any Temporary Works designed by the Engineer. The Contractor shall exercise all reasonable skill care and diligence in designing any part of the Permanent Works for which he is responsible.

(3) The Contractor shall take full responsibility for the adequacy stability and safety of all site operations and methods of construction."

Sub-clause (1) repeats obligations contained in the tender to "construct and complete the whole of the said works"; and in the form of agreement, to "construct and complete the works." The question of the contractor's design responsibility is covered also in other clauses:

"7(2) Where sub-clause (6) of this Clause applies the Engineer may require the Contractor to supply such further documents as shall in his opinion be necessary for the purpose of the proper and adequate construction completion and maintenance of the Works and when approved by the Engineer the Contractor shall carry out and be bound by the same.

(6) Where the Contract expressly provides that part of the Permanent Works shall be designed by the Contractor he shall submit to the Engineer for approval:

(a) such drawings specifications calculations and other information as shall be necessary to satisfy the Engineer as to the suitability and adequacy of the design and

(b) operation and maintenance manuals together with as completed drawings of that part of the Permanent Works in sufficient detail to enable the Employer to operate maintain dismantle reassemble and adjust the Permanent Works incorporating that design. No certificate under Clause 48 covering any part of the Permanent Works designed by the Contractor shall be issued until manuals and drawings in such detail have been submitted to and approved by the Engineer.

(7) Approval by the Engineer in accordance with sub-clause (6) of this Clause shall not relieve the Contractor of any of his responsibilities under the Contract. The Engineer shall be responsible for the integration and co-ordination of the Contractor's design with the rest of the Works."

Clause 58(3) deals with design responsibility where a nominated sub-contractor is involved. Clause 20 makes the contractor generally responsible for the works with exceptions (the "Excepted Risks") which include the engineer's design. The contractor's design responsibility may therefore not be limited to reasonable skill as suggested by clause 8(2). The meaning of "Temporary" and "Permanent" work is as follows:

"1(1)(n) 'Permanent Works' means the permanent works to be constructed and completed in accordance with the Contract.

(o) 'Temporary Works' means all temporary works of every kind required in or about the construction and completion of the Works.

(p) 'Works' means the Permanent Works together with the Temporary Works."

These definitions are not at all precise. It is often difficult to decide whether work is permanent or temporary, for example, a cofferdam which is intended to be left in position, after temporary use to facilitate excavation. Temporary works are not mentioned in the JCT contract. They are dealt with in these conditions because such works are often very costly and are frequently designed by the contractor, who assumes responsibility therefor.

The contractor's responsibility for methods of construction is dealt with further below. In addition to his responsibility for the site operations and methods of construction and for the works themselves (clause 20(2), see below), the contractor is required to take the risk of the site and the sub-soil:

"**11(1)** The Employer shall be deemed to have made available to the Contractor before the submission of the Tender all information on

    (a) the nature of the ground and sub-soil including hydrological conditions, and

    (b) pipes and cables in on or over the ground

obtained by or on behalf of the Employer from investigations undertaken relevant to the Works.
The Contractor shall be responsible for the interpretation of all such information for the purposes of constructing the Works and for any design which is the Contractor's responsibility under the Contract.

    (2) The Contractor shall be deemed to have inspected and examined the Site and its surroundings and information available in connection therewith and to have satisfied himself so far as is practicable and reasonable before submitting his tender as to

    (a) the form and nature thereof including the ground and sub-soil

    (b) the extent and nature of work and materials necessary for constructing and completing the Works and

    (c) the means of communication with and access to the Site and the accommodation he may require

and in general to have obtained for himself all necessary information as to risks contingencies and all other circumstances which may influence or affect his tender.

    (3) The Contractor shall be deemed to have

    (a) based his tender on the information made available by the Employer and on his own inspection and examination all as aforementioned and

    . . . . .

    (c) satisfied himself before submitting his tender as to the correctness and sufficiency of the rates and prices stated by him in the Bill of Quantities which shall (unless otherwise provided in the Contract) cover all his obligations under the Contract."

The contractor's liability is subject to two exceptions. First, clause 11(2) expressly permits the contractor to take account of any sub-soil information provided, *i.e.* any site investigation data. The effect of this provision is obscure, but it may permit the contractor to bring proceedings under the Misrepresentation Act 1967, if the data is misleading. Secondly, clause 12 may entitle the contractor to additional payment if unforeseeable physical conditions or artificial obstructions are encountered in the sub-soil (see below). There will be a further exception where the employer has failed to disclose relevant sub-soil information. Since the tender is deemed to be based on the information made available (clause 11(3)(a)) there may be a claim for re-pricing. Other clauses which bear on the contractor's general responsibility are as follows:

"**13(1)** Save in so far as it is legally or physically impossible the Contractor shall construct and complete the Works in strict accordance with the Contract to the satisfaction of the Engineer and shall comply with and adhere strictly to the Engineer's instructions on any matter connected therewith (whether mentioned in the Contract or not). The Contractor shall take instructions only from the Engineer or (subject to the limitations referred to in Clause 2) from the Engineer's Representative."

This clause gives the engineer very wide powers to give instructions. This is important in regard to claims, as to which see clause 13(3), referred to below. The contractor is not obliged to carry out the work to the extent it is legally or physically impossible nor is the employer obliged to pay for work omitted on this ground. The degree of impossibility required to absolve the contractor from further performance was considered in *Turriff v. Welsh Water Authority*,[1] where it was held sufficient that the work was commercially impossible in a practical sense. The contractor was attempting to join rectangular pre-cast sections of culvert, whose design tolerances prevented a seal being achieved. It was held the contractor was not under a duty to re-design the work to render it capable of being constructed.

Subject to specified exceptions the contractor is required to assume responsibility for the works, irrespective of fault:

"**20(1)(a)** The Contractor shall save as in paragraph (b) hereof and subject to sub-clause (2) of this Clause take full respon-

---

[1] (1979) [1995] Con. L.Yb. 122.

sibility for the care of the Works and materials plant and equipment for incorporation therein from the Works Commencement Date until the date of issue of a Certificate of Substantial Completion for the whole of the Works when the responsibility for the said care shall pass to the Employer.

(b) If the Engineer issues a Certificate of Substantial Completion for any Section or part of the Permanent Works the Contractor shall cease to be responsible for the care of that Section or part from the date of issue of such Certificate of Substantial Completion when the responsibility for the care of that Section or part shall pass to the Employer.

(c) The Contractor shall take full responsibility for the care of any outstanding work and materials plant and equipment for incorporation therein which he undertakes to finish during the Defects Correction Period until such outstanding work has been completed.

(2) The Excepted Risks for which the Contractor is not liable are loss or damage to the extent that it is due to

(a) the use or occupation by the Employer his agents servants or other Contractors (not being employed by the Contractor) of any part of the Permanent Works

(b) any fault defect error or omission in the design of the Works (other than a design provided by the Contractor pursuant to his obligations under the Contract)

(c) riot war invasion act of foreign enemies or hostilities (whether war be declared or not)

(d) civil war rebellion revolution insurrection or military or usurped power

(e) ionising radiations or contamination by radio-activity from any nuclear fuel or from any nuclear waste from the combustion of nuclear fuel radio-active toxic explosive or other hazardous properties of any explosive nuclear assembly or nuclear component thereof and

(f) pressure waves caused by aircraft or other aerial devices travelling at sonic or supersonic speeds.

(3)(a) In the event of any loss or damage to

(i) the Works or any Section or part thereof or
(ii) materials plant or equipment for incorporation therein

while the Contractor is responsible for the care thereof (except as provided in sub-clause (2) of this Clause) the Contractor shall at his own cost rectify such loss or damage so that the Permanent Works conform in every respect with the provisions of the Contract and the Engineer's (a) instructions. The Contractor shall also be liable for any loss or damage to the Works occasioned by him in the course of any operations carried out by him for the purpose of complying with his obligations under Clauses 49 and 50.

(b) Should any such loss or damage arise from any of the Excepted Risks defined in sub-clause (2) of this Clause the

Contractor shall if and to the extent required by the Engineer rectify the loss or damage at the expense of the Employer.

(c) In the event of loss or damage as a result of an Excepted Risk and a risk for which the Contractor is responsible under sub-clause (1)(a) of this Clause then the Engineer shall when determining the expense to be borne by the Employer under the Contract apportion the cost of rectification into that part caused by the Excepted Risk and that part which is the responsibility of the Contractor."

This is a most important general obligation, which can override other specific responsibilities. The contractor must make good loss to the works from any cause, occurring before completion. The cost is to be met from the insurance provided for under clause 21 (below). If the loss is due to an "Excepted Risk" an instruction is required and the employer must bear the cost.

Clauses 21 to 25 contain important requirements as to insurance and liability for losses. Clause 21 requires the contractor to insure the works and his materials and plant against loss, from any cause other than the excepted risks, so as to cover his liability under clause 20. This insurance is required to be in the joint names of employer and contractor. It is important for the employer to be aware that the works are not insured against damage caused by any fault in the engineer's design. The employer's ability to recover for such loss will depend on the limit of the engineer's professional indemnity policy (see Chapter 7).

Clause 22 apportions liability for third party claims which arise out of or in consequence of the work; and clause 23 requires the contractor to insure his own liability. Clause 24 deals with injuries to workmen. Clause 25 gives the employer the right to effect any insurance which the contractor fails to take out.

The contractor's methods of work

Prima facie, the contractor's methods of carrying out the work are his responsibility and his choice. The contractor is responsible for the works and for their safety (clauses 20 and 8(3)). There is an exception if the contract designates the method of construction and that method becomes impossible.[2] Other clauses which bear on the choice and responsibility for the method are as follows:

"**13(2)** The whole of the materials plant and labour to be provided by the Contractor under Clause 8 and the mode manner and speed

---

[2] See *McAlpine v. Yorkshire Water Authority* (1985) 32 B.L.R. 114 and Chap. 9.

of construction of the Works are to be of a kind and conducted in a manner acceptable to the Engineer.

14(1) (a) Within 21 days after the award of the Contract the Contractor shall submit to the Engineer for his acceptance a programme showing the order in which he proposes to carry out the Works having regard to the provisions of Clause 42(1).

(b) At the same time the Contractor shall also provide in writing for the information of the Engineer a general description of the arrangements and methods of construction which the Contractor proposes to adopt for the carrying out of the Works.

(c) Should the Engineer reject any programme under sub-clause (2)(b) of this Clause the Contractor shall within 21 days of such rejection submit a revised programme.

(2) The Engineer shall within 21 days after receipt of the Contractor's programme

(a) accept the programme in writing or

(b) reject the programme in writing with reasons or

(c) request the Contractor to supply further information to clarify or substantiate the programme or to satisfy the Engineer as to its reasonableness having regard to the Contractor's obligations under the Contract.

Provided that if none of the above actions is taken within the said period of 21 days the Engineer shall be deemed to have accepted the programme as submitted.

(3) The Contractor shall within 21 days after receiving from the Engineer any request under sub-clause (2)(c) of this Clause or within such further period as the Engineer may allow provide the further information requested failing which the relevant programme shall be deemed to be rejected.

Upon receipt of such further information the Engineer shall within a further 21 days accept or reject the programme in accordance with sub-clauses (2)(a) or (2)(b) of this Clause.

(4) Should it appear to the Engineer at any time that the actual progress of the work does not conform with the accepted programme referred to in sub-clause (1) of this Clause, the Engineer shall be entitled to require the Contractor to produce a revised programme showing such modifications to the original programme as may be necessary to ensure completion of the Works or any Section within the time for completion as defined in Clause 43 or extended time granted pursuant to Clause 44. In such event the Contractor shall submit his revised programme within 21 days or within such further period as the Engineer shall allow. Thereafter the provisions of sub-clauses (2) and (3) of this Clause shall apply.

(5) The Engineer shall provide to the Contractor such design criteria relevant to the Permanent Works or any Temporary Works design supplied by the Engineer as may be necessary to enable the Contractor to comply with sub-clauses (6) and (7) of this Clause.

(6) If requested by the Engineer the Contractor shall submit at such times and in such further detail as the Engineer may reasonably require information pertaining to the methods of construction (including Temporary Works and the use of Contractor's Equipment) which the Contractor proposes to adopt or use and calculations of stresses strains and deflections that will arise in the Permanent Works or any parts thereof during construction so as to enable the Engineer to decide whether, if these methods are adhered to the Works can be constructed and completed in accordance with the Contract and without detriment to the Permanent Works when completed.

(7) The Engineer shall inform the Contractor in writing within 21 days after receipt of the information submitted in accordance with sub-clauses (1)(b) and (6) of this Clause either

   (a) that the Contractor's proposed methods have the consent of the Engineer or

   (b) in what respects in the opinion of the Engineer they fail to meet the requirements of the Contract or will be detrimental to the Permanent Works.

In the latter event the Contractor shall take such steps or make such changes in the said methods as may be necessary to meet the Engineer's requirements and to obtain his consent. The Contractor shall not change the methods which have received the Engineer's consent without the further consent in writing of the Engineer which shall not be unreasonably withheld.

(8) If the Contractor unavoidably incurs delay or cost because

   (a) the Engineer's consent to the proposed methods of construction is unreasonably delayed or

   (b) the Engineer's requirements pursuant to sub-clause (7) of this clause or any limitations imposed by any of the design criteria supplied by the Engineer pursuant to sub-clause (5) of this Clause could not reasonably have been foreseen by an experienced Contractor at the time of tender

   the Engineer shall take such delay into account in determining any extension of time to which the Contractor is entitled under Clause 44 and the Contractor shall subject to Clause 52(4) be paid in accordance with Clause 60 such sum in respect of the cost incurred as the Engineer considers fair in all the circumstances. Profit shall be added thereto in respect of any additional permanent or temporary work.

(9) Acceptance by the Engineer of the Contractor's programme in accordance with sub-clauses (2) (3) or (4) of this Clause and the consent of the Engineer to the Contractor's proposed methods of construction in accordance with sub-clause (7) of this Clause shall not relieve the Contractor of any of his duties or responsibilities under the Contract."

Clause 14 deals both with programme and method. By sub-clause (7) the engineer is required to respond to the contractor's method information. In most circumstances the contractor will remain responsible for the method, but it is to be noted that clause 51(1) includes a method change within the definition of a variation.

In addition to the general obligation as to safety under clause 8(3), the contractor is specifically required to have regard to the safety of persons on the site:

"19(1) The Contractor shall throughout the progress of the Works have full regard for the safety of all persons entitled to be upon the Site and shall keep the Site (so far as the same is under his control) and the Works (so far as the same are not completed or occupied by the Employer) in an orderly state appropriate to the avoidance of danger to such persons, and shall inter alia in connection with the Works provide and maintain at his own cost all rights, guards, fencing, warning signs and watching when and where necessary or required by the Engineer, or the Engineer's Representative, or by any competent statutory or other authority for the protection of the Works or for the safety and convenience of the public or others.

(2) If under Clause 31 the Employer shall carry out work on the Site with his own workmen he shall in respect of such work

(a) have full regard to the safety of all persons entitled to be upon the Site and

(b) keep the Site in an orderly state appropriate to the avoidance of danger to such persons."

If under Clause 31 the Employer shall employ other Contractors on the Site he shall require them to have the same regard for safety and avoidance of danger.

Clause 22 provides for indemnities in respect of personal injury or damage to property. Every contractor must also comply with regulations as to site safety, under the Health and Safety at Work Act, etc., 1974 (see Chapter 16).

Workmanship and materials

The following clauses should be read in the light of the general obligations to comply with the contract (clause 13(1), tender and form of agreement):

"36(1) All materials and workmanship shall be of the respective kinds described in the Contract and in accordance with the Engineer's instructions and shall be subjected from time to time to such tests as the Engineer may direct at the place of manufacture or fabrication or on the Site or such other place or places as may be specified in the Contract. The Contractor shall provide such assistance instruments machines labour and materials as are

> normally required for examining measuring and testing any work and the quality weight or quantity of any materials used and shall supply samples of materials before incorporation in the Works for testing as may be selected and required by the Engineer."

The "instructions" referred to must arise under some express power in the contract: this clause does not empower the engineer to change the specification without giving an instruction to do so, which must be paid for under clauses 51(1) or 13(1). Clause 36(2) and (3) stipulates how the cost of tests and samples is to be borne.

> "**38(1)** No work shall be covered up or put out of view without the consent of the Engineer and the Contractor shall afford full opportunity for the Engineer to examine and measure any work which is about to be covered up or put out of view and to examine foundations before permanent work is placed thereon. The Contractor shall give due notice to the Engineer whenever any such work or foundations is or are ready or about to be ready for examination and the Engineer shall without unreasonable delay unless he considers it unnecessary and advises the Contractor accordingly attend for the purpose of examining and measuring such work or of examining such foundations."

Clause 38(2) permits the engineer to order uncovering of work and provides for apportioning the cost.

> "**39(1)** The Engineer shall during the progress of the Works have power to instruct in writing the
>
> > (**a**) removal from the Site within such time or times specified in the instruction of any materials which in the opinion of the Engineer are not in accordance with the Contract
> > (**b**) substitution with materials in accordance with the Contracts and
> > (**c**) removal and proper re-execution, notwithstanding any previous test thereof or interim payment therefor, of any work which in respect of
> >
> > > (i) material or workmanship or
> > > (ii) design by the Contractor or for which he is responsible
> >
> > is not in the opinion of the Engineer in accordance with the Contract.
>
> (**2**) In case of default on the part of the Contractor in carrying out such instruction the Employer shall be entitled to employ and pay other persons to carry out the same and all costs consequent thereon or incidental thereto as determined by the Engineer shall be recoverable from the Contractor by the Employer and may be deducted by the Employer from any monies due or to become due to him and the Engineer shall notify the Contractor accordingly with a copy to the Employer.

(3) Failure of the Engineer or any person acting under him pursuant to Clause 2 to disapprove any work or materials shall not prejudice the power of the Engineer or any such person subsequently to take action under this Clause."

This clause is of importance. Without it the engineer's only remedy for defective work would be determination (under clause 63) or refusal to certify payment on completion, the latter carrying the sanction of liquidated damages. Work is not in accordance with the contract if it does not comply with the drawings or the specification (clause 1(1)(e)).

The contractor is responsible for setting-out errors, unless based on incorrect data supplied:

"17(1) The Contractor shall be responsible for the true and proper setting-out of the Works and for the correctness of the position levels dimensions and alignment of all parts of the Works and for the provision of all necessary instruments appliances and labour in connection therewith.

(2) If at any time during the progress of the Works any error shall appear or arise in the position levels dimensions or alignment of any part of the Works the Contractor on being required so to do by the Engineer shall at his own costs rectify such error to the satisfaction of the Engineer unless such error is based on incorrect data supplied in writing by the Engineer or the Engineer's Representative in which case the cost of rectifying the same shall be borne by the Employer.

(3) The checking of any setting-out or of any line or level by the Engineer or the Engineer's Representative shall not in any way relieve the Contractor of his responsibility for the correctness thereof and the Contractor shall carefully protect and preserve all bench-marks sight rails pegs and other things used in setting out the Works."

Independent of the contract, the work must, by law, comply with various statutory requirements (see Chapter 16). Difficult questions can arise in the event of conflict between these requirements and the engineer's design. This is dealt with as follows:

"26(1) The Contractor shall give all notices and pay all fees required to be given or paid by any Act of Parliament or any Regulation or By-law of any local or other statutory authority in relation to the construction and completion of the Works and by the rules and regulations of all public bodies and companies whose property or rights are or may be affected in any way by the Works.

(2) The Employer shall repay or allow to the Contractor all such sums as the Engineer shall certify to have been properly payable and paid by the Contractor in respect of such fees and also all rates and taxes paid by the Contractor in respect of the Site or any part thereof, or anything constructed or erected thereon, or

on any part thereof or any temporary structure situated elsewhere but used exclusively for the purposes of the Works or any structures used temporarily and exclusively for the purposes of the Works.

(3) The Contractor shall ascertain and conform in all respects with the provisions of any general or local Act of Parliament and the Regulations and By-laws of any local or other statutory authority which may be applicable to the Works and with such rules and regulations of public bodies and companies as aforesaid and shall keep the Employer indemnified against all penalties and liability of every kind for breach of any such Act Regulation or By-Law. Provided always that

    (a) the Contractor shall not be required to indemnify the Employer against the consequences of any such breach which is the unavoidable result of complying with the Contract or instructions of the Engineer

    (b) if the Contract or instructions of the Engineer shall at any time be found not to be in conformity with any such Act Regulation or By-law the Engineer shall issue such instructions including the ordering of a variation under Clause 51 as may be necessary to ensure conformity with such Act Regulation or By-law

    (c) the Contractor shall not be responsible for obtaining any planning permission which may be necessary in respect of the Permanent Works or any Temporary Works design supplied by the Engineer and the Employer hereby warrants that all the said permissions have been or will in due time be obtained."

If the contractor, in following the engineer's design, contravenes any statute or by-law, etc., the employer is not entitled to indemnity because of proviso (a). But the contractor may not be entitled to payment for work carried out contrary to byelaws, etc., since the work is in breach of contract by virtue of sub-clause (2).

Clause 27 as re-issued in the 1993 deals with the New Roads and Street Works Act 1991. Essentially the employer is to obtain licences and consents for the permanent works and the contractor is to give notices (see further Chapter 16).

Completion and defects correction

The contractor's obligations fall into two separate periods: the period up to the completion certificate, when the work is carried out; and the period during which the contractor must repair any defects. These are dealt with in the following provisions. It is important to note that these clauses do not limit the contractor's continuing liability for any failure to comply with the contract.

"**48(1)** When the Contractor considers that

(a) the whole of the Works or

(b) any Section in respect of which a separate time for completion is provided in the Appendix to the Form of Tender.

has been substantially completed and has satisfactorily passed any final test that may be prescribed by the Contract he may give notice in writing to that effect to the Engineer or to the Engineer's Representative. Such notice shall be accompanied by an undertaking to finish any outstanding work in accordance with the provisions of Clause 49(1).

(2) The Engineer shall within 21 days of the date of delivery of such notice either

(a) issue to the Contractor (with a copy to the Employer) a Certificate of Substantial Completion stating the date on which in his opinion the Works were or the Section was substantially completed in accordance with the Contract or

(b) give instructions in writing to the Contractor specifying all the work which in the Engineer's opinion requires to be done by the Contractor before the issue of such certificate.

If the Engineer gives such instructions the Contractor shall be entitled to receive a Certificate of Substantial Completion within 21 days of completion to the satisfaction of the Engineer of the work specified in the said instructions.

(3) If any substantial part of the Works has been occupied or used by the Employer other than as provided in the Contract the Contractor may request in writing and the Engineer shall issue a Certificate of Substantial Completion in respect thereof. Such certificate shall take effect from the date of delivery of the Contractor's request and upon the issue of such certificate the Contractor shall be deemed to have undertaken to complete any outstanding work in that part of the Works during the Defects Correction Period.

(4) If the Engineer considers that any part of the Works has been substantially completed and has passed any final test that may be prescribed by the Contract he may issue a Certificate of Substantial Completion in respect of that part of the Works before completion of the whole of the Works and upon the issue of such certificate the Contractor shall be deemed to have undertaken to complete any outstanding work in that part of the Works during the Defects Correction Period.

(5) A Certificate of Substantial Completion given in respect of any Section or part of the Works before completion of the whole shall not be deemed to certify completion of any ground or surfaces requiring reinstatement unless such certificate shall expressly so state."

Note the engineer's useful power to certify completion despite the existence of outstanding work. This must be such that the employer may still take-over and use the works; otherwise the engineer should certify in respect of completed parts. The employer must take over any parts which are usable in order to mitigate damages for delay. The later discovery of hidden defects does not invalidate the completion certificate (see Chapter 9).

"**49(1)** The undertaking to be given under Clause 48(1) may after agreement between the Engineer and the Contractor specify a time or times within which the outstanding work shall be completed. If no such times are specified any outstanding work shall be completed as soon as practicable during the Defects Correction Period.

(2) The Contractor shall deliver up to the Employer the Works and each Section and part thereof at or as soon as practicable after the expiry of the relevant Defects Correction Period in the condition required by the Contract (fair wear and tear excepted) to the satisfaction of the Engineer. To this end the Contractor shall as soon as practicable execute all work of repair amendment reconstruction rectification and making good of defects of whatever nature as may be required of him in writing by the Engineer during the relevant Defects Correction Period or within 14 days after its expiry as a result of an inspection made by or on behalf of the Engineer prior to its expiry.

(3) All work required under sub-clause (2) of this Clause shall be carried out by the Contractor at his own expense if in the Engineer's opinion it is necessary due to the use of materials or workmanship not in accordance with the Contract or to neglect or failure by the Contractor to comply with any of his obligations under the Contract. In any other event the value of such work shall be ascertained and paid for as if it were additional work.

(4) If the Contractor fails to do any such work as aforesaid the Employer shall be entitled to carry out such work by his own workpeople or by other Contractors and if such work is work which the Contractor should have carried out at his own expense the Employer shall be entitled to recover the cost thereof from the Contractor and may deduct the same from any monies that are or may become due to the Contractor."

The term "Maintenance Period" appearing in previous editions of the form is now replaced by "Defects Correction Period," which more accurately reflects the effect of the clause. The contractor is obliged to put right any defects, whether or not they are due to his failure to comply with the contract but clause 49(3) entitles the contractor to payment if the defect is not due to his default. The engineer has specific power to order tests. This may be exercised after completion, and is not, apparently, limited to the maintenance period:

"**50** The Contractor shall if required by the Engineer in writing carry out such searches tests or trials as may be necessary to determine the cause of any defect imperfection or fault under the directions of the Engineer. Unless such defect imperfection or fault shall be one for which the Contractor is liable under the Contract the cost of the work carried out by the Contractor as aforesaid shall be borne by the Employer. But if such defect imperfection or fault shall be one for which the Contractor is liable the cost of the work carried out as aforesaid shall be borne by the Contractor and he shall in such case repair rectify and make good such defect imperfection or fault at his own expense in accordance with Clause 49."

If defects are discovered after the defects correction period the contractor remains liable (within the period of limitation) and is not entitled to insist on carrying out remedial work himself. But, in practice, it is often to the advantage of both parties to agree upon remedial works to be carried out by the contractor.

Upon completion the contractor must clear the site:

"33 On the completion of the Works the Contractor shall clear away and remove from the Site all Contractor's Equipment surplus material rubbish and Temporary Works of every kind and leave the whole of the Site and Permanent Works clean and in a workmanlike condition to the satisfaction of the Engineer."

Time

The speed with which the contractor carries out the work is an important element in the performance of the contract. The appendix specifies the time for completion, but the contract contains extensive provisions under which the engineer may extend the completion date.

"41(1) The Works Commencement Date shall be

(a) the date specified in the Appendix to the Form of Tender or if no date is specified

(b) a date within 28 days of the award of the Contract to be notified by the Engineer in writing or

(c) such other date as may be agreed between the parties.

(2) The Contractor shall start the Works on or as soon as is reasonably practicable after the Works Commencement Date. Thereafter the Contractor shall proceed with the Works with due expedition and without delay in accordance with the Contract."

By clause 42 the contractor must be given possession of the site so far as necessary. Sub-clause (3) deals with the consequences of failure to give possession:

"42(1) The Contract may prescribe

(a) the extent of portions of the Site of which the Contractor is to be given possession from time to time

(b) the order in which such portions of the Site shall be made available to the Contractor

(c) he availability and the nature of the access which is to be provided by the Employer

(d) the order in which the Works shall be constructed.

(2) (a) Subject to sub-clause (1) of this Clause the Employer shall give to the Contractor on the Works Commencement Date possession of so much of the Site and access thereto as may be required to enable the Contractor to commence and proceed with the construction of the works.

(b) Thereafter the Employer shall during the course of the Works give to the Contractor possession of such further portions of the Site as may be required in accordance with the programme which the Engineer has accepted under Clause 14 and such further access as is necessary to enable the Contractor to proceed with the construction of the Works with due despatch.

(3) If the Contractor suffers delay and/or incurs additional cost from failure on the part of the Employer to give possession in accordance with the terms of this Clause the Engineer shall determine

(a) any extension of time to which the Contractor is entitled under Clause 44 and

(b) subject to Clause 52(4) the amount of any additional cost to which the Contractor may be entitled. Profit shall be added thereto in respect of any additional permanent or temporary works.

The Engineer shall notify the Contractor accordingly with a copy to the Employer.

(4) The Contractor shall bear all costs and charges for any access required by him additional to those provided by the Employer. The Contractor shall also provide at his own cost any additional facilities outside the Site required by him for the purposes of the Works."

Except where the contract lays down the parties' intention, the contractor's right to possession depends on the provisions of the programme under clause 14, which requires the engineer's approval. The definition of the site, under clause 1(1), is imprecise:

"(v) 'Site' means the lands and other places on under in or through which the Works are to be executed and any other lands or places provided by the Employer for the purposes of the Contract together with such other places as may be designated in the Contract of subsequently agreed by the Engineer as forming part of the Site."

It is advisable, therefore, to define in the contract both the extent of the site and the times at which it will be released to the contractor.

During the course of the work the engineer has powers to require expedition of progress under clause 14(4) (see above), and under clause 46 (see below) where the contractor is in default. Repeated failure to proceed at an adequate rate may lead to determination under clause 63. Conversely the engineer may, at the employer's cost, order suspension of the work:

"**40(1)** The Contractor shall on the written order of the Engineer suspend the progress of the Works or any part thereof for such time or times and in such manner as the Engineer may consider necessary and shall during such suspension properly protect and secure the work so far as is necessary in the opinion of the Engineer. Subject to Clause 52(4) the Contractor shall be paid in accordance with Clause 60 the extra cost (if any) incurred in giving effect to the Engineer's instructions under this Clause except to the extent that such suspension is

    (a) otherwise provided for in the Contract or
    (b) necessary by reason of weather conditions or by some default on the part of the Contractor or
    (c) necessary for the proper execution or for the safety of the Works or any part thereof in as much as such necessity does not arise from any act or default of the Engineer or the Employer or from any of the Excepted Risks defined in Clause 20(2).

Profit shall be added thereto in respect of any additional permanent or temporary work.
The Engineer shall take any delay occasioned by a suspension ordered under this Clause (including that arising from any act or default of the Engineer or the Employer) into account in determining any extension of time to which the Contractor is entitled under Clause 44 except when such suspension is otherwise provided for in the Contract or is necessary by reason of some default on the part of the Contractor.

(2) If the progress of the Works or any part thereof is suspended on the written order of the Engineer and if permission to resume work is not given by the Engineer within a period of three months from the date of suspension then the Contractor may unless such suspension is otherwise provided for in the Contract or continues to be necessary by reason of some default on the part of the Contractor serve a written notice on the Engineer requiring permission within 28 days from the receipt of such notice to proceed with the Works or that part thereof in regard to which progress is suspended. If within the said 28 days the Engineer does not grant such permission the Contractor by a further written notice so served may (but is not bound to) elect to treat the suspension where it affects part only of the Works as an omission of such part under Clause 51 or where it affects the whole Works as as an abandonment of the Contract by the Employer."

This is the only power given to the contractor to determine the contract (compare clause 28 of the JCT form) outside common law rights (see Chapter 6).

"43 The whole of the Works and any Section required to be completed within a particular time as stated in the Appendix to the Form of Tender shall be substantially completed within the time so stated (or such extended time as may be allowed under Clause 44) calculated from the Works Commencement Date notified under Clause 41.

44(1) Should the Contractor consider that

    (a) any variation ordered under Clause 51(1) or
    (b) increased quantities referred to in Clause 51(4) or
    (c) any cause of delay referred to in these Conditions or
    (d) exceptional adverse weather conditions or
    (e) any delay impediment prevention or default by the Employer or
    (f) other special circumstances of any kind whatsoever which may occur

be such as to entitle him to an extension of time for the substantial completion of the Works or any Section thereof he shall within 28 days after the cause of any delay has arisen or as soon thereafter as is reasonable deliver to the Engineer full and detailed particulars in justification of the period of extension claimed in order that the claim may be investigated at the time.

(2) (a) The Engineer shall upon receipt of such particulars consider all the circumstances known to him at that time and make an assessment of the delay (if any) that has been suffered by the Contractor as a result of the alleged cause and shall so notify the Contractor in writing.

(b) The Engineer may in the absence of any claim make an assessment of the delay that he considers has been suffered by the Contractor as a result of any of the circumstances listed in sub-clause (1) of this Clause and shall so notify the Contractor in writing.

(3) Should the Engineer consider that the delay suffered fairly entitles the Contractor to an extension of the time for the substantial completion of the Works or any Section thereof such interim extension shall be granted forthwith and be notified to the Contractor in writing. In the event that the Contractor has made a claim for an extension of time but the Engineer does not consider the Contractor entitled to an extension of time he shall so inform the Contractor without delay.

(4) The Engineer shall not later than 14 days after the due date or extended date for completion of the Works or any Section thereof (and whether or not the Contractor shall have made any claim for an extension of time) consider all the circumstances known to him at that time and take action similar to that provided for in sub-clause (3) of this Clause. Should the

Engineer consider that the Contractor is not entitled to an extension of time he shall so notify the Employer and the Contractor.

(5) The Engineer shall within 14 days of the issue of the Certificate of Substantial Completion for the Works or for any Section thereof review all the circumstances of the kind referred to in sub-clause (1) of this Clause, and shall finally determine and certify to the Contractor with a copy to the Employer the overall extension of time (if any) to which he considers the Contractor entitled in respect of the Works or the relevant Section. No such final review of the circumstances shall result in a decrease in any extension of time already granted by the Engineer pursuant to sub-clauses (3) or (4) of this Clause."

An interim extension under clause 44(3) may be allowed where the contractor has not given notice under sub-clause (1). The engineer is bound to review the grounds for extension under sub-clauses (4) and (5) at the contractual date for completion and again at the actual date, whether or not extensions have been requested. This allows the engineer to grant extensions on what he considers the true grounds of delay, when the contractor may have applied on other grounds. The grounds which entitle the contractor to extensions, other than those mentioned in clause 44(1), are, principally: late instructions (clause 7), unforeseen adverse conditions (clause 12), instructions under clauses 5 or 13, delayed or unforeseen requirements (clause 14), other contractors (clause 31), suspension of work (clause 40), and non-possession (clause 42).

The contractor's failure to complete the work by the date or extended date for completion entitles the employer to deduct liquidated damages as specified in the Appendix. Clause 47 contains extensive provisions governing the deduction of liquidated damages for the whole or for specified sections of the work. In the event of the work being delayed by a default on the part of the employer which does not permit an extension to be granted, the liquidated damages clause becomes unenforceable (see Chapter 9). While the general words "other special circumstances" may not allow extension, a new sub-clause 44(1)(e) now expressly covers delay due to the employer's default.

POWERS AND REMEDIES OF THE ENGINEER AND EMPLOYER

Many of these have already been dealt with. The engineer has a sweeping power under clause 13(1) to give instructions and directions on any matter connected with the works. Under clauses

13(2) and 14 the engineer has powers to control the contractor's methods of work. Clause 14 also deals with the question of programming the works; sub-clauses (1) to (4) require the contractor to provide a programme acceptable to the engineer, and empowers the engineer to require a revision should the work fall into delay.

Note that the programme is not a contract document and does not strictly bind either party. It is intended to monitor performance of other obligations required to be carried out timeously. As an additional remedy for slow progress, the engineer may require the contractor to specify steps to expedite the work under clause 46, which also contains provisions for agreed acceleration:

> "**46(1)** If for any reason which does not entitle the Contractor to an extension of time the rate of progress of the Works or any Section is at any time in the opinion of the Engineer too slow to ensure substantial completion by the time or extended time for completion prescribed by Clauses 43 and 44 as appropriate, or the revised time for completion agreed under sub-clause (3) of this clause, the Engineer shall notify the Contractor in writing and the Contractor shall thereupon take such steps as are necessary and to which the Engineer may consent to expedite the progress so as substantially to complete the Works or such Section by that prescribed time or extended time. The Contractor shall not be entitled to any additional payment for taking such steps.
>
> **(2)** If as a result of any notice given by the Engineer under sub-clause (1) of this Clause the Contractor shall seek the Engineer's permission to do any work on Site at night or on Sundays such permission shall not be unreasonably refused.
>
> **(3)** If the Contractor is requested by the Employer or the Engineer to complete the Works or any Section within a revised time being less than the time or extended time for completion prescribed by Clauses 43 and 44 as appropriate and the Contractor agrees so to do then any special terms and conditions of payment shall be agreed between the Contractor and the Employer before any such action is taken."

In default of compliance by the contractor the employer's positive powers are limited to deduction of liquidated damages or determination (clauses 47, 63).

The engineer's most important express power under the contract is to vary the works:

> "**51(1)** The Engineer
>
> **(a)** shall order any variation to any part of the Works that is in his opinion necessary for the completion of the Works and
>
> **(b)** may order any variation that for any other reason shall in his opinion be desirable for the completion and/or improved functioning of the Works.

Such variations may include additions omissions substitutions alterations changes in quality form character kind position dimension level or line and changes in any specified sequence method or timing of construction required by the Contract and may be ordered during the Defects Correction Period.

(2) All variations shall be ordered in writing but the provisions of Clause 2(6) in respect of oral instructions shall apply.

(3) No variation ordered in accordance with sub-clauses (1) and (2) of this Clause shall in any way vitiate or invalidate the Contract but the value (if any) of all such variations shall be taken into account in ascertaining the amount of the Contract Price except to the extent that such variation is necessitated by the Contractor's default.

(4) No order in writing shall be required for increase or decrease in the quantity of any work where such increase or decrease is not the result of an order given under this Clause but is the result of the quantities exceeding or being less than those stated in the Bill of Quantities."

The exception at the end of sub-clause (3) is new to the sixth edition and is intended to prevent a contractor who is in default taking advantage of a chance instruction which could be construed as a variation order. Note that the engineer's powers include varying the works and also ordering changes in the specified sequence, method or timing of construction required by the contract. The obligation to order a variation under sub-clause (1)(a) is of limited ambit. It may apply where the work becomes impossible to carry out in accordance with express requirements of the contract, such as an incorporated method of working (see above and Chapter 9). Valuation of variations and of changes in quantities of the work are dealt with in clauses 52 and 56(2).

Sub-contractors

Clauses 3 and 4 deal with the question of vicarious performance. Clause 3 prevents assignment (by either party) without consent. Clause 4 permits sub-contracting of any parts of the work.

"3 Neither the Employer nor the Contractor shall assign the Contract or any part thereof or any benefit or interest therein or thereunder without the prior written consent of the other party which consent shall not unreasonably be withheld.

4(1) The Contractor shall not sub-contract the whole of the Works without the prior written consent of the Employer.

(2) Except where otherwise provided the Contractor may sub-contract any part of the Works or their design. The extent of the work to be sub-contracted and the name and address of the sub-contractor must be notified in writing to the Engineer prior to

the sub-contractor's entry on to the Site or in the case of design on appointment.

(3) The employment of labour only sub-contractors does not require notification to the Engineer under sub-clause (2) of this Clause.

(4) The Contractor shall be and remain liable under the Contract for all work sub-contracted under this Clause and for acts defaults or neglects of any sub-contractor his agents servants or workpeople.

(5) The Engineer shall be at liberty after due warning in writing to require the Contractor to remove from the Works any sub-contractor who mis-conducts himself or is incompetent or negligent in the performance of his duties or fails to conform with any particular provisions with regard to safety which may be set out in the Contract or persists in any conduct which is prejudicial to safety or health and such sub-contractor shall not be again employed upon the Works without the permission of the Engineer."

The contractor is to remain fully liable for domestic sub-contractors. Where the contract requires specific work to be carried out by a nominated sub-contractor, clauses 58 and 59 lay down extensive provisions governing the parties' rights, particularly in regard to default by the sub-contractor. Such work will be designated as either a provisional sum or a prime cost item. In regard to these the engineer's powers are as follows:

"58(1) In respect of every Provisional Sum the Engineer may order either or both of the following

(a) work to be executed or goods materials or services to be supplied by the Contractor the value thereof being determined in accordance with Clause 52 and included in the Contract Price

(b) work to be executed or goods materials or services to be supplied by a Nominated Sub-Contractor in accordance with Clause 59.

(2) In respect of every Prime Cost Item the Engineer may order either or both of the following

(a) subject to Clause 59 that the Contractor employ a sub-contractor nominated by the Engineer for the execution of any work or the supply of any goods materials or services included therein

(b) with the consent of the Contractor that the Contractor himself execute any such work or supply any such goods, materials or services in which event the Contractor shall be paid in accordance with the terms of a quotation submitted by him and accepted by the Engineer or in the absence thereof the value shall be determined in accordance with Clause 52 and included in the Contract Price.

(3) If in connection with any Provisional Sum or Prime Cost Item the services to be provided include any matter of design or

specification of any part of the Permanent Works or of any equipment or plant to be incorporated therein such requirement shall be expressly stated in the Contract and shall be included in any Nominated Sub-contract. The obligation of the Contractor in respect thereof shall be only that which has been expressly stated in accordance with this sub-clause."

With regard to design, the contract must contain an express provision to make the contractor liable for the design of the permanent works (clause 8(2)). It frequently happens that sub-contractors are chosen for nomination because they can carry out specialist designs. Clause 58(3) requires this obligation to appear also in the sub-contract before the main contractor can be made liable. The employer should, in such a case, also consider taking a direct warranty from the sub-contractor (see Chapter 9).

Clause 59 contains extensive provisions in regard to nominated sub-contractors, covering the following matters:

(1) terms which the contractor is entitled to have in any nominated sub-contract;
(2) powers available to the engineer if the parties to the sub-contract cannot agree terms;
(3) payments in respect of nominated sub-contract work, including the power to make direct payment to the sub-contractor;
(4) consequences of termination of a nominated sub-contract;
(5) limitation of the contractor's liability in the event of default by a nominated sub-contractor.

The last two of these matters are of great concern to the employer, and potential benefit to the contractor. Clause 59 provides that, where a nominated sub-contract is terminated, the engineer must either make a re-nomination or order a variation, in the first instance at the employer's expense. Where the termination was with the engineer's consent, clause 59(4) provides for the contractor to attempt recovery of the employer's loss, subject to the employer reimbursing any unrecovered expenses. These provisions are based on clauses in the JCT management contract (particularly clause 3.21). It is questionable whether they provide any real benefit to the employer.

In regard to default by a nominated sub-contractor which does not lead to termination, the main contractor is fully liable:

"59(3) Except as otherwise provided in Clause 58(3) the Contractor shall be as responsible for the work executed or goods materials or services supplied by a Nominated Sub-Contractor employed by him as if he had himself executed such work or supplied such goods materials or services."

There is, however, a restriction on withholding sums on account of default by a nominated sub-contractor:

"**60(8)**   The Engineer shall have power to omit from any certificate the value of any work done goods or materials supplied or services rendered with which he may for the time being be dissatisfied and for that purpose or for any other reason which to him may seem proper may by any certificate delete correct or modify any sum previously certified by him. Provided that

(a)   the Engineer shall not in any interim certificate delete or reduce any sum previously certified in respect of work done goods or materials supplied or services rendered by a Nominated Sub-Contractor if the Contractor shall have already paid or be bound to pay that sum to the Nominated Sub-Contractor;

(b)   if the Engineer in the final certificate shall delete or reduce any sum previously certified in respect of work done goods or materials supplied or services rendered by a Nominated Sub-Contractor which sum shall have been already paid by the Contractor to the Nominated Sub-Contractor the Employer shall reimburse to the Contractor the amount of any sum overpaid by the Contractor to the Sub-Contractor in accordance with the certificates issued under sub-clause (2) of this Clause which the Contractor shall be unable to recover from the Nominated Sub-Contractor together with interest thereon at the rate stated in sub-clause (7) of this Clause from 28 days after the date of the final certificate issued under sub-clause (4) of this Clause until the date of such reimbursement."

Clause 60(1) to (4) is set out below. Clause 60(7) (not printed) provides for compound interest on unpaid sums at two per cent above base lending rate.

Default by the contractor

In the event that the contractor fails to comply with the contract, the employer has (in addition to the sanctions discussed below) some degree of security against loss. As to the work done and materials supplied, the employer withholds a percentage of the value of work done until completion (see under "Payment" below). The employer also has potentially valuable rights over the contractor's plant and materials on the site.

"**53(1)**   All Contractor's Equipment Temporary Works materials for Temporary Works or other goods or materials owned by the Contractor shall when on Site be deemed to be the property of the Employer and shall not be removed therefrom without the written consent of the Engineer which consent shall not unreasonably be withheld where the items in question are no longer immediately required for the purposes of the completion of the Works.

(2) The Employer shall not at any time be liable save as mentioned in Clauses 22 and 65 for the loss of or damage to any Contractor's Equipment Temporary Works goods or materials.

(3) If the Contractor fails to remove any of the said Contractor's Equipment Temporary Works goods or materials as required by Clause 33 within such reasonable time after completion of the Works as the Engineer may allow then the Employer may sell or otherwise dispose of such items. From the proceeds of the sale of any such items the Employer shall be entitled to retain any costs or expenses incurred in connection with their sale and disposal before paying the balance (if any) to the Contractor."

The deemed vesting of the contractor's plant and materials was held in *Re Cosslett (Contractors) Ltd*[3] to be ineffective to pass title to the employer. However, upon the removal of the contractor from site, clause 63 was held effective in giving the employer the right to retain and use the plant to complete the work. As to the effect of insolvency see Chapter 10.

Some of the sanctions exercisable on the default of the contractor are discussed above. If the contractor does not make adequate progress the engineer may require steps to be taken to expedite progress under clause 46, and require a revised programme under clause 14. If the works are not completed by the date or extended date for completion the employer may deduct liquidated damages. Where defective work is done the engineer has important powers under clause 39 to order its removal and proper replacement. In respect of all these powers, the choice remains with the contractor to obey the engineer's instructions or pay for his default.

Where the contractor's default is more serious, or when he becomes unable to perform the contract, the employer may become entitled to determine the contractor's employment.

"63(1) If

  (a) the Contractor shall be in default in that he

      (i) becomes bankrupt or has a receiving order or administration order made against him or presents his petition in bankruptcy or makes an arrangement with or assignment in favour of his creditors or agrees to carry out the Contract under a committee of inspection of his creditors or (being a corporation) goes into

---

[3] [1998] Ch. 495.

liquidation (other than a voluntary liquidation for the purposes of amalgamation or reconstruction) or

(ii) assigns the Contract without the consent in writing of the Employer first obtained or

(iii) has an execution levied on his goods which is not stayed or discharged within 28 days or

(b) the Engineer certifies in writing to the Employer with a copy to the Contractor that in his opinion the Contractor

(i) has abandoned the Contract or

(ii) without reasonable excuse has failed to commence the Works in accordance with Clause 41 or has suspended the progress of the Works for 14 days after receiving from the Engineer written notice to proceed or

(iii) has failed to remove goods or materials from the Site or to pull down and replace work for 14 days after receiving from the Engineer written notice that the said goods materials or work have been condemned and rejected by the Engineer or

(iv) despite previous warnings by the Engineer in writing is failing to proceed with the Works with due diligence or is otherwise persistently or fundamentally in breach of his obligations under Contract

Then the Employer may after giving seven days' notice in writing to the Contractor specifying the default enter upon the Site and the Works and expel the Contractor therefrom without thereby avoiding the Contract or releasing the Contractor from any of his obligations or liabilities under the Contract. Provided that the Employer may extend the period of notice to give the Contractor opportunity to remedy the default.

Where a notice of determination is given pursuant to this sub-clause it shall be given as soon as is reasonably possible after receipt of the Engineer's certificate.

(2) Where the Employer has entered upon the Site and the Works as hereinbefore provided he may himself complete the Works or may employ any other Contractor to complete the Works and the Employer or such other Contractor may use for such completion so much of the Contractor's Equipment Temporary Works goods and materials which have been deemed to become the property of the Employer under Clauses 53 and 54 as he or they may think proper and the Employer may at any time sell any of the said Contractor's Equipment Temporary Works and unused goods and materials and apply the proceeds of sale in or towards the satisfaction of any sums due or which may become due to him from the Contractor under the Contract."

Note that the operation of this clause does not determine the contract. Clause 63(4), which remains binding on the contractor, entitles the employer to be paid the additional costs of completing the work by another contractor. Sub-clause (3) requires the contractor to assign sub-contracts to the employer after determination.

This clause is, in practice, difficult to operate. If the employer expels the contractor from the site without complying precisely with sub-clause (1), he is likely to have repudiated the contract, rendering himself liable to the contractor in damages (see Chapter 6). The clause is an example of the unnecessary obscurity which pervades much of the contract. It is not clear to what extent these remedies supersede other remedies based on the same grounds. For example, if the engineer exercises the power under clause 39(2) to bring in another contractor to re-execute defective work (see above) can notice also be given under clause 63(1)(b)(iii)?

CERTIFICATION AND PAYMENT

These provisions are vital to the contractor, who will often, during the course of the work, lay out sums of money or incur liabilities exceeding the value of his assets. The provisions for interim payment make this possible. Cash flow has been referred to by the courts as the life-blood of the industry.

Clause 60 lays down the basic monthly accounting procedure:

60(1) Unless otherwise agreed the Contractor shall submit to the Engineer at monthly intervals commencing within one month after the Works Commencement Date a statement (in such form if any as may be prescribed in the Specification) showing:

(a) the estimated contract value of the Permanent Works executed up to the end of that month

(b) a list of any goods or materials delivered to the Site for but not yet incorporated in the Permanent Works and their value

(c) a list of any of those goods or materials identified in the Appendix to the Form of Tender which have not yet been delivered to the Site but of which the property has vested in the Employer pursuant to Clause 54 and their value

(d) the estimated amounts to which the Contractor considers himself entitled in connection with all other matters for which provision is made under the Contract including any Temporary Works or Contractor's Equipment for which separate amounts are included in the Bill of Quantities

unless in the opinion of the Contractor such values and amounts together will not justify the issue of an interim certificate.

Amounts payable in respect of Nominated Sub-contracts are to be listed separately.

(2) Within 25 days of the date of delivery of the Contractor's monthly statement to the Engineer or the Engineer's Representative in accordance with sub-clause (1) of this clause the Engineer shall certify and within 28 days of the same date the Employer shall pay to the Contractor (after deducting any previous payments on account)

   (a) the amount which in the opinion of the Engineer on the basis of the monthly statement is due to the Contractor on account of sub-clauses (1)(a) and (1)(d) of this Clause less a retention as provided in sub-clause (5) of this Clause and

   (b) such amounts (if any) as the Engineer may consider proper (but in no case exceeding the percentage of the value stated in the Appendix to the Form of Tender) in respect of sub-clauses (1)(b) and (1)(c) of this Clause.

The payments become due on certification with the final date for payment being 28 days after the date of delivery of the Contractor's monthly statement.

The amounts certified in respect of Nominated Sub-contracts shall be shown separately in the certificate."

Note that retention is not deducted on the value of unfixed goods. The contractor is entitled only to the percentage of their value specified in the appendix. Clause 54 lays down conditions to be satisfied if the contractor wishes to obtain payment for materials (to be specified in the appendix) before delivery to the site. Such goods must become the property of the employer.

Clause 60(3) provides for a minimum amount for certificates. Clause 60(4) deals with the final account. Clause 60(5) regulates the deduction of retention, which is recommended not to exceed five per cent. Clause 60(6) provides for payment of the retention. Subject to reductions for sectional completion, the money is to be paid to the contractor as to half on the certificate of completion of the whole of the works, and half on expiry of the period of maintenance, subject to deduction for outstanding work. Clause 60(7) provides for payment of compound interest on certificates withheld or unpaid, at a rate of two per cent over base lending rate. Clause 60 has been amended to comply with the Housing Grants, etc., Act 1996.

The measurement of the work is dealt with in the following group of clauses:

"55(1) The quantities set out in the Bill of Quantities are the estimated quantities of the work but they are not to be taken as the actual and correct quantities of the Works to be executed by the Contractor in fulfilment of his obligations under the Contract.

(2) Any error in description in the Bill of Quantities or omission therefrom shall not vitiate the Contract nor release the Contractor from the execution of the whole or any part of the Works according to the Drawings and Specification or from any of his obligations or liabilities under the Contract. Any such error or ommission shall be corrected by the Engineer and the value of the work actually carried out shall be ascertained in accordance with Clause 52. Provided that there shall be no rectification of any errors omissions or wrong estimates in the descriptions rates and prices inserted by the Contractor in the Bill of Quantities.

56(1) The Engineer shall except as otherwise stated ascertain and determine by admeasurement the value in accordance with the Contract of the work done in accordance with the Contract.

(2) Should the actual quantities executed in respect of any item be greater or less than those stated in the Bill of Quantities and if in the opinion of the Engineer such increase or decrease of itself shall so warrant the Engineer shall after consultation with the Contractor determine an appropriate increase or decrease of any rates or prices rendered unreasonable or inapplicable in consequence thereof and shall notify the Contractor accordingly.

(3) The Engineer shall when he requires any part or parts of the work to be measured give reasonable notice to the Contractor who shall attend or send a qualified agent to assist the Engineer or the Engineer's Representative in making such measurement and shall furnish all particulars required by either of them. Should the Contractor not attend or neglect or omit to send such agent then the measurement made by the Engineer or approved by him shall be taken to be the correct measurement of the work.

Sub-clause 56(4) deals with dayworks.

57 Unless otherwise provided in the Contract or unless general or detailed description of the work in the Bill of Quantities or any other statement clearly shows to the contrary the Bill of Quantities shall be deemed to have been prepared and measurements shall be made according to the procedure set out in the 'Civil Engineering Standard Method of Measurement Second Edition 1985' approved by the Institution of Civil Engineers and the Federation of Civil Engineering Contractors in association with the Association of Consulting Engineers or such later or amended edition thereof as may be stated in the Appendix to the Form of Tender to have been adopted in its preparation."

These clauses mean that the contract is subject to remeasurement, *i.e.* the contractor is to be paid for the actual quantities of work executed at the contract rates, which may themselves be varied under clause 56(2). The effect of clauses 57 and 55(2) is that work shown on the drawings or in the specification but omitted from the bill, contrary to the Standard Method, is to be paid for as an extra.

The amounts to which the contractor is entitled under clause 60(1)(d) "in connection with all other matters for which provision is made under the contract" depend on many clauses including 7(3), 12(3), 13(3), 14(6), 31(2), 40(1), 42(2) and 52. The most important is the provision for valuing variations.

> "**52(1)** The value of all variations ordered by the Engineer in accordance with Clause 51 shall be ascertained by the Engineer after consultation with the Contractor in accordance with the following principles
>
>   (a) where work is of similar character and executed under similar conditions to work priced in the Bill of Quantities it shall be valued at such rates and prices contained therein as may be applicable
>
>   (b) where work is not of a similar character or is not executed under similar conditions or is ordered during the Defects Correction Period the rates and prices in the Bill of quantities shall be used as the basis for valuation so far as may be reasonable failing which a fair valuation shall be made.
>
> Failing agreement between the Engineer and the Contractor as to any rate or price to be applied in the valuation of any variation the Engineer shall determine the rate or price in accordance with the foregoing principles and he shall notify the Contractor accordingly.
>
>   (2) If the nature or amount of any variation relative to the nature or amount of the whole of the contract work or to any part thereof shall be such that in the opinion of the Engineer or the Contractor any rate or price contained in the Contract for any item of work is by reason of such variation rendered unreasonable or inapplicable either the Engineer shall give to the Contractor or the Contractor shall give the Engineer notice before the varied work is commenced or as soon thereafter as is reasonable in all the circumstances that such rate or price should be varied and the Engineer shall fix such rate or price as in the circumstances he shall think reasonable and proper."

Note that the engineer's power to vary the contract rates under sub-clause (2) applies "to any rate or price contained in the contract for any item of work" and is not limited to the work which is varied. A claim under clause 52(2) is the nearest equivalent to a claim for loss and/or expense under clause 26 of the JCT form. The operation of clause 52(1) and (2) was considered recently in *Henry Boot v. Alstom*.[4] Clause 52(3) enables the engineer to order additional or substituted work to be executed on a daywork basis.

---

[4] [1999] B.L.R. 123.

Clause 52(4) provides for notices to be given of claims under any clause of the contract (see below). In addition to sums due for varied work, there are many other provisions which may entitle the contractor to claim further payment.

In addition to payment at the rates and prices fixed under the contract, there may be incorporated optional fluctuations clauses. Separate clauses are available for use where the contract consists primarily of civil engineering work (the C.E. clause) or of fabricated structural steelwork (the FSS clause). Both may be used, together with a regulating clause referred to as CE/FSS. These clauses operate exclusively on formula adjustments. The amounts payable therefore depend only on the net value of works executed and index changes, and are independent of the actual constituents of the work. The current index figures, from which payments are calculated, are defined as:

> "(2)(c) 'Current Index Figure' shall mean the appropriate Final Index Figure to be applied in respect of any certificate issued or due to be issued by the Engineer pursuant to Clause 60 and shall be the appropriate Final Index Figure applicable to the date 42 days prior to
>> (i)   the due date (or extended date) for completion or
>> (ii) the date certified pursuant to Clause 48 of completion of the whole of the Works or
>> (iii) the last day of the period to which the certificate relates; whichever is the earliest."

Payments are thus frozen at the prices applying when the work should be completed. There are no requirements such as those of clause 39.5.8 of the JCT form (see Chapter 12).

Claims

Additional payments which may be due to the contractor under provisions other than those covering valuation of the work done are often referred to as "claims". The term may also include damages for breach of contract; but this section is limited to consideration of sums due under the contract. The principal claims available to the contractor are set out below.

Clause 7(4) entitles the contractor to payment in respect of the late issue of drawings or instructions:

> "7(4)(a) If by reason of any failure or inability of the Engineer to issue at a time reasonable in all the circumstances Drawings Specifica-

tions or instructions requested by the Contractor and considered necessary by the Engineer in accordance with sub-clause (1) of this Clause the Contractor suffers delay or incurs cost then the Engineer shall take such delay into account in determining any extension of time to which the Contractor is entitled under Clause 44 and the Contractor shall subject to Clause 52(4) be paid in accordance with Clause 60 the amount of such cost as may be reasonable.

(b) If the failure of the Engineer to issue any Drawing Specification or instruction is caused in whole or in part by the failure of the Contractor after due notice in writing to submit drawings specifications or other documents which he is required to submit under the Contract the Engineer shall take into account such failure by the Contractor in taking any action under sub-clause (4)(a) of this Clause."

Clause 12 allows the contractor to claim additional payment for work to overcome unforeseen physical conditions or artificial obstructions:

"**12(1)** If during the execution of the Works the Contractor shall encounter physical conditions (other than weather condition or conditions due to weather conditions) or artificial obstructions which conditions or obstructions could not in his opinion reasonably have been foreseen by an experienced Contractor the Contractor shall as early as practicable give written notice thereof to the Engineer.

(2) If in addition the Contractor intends to make any claim for additional payment or extension of time arising from such condition or obstruction he shall at the same time or as soon thereafter as may be reasonable inform the Engineer in writing pursuant to Clause 52(4) and/or Clause 44(1) as may be appropriate specifying the condition or obstruction to which the claim relates.

(3) When giving notification in accordance with sub-clauses (1) and (2) of this Clause or as soon as practicable thereafter the Contractor shall give details of any anticipated effects of the condition or obstruction the measures he has taken is taking or is proposing to take their estimated cost and the extent of the anticipated delay in or interference with the execution of the Works.

(4) Following receipt of any notification under sub-clauses (1) or (2) or receipt of details in accordance with sub-clause (3) of this Clause the Engineer may if he thinks fit, *inter alia,*

   (a) require the Contractor to investigate and report upon the practicality cost and timing of alternative measures which may be available

   (b) give written consent to measures notified under sub-clause (3) of this Clause with or without modification

   (c) give written instructions as to how the physical conditions or artificial obstructions are to be dealt with

(d) order a suspension under Clause 40 or a variation under Clause 51.

(5) If the Engineer shall decide that the physical conditions or artificial obstructions could in whole or in part have been reasonably foreseen by an experienced Contractor he shall so inform the Contractor in writing as soon as he shall have reached that decision but the value of any variation previously ordered by him pursuant to sub-clause (4)(d) of this Clause shall be ascertained in accordance with Clause 52 and included in the Contract Price.

(6) Where an extension of time or additional payment is claimed pursuant to sub-clause (2) of this Clause the Engineer shall if in his opinion such conditions or obstructions could not reasonably have been foreseen by an experienced Contractor determine the amount of any costs which may reasonably have been incurred by the Contractor by reason of such conditions or obstructions together with a reasonable percentage addition thereto in respect of profit and any extension of time to which the Contractor may be entitled and shall notify the Contractor accordingly with a copy to the Employer. The contractor shall subject to clause 52(4) be paid in accordance with clause 60 the amount so determined."

Clause 12 is widely used as a vehicle for presenting claims for additional cost arising from constructional difficulties. Claims usually relate to sub-soil conditions, but the terms of sub-clause (1) are not limited to any particular sort of condition or obstruction. The Court of Appeal upheld an award in which a transient combination of soil strength and stress was held to constitute a physical condition within clause 12.[5] A further claim provision of general application is clause 13(3):

"**13**(3) If in pursuance of Clause 5 or sub-clause (1) of this Clause the Engineer shall issue instructions which involve the Contractor in delay or disrupt his arrangements or methods of construction so as to cause him to incur cost beyond that reasonably to have been foreseen by an experienced Contractor at the time of tender then the Engineer shall take such delay into account in determining any extension of time to which the Contractor is entitled under Clause 44 and the Contractor shall subject to Clause 52(4) be paid in accordance with Clause 60 the amount of such cost as may be reasonable except to the extent that such delay and extra cost result from the Contractor's default. Profit shall be added thereto in respect of any additional permanent or temporary work. If such instructions require any variation to any part of the Works the same shall be deemed to have been given pursuant to Clause 51."

---

[5] *Humber Oil v. Harbour & General* (1991) 59 B.L.R. 1; and see [1995] Con. L.Yb. 98.

An instruction under clause 5 is one given to explain and adjust an ambiguity in the contract. Clause 13(1) refers to instructions on any matter connected with the works (whether mentioned in the contract or not). In practice, instructions are often given without specifying any clause. It is thought that an instruction is given in pursuance of clause 13(1) only if it cannot be given under any other clause. But the contrary is arguable, *i.e.* that any instruction gives rise to a claim under clause 13(3). An instruction under clause 13(1) need not be in writing.

Clause 14(8) allows claims in respect of the engineer's requirements in regard to the methods of construction (printed above). Clause 31 requires the contractor to give facilities for other contractors and provides for payment of unforeseen cost:

> "**31(1)** The Contractor shall to accordance with the requirements of the Engineer or Engineer's Representative afford all reasonable facilities for any other Contractors employed by the Employer and their workmen and for the workmen of the Employer and of any other properly authorised authorities or statutory bodies who may be employed in the execution on or near the Site of any work not in the Contract or of any contract which the Employer may enter into in connection with or ancillary to the Works.
>
> **(2)** If compliance with sub-clause (1) of this Clause shall involve the Contractor in delay or cost beyond that reasonably to be foreseen by an experienced Contractor at the time of tender then the Engineer shall take such delay into account in determining any extension of time to which the Contractor is entitled under Clause 44 and the Contractor shall subject to Clause 52(4) be paid in accordance with Clause 60 the amount of such cost as may be reasonable. Profit shall be added thereto in respect of any additional permanent or temporary work."

Claims for extra cost arising from suspension of the works or non-possession are provided by clauses 40(1) and 42(1), set out above. In respect of all such claims, the cost recoverable may include overheads and interest incurred:

> "**1(5)** The word 'cost' when used in the Conditions of Contract means all expenditure properly incurred or to be incurred whether on or off the Site including overhead finance and other charges properly allocatable thereto but does not include any allowance for profit."

The addition of profit is dealt with in individual clauses: see clauses 12(6), 13(3) and 31(2) above. All the above claims (and others) are subject to clause 52(4) which requires the contractor to give notice in writing "as soon as reasonably possible" (note that stricter notices are required for claims under clauses 52 and 56). The contractor's right to final and interim payments in respect of claims are governed by paragraphs (e) and (f):

"52(4)(a) If the Contractor intends to claim a higher rate or price than the one notified to him by the Engineer pursuant to sub-clauses (1) and (2) of this Clause or Clause 56(2) the Contractor shall within 28 days after such notification give notice in writing of his intention to the Engineer.

(b) If the Contractor intends to claim any additional payment pursuant to any Clause of these Conditions other than sub-clauses (1) and (2) of this Clause or Clause 56(2) he shall give notice in writing of his intention to the Engineer as soon as may be reasonable and in any event within 28 days after the happening of the events giving rise to the claim. Upon the happening of such events the Contractor shall keep such contemporary records as may reasonably be necessary to support any claim he may subsequently wish to make.

(c) Without necessarily admitting the Employer's liability the Engineer may upon receipt of a notice under this Clause instruct the Contractor to keep such contemporary records or further contemporary records as the case may be as are reasonable and may be material to the claim of which notice has been given and the Contractor shall keep such records. The Contractor shall permit the Engineer to inspect all records kept pursuant to this Clause and shall supply him with copies thereof as and when the Engineer shall so instruct.

(d) After the giving of a notice to the Engineer under this Clause the Contractor shall as soon as is reasonable in all the circumstances send to the Engineer a first interim account giving full and detailed particulars of the amount claimed to that date and of the grounds upon which the claim is based. Thereafter at such intervals as the Engineer may reasonably require the Contractor shall send to the Engineer further up to date accounts giving the accumulated total of the claim and any further grounds upon which it is based.

(e) If the Contractor fails to comply with any of the provisions of this Clause in respect of any claim which he shall seek to make then the Contractor shall be entitled to payment in respect thereof only to the extent that the Engineer has not been prevented from or substantially prejudiced by such failure in investigating the said claim.

(f) The Contractor shall be entitled to have included in any interim payment certified by the Engineer pursuant to Clause 60 such amount in respect of any claim as the Engineer may consider due to the Contractor provided that the Contractor shall have supplied sufficient particulars to enable the Engineer to determine the amount due. If such particulars are insufficient to substantiate the whole of the claim the Contractor shall be entitled to payment in respect of such part of the claim as the particulars may substantiate to the satisfaction of the Engineer."

Note that paragraph (e) sets no time limit on the engineer's investigation. The contractor may re-submit his claim if further information comes to light.

Final accounting

After the completion of the work and correction of notified defects and omissions, the engineer is required to issue the Defects Correction Certificate. This is followed by vetting the contractor's accounts and issuing a final certificate.

"**61(1)** Upon the expiry of the Defects Correction Period or where there is more than one such period upon the expiration of the last of such periods and when all outstanding work referred to under Clause 48 and all work of repair amendment reconstruction rectification and making good of defects imperfections shrinkages and other faults referred to under Clauses 49 and 50 shall have been completed the Engineer shall issue to the Employer (with a copy to the Contractor) a Defects Correction Certificate stating the date on which the Contractor shall have completed his obligations to construct and complete the Works to the Engineer's satisfaction.

(2) The issue of the Defects Correction Certificate shall not be taken as relieving either the Contractor or the Employer from any liability the one towards the other arising out of or in any way connected with the performance of their respective obligations under the Contract.

60(3) Until the whole of the Works has been certified as substantially complete in accordance with clause 48 the Engineer shall not be bound to issue an interim certificate for a sum less than that stated in the Appendix to the Form of Tender but thereafter he shall be bound to do so and the certification and payment of amounts due to the Contractor shall be in accordance with the time limits contained in this clause.

(4) Not later than three months after the date of the Defects Correction Certificate the Contractor shall submit to the Engineer a statement of final account and supporting documentation showing in detail the value in accordance with the Contract of the Works executed together with all further sums which the Contractor considers to be due to him under the Contract up to the date of the Defects Correction Certificate.

Within three months after receipt of this final account and of all information reasonably required for its verification the Engineer shall issue a certificate stating the amount which in his opinion is finally due under the Contract from the Employer to the Contractor or from the Contractor to the Employer as the case may be up to the date of the Defects Correction Certificate and after giving credit to the Employer for all amounts previously paid by the Employer and for all sums to which the Employer is entitled under the Contract.

Such amount shall subject to Clause 47 be paid to or by the Contractor as the case may require. The paymnet becomes due on certification. The final date for payment is 28 days later."

Neither the Defects Correction certificate nor the final certificate constitutes a binding approval of the work. This should be compared to the final certificate under clause 30.8 of the JCT form.

DISPUTES

The ICE Conditions have traditionally provided a multi-tier system for dispute resolution, including mandatory reference of a dispute to the Engineer and optional reference to conciliation. As a result of the Housing Grants, etc., Act 1996, it has been necessary to add the further tier of Adjudication. At first sight, it would seem doubtful that the mandatory reference to the Engineer could be maintained in the light of the statutory entitlement under clause 108 of the Act to refer a dispute for adjudication "at any time". The way in which the drafting body has sought to avoid this difficulty is to re-define "dispute" as a matter in respect of which the Engineer has given a decision which is unacceptable to one party (or in respect of which an adjudicator has given a decision which has not been given accepted). Clause 66(2) re-defines what was formerly a dispute as a "matter of dissatisfaction", which must be referred to the Engineer (in the traditional way) for his decision. Thereafter, either party may initiate optional conciliation under clause 66(5). Alternatively, (or after an unsuccessful conciliation), either party may refer the dispute to adjudication.

After the decision of the adjudicator a dispute may be referred to arbitration in the traditional way (with no further time limit). If adjudication is not invoked arbitration may follow the decision of the engineer or conciliator. The arbitration may be conducted optionally in accordance with the ICE Arbitration Procedure 1997 or the Construction Industry Model Arbitration Rules (see Chapter 3). The arbitration provision in clause 66(9) does not apply to the failure to give effect to a decision of an adjudicator, which may accordingly be the subject of an application to the court for summary judgment.[6] Some doubt must exist as to whether a "matter of dissatisfaction" may not also constitute a dispute within the Housing Grants, etc., Act, which would allow the party to proceed immediately to adjudication without waiting for the decision of the Engineer. The full clause is printed below.

Avoidance of disputes

"(1) In order to overcome where possible the causes of disputes and in those cases where disputes are likely still to arise to facilitate their clear definition and early resolution (whether by agreement or otherwise) the following procedure shall apply for the avoidance and settlement of disputes.

(2) If at any time

(a) the Contractor is dissatisfied with any act or instruction of the Engineer's Representative or any other person responsible to the Engineer or

(b) The Employer or the Contractor is dissatisfied with any decision opinion instruction direction certificate or valuation of the Engineer or with any other matter arising under or in connection with the Contract or the carrying out of the Works.
The matter of dissatisfaction shall be referred to the Engineer who shall notify his written decision to the Employer and the Contractor within one month of the reference to him.

(3) The Employer and the Contractor agree that no matter shall constitute nor be said to give rise to a dispute unless and until in respect of that matter

(a) the time for the giving of a decision by the Engineer on a matter of dissatisfaction under Clause 66(2) has expired or the decision given is unacceptable or has not been implemented and in consequence the Employer or the Contractor has served on the other and on the Engineer a notice in writing (hereinafter called the Notice of Dispute) or

(b) and adjudicator has given a decision on a dispute under Clause 66(6) and the Employer or the Contractor is not giving effect to the decision, and in consequence the other has served on him and the Engineer a Notice of Dispute
and the dispute shall be that stated in the Notice of Dispute. For the purposes of all matters arising under or connection with the Contract or the carrying out of the Works the word 'dispute' shall be construed accordingly and shall include any difference.

(4) (a) Notwithstanding the existence of a dispute following the service of a Notice under Clause 66(3) and unless the

---

[6] See *Stay of Proceedings*, Chap.3.

Contract has already been determined or abandoned the Employer and the Contractor shall continue to perform their obligations.

(b) The Employer and the Contractor shall give effect forthwith to every decision of

    (i) the Engineer on a matter of dissatisfaction given under Clause 66(2) and

    (ii) the adjudicator on a dispute given under Clause 66(6) unless and until that decision is revised by agreement of the Employer and Contractor or pursuant to Clause 66.

(5) (a) The Employer or the Contractor may at any time before service of a Notice to Refer to arbitration under Clause 66(9) by notice in writing seek the agreement of the other for the dispute to be considered under the Institution of Civil Engineers' Conciliation Procedure (1994) or any amendment or modification thereof being in force at the date of such notice.

(b) If the other party agrees to this procedure any recommendation of the conciliator shall be deemed to have been accepted as finally determining the dispute by agreement so that the matter is no longer in dispute unless a Notice of Adjudication under Clause 66(6) or a Notice to Refer to arbitration under Clause 66(9) has been served in respect of that dispute not later than 1 month after receipt of the recommendation by the dissenting party.

(6) (a) The Employer and the Contractor each has the right to refer a dispute as to a matter under the Contract for adjudication and either party may give notice in writing (hereinafter called the Notice of Adjudication) to the other at any time of his intention so to do. The adjudication shall be conducted under the Institution of Civil Engineers' Adjudication Procedure (1997) or any amendment of modification thereof being in force at the time of the said Notice.

(b) Unless the adjudicator has already been appointed he is to be appointed by a timetable with the object of securing his appointment and referral of the dispute to him within 7 days of such notice.

(c) The adjudicator shall reach a decision within 28 days of referral or such longer period as is agreed by the parties after the dispute has been referred.

(d) The adjudicator may extend the period of 28 days by up to 14 days with the consent of the party by whom the dispute was referred.

(e) The adjudicator shall act impartially.

(f) The adjudicator may take the initiative in ascertaining the facts and the law.

(7) The decision of the adjudicator shall be binding until the dispute is finally determined by legal proceedings or any arbitration (if the Contract provides for arbitration or the parties otherwise agreed to arbitration) or by agreement.

(8) The adjudicator is not liable for anything done or omitted in the discharge or purported discharge of his functions as adjudicator unless the act or omission is in bad faith and any employee or agent of the adjudicator is similarly not liable.

(9) (a) All disputes arising under or in connection with the Contract or the carrying out of the Works other than failure to give effect to a decision of an adjudicator shall be finally determined by reference to arbitration. The Party seeking arbitration shall serve on the on the other party a notice in writing (called the Notice to Refer) to refer the dispute to arbitration.

(b) Where an adjudicator has given a decision under Clause 66(6) in respect of the particular dispute the Notice to Refer must be served within three months of the giving of the decision otherwise it shall be final as well as binding.

(10) (a) The arbitrator shall be a person appointed by agreement of the parties.

(b) If the parties fail to appoint an arbitrator within one month of either party serving on the other party a notice in writing (hereinafter called the Notice to Concur) to concur in the appointment of an arbitrator the dispute shall be referred to a person to be appointed on the application of either party by the President for the time being of the Institution of Civil Engineers.

(c) If an arbitrator declines the appointment of after appointment is removed by order of a competent court or is capable of acting or dies and the parties do not within one month of the vacancy arising fill the vacancy then either party may apply to the President for the time being of the Institution of Civil Engineers to appoint another arbitrator to fill the vacancy.

(d) In any case where the President for the time being of the Institution of Civil Engineers is not able to exercise the functions conferred on him by this Clause the said functions shall be exercised on his behalf by a Vice-President for the time being of the said Institution.

(11) (a) Any reference to arbitration under this Clause shall be deemed to be a submission to arbitration within the meaning of the Arbitration Act 1996 or any statutory re-enactment or amendment thereof for the time being in force. The reference shall be conducted in accordance with the procedure set out in the Appendix to the Form of Tender or any amendment or modification thereof being in force at the time of the appointment of the arbitrator. Such arbitrator shall have full power to open up review and revise any decision opinion instruction direction certificate or valuation of the Engineer or an adjudicator.

(b) Neither party shall be limited in the arbitration to the evidence or arguments put to the Engineer or any adjudicator pursuant to Clause 66(2) or 66(6) respectively.

(c) The award of the arbitrator shall be binding on all parties.

(d) Unless the parties otherwise agree in writing any reference to arbitration may proceed notwithstanding that the Works are not then complete or alleged to be complete.

(12) (a) No decision opinion instruction direction certificate or valuation given by the Engineer shall disqualify him from being called as a witness and giving evidence before a conciliator adjudicator or arbitrator on any matter whatsoever relevant to the dispute.

(b) All matters and information placed before a conciliator pursuant to a reference under sub-clause (5) of this Clause shall be deemed to be submitted to him without prejudice and the conciliator shall not be called as witness by the parties or anyone claiming through them in connection with any adjudication arbitration or other legal proceedings arising out of or connected with any matter so referred to him."

The revised clause removes the potentially binding effect of the Engineer's decision, which has applied in all previous drafts of the clause. Under clause 66(4)(b) the parties are merely bound to "give effect forthwith" to the Engineer's decisions, which may be revised by conciliation, adjudication or arbitration at any time following the decision. Further, either party remains free to bring court proceedings, subject to prior reference to the Engineer. The other party, however, would be entitled to a mandatory stay for arbitration,[7] or alternatively would appear to have the right to require an adjudication decision irrespective of the existence of other proceedings. The only condition precedent to challenging the Engineer's decision is service of a Notice of Dispute which may, however, be served at any time.

Scotland and Northern Ireland

The form is intended for use in any part of the United Kingdom. A separate clause is included for Scotland and Northern Ireland (clause 67, as amended), which have different legal systems. In the case of Scotland, differences in the law applicable to arbitration are long-standing and permanent, in view of the adoption in Scotland of the UNCITRAL model law (see Chapter 3). Differences in relation to Northern Ireland are of less moment, particularly as the 1996 Arbitration Act now applies.

---

[7] Arbitration Act 1996, s. 9 and see Chap. 3.

CHAPTER 14

LAW OF TORT

THE law of tort is mostly to be found in the common law, but there are some important statutes. Tort can be defined as a civil wrong independent of contract; or as breach of a legal duty owed to persons generally. The practical consequences of the law of tort are concerned with the adjustment of losses. Where the elements of fault and damage exist, the law determines who should bear the resulting financial loss.

There is no complete body of general principles which applies to all torts, in the way that all contracts are governed by the same general principles. Some jurists view torts as a series of separate civil wrongs. For more practical reasons the torts discussed in this chapter are set out in separate sections. There are, however, some principles common to all or most torts, and these are discussed by way of introduction. This chapter covers those specific torts which are most relevant to the construction industry and to those professionally involved in it.

NEGLIGENCE

Negligence is by far the most important of torts for several reasons. It forms the cause of action in the majority of cases brought in tort; its scope is very wide; and it may also be an element in liability for other torts. The term negligence is also found in the context of breach of contract, for example, where an architect is alleged to have carried out negligent design or supervision. A common type of action in negligence heard in the courts is that between two or more drivers involved in a road accident. In such cases it is not infrequent for all parties to be held to be negligent in some degree.

Element of liability

The plaintiff in an action for negligence must show: (1) that the defendant owed him a duty of care; (2) that there was a breach of that duty; and (3) that recoverable damage was thereby caused. Considering the first of these elements, it is necessary to decide whether in the particular circumstances one person (the defendant) owed a duty of care to the other (the plaintiff). The classic test as to when a duty of care might arise was stated in the leading case of *Donoghue v. Stevenson*.[1] The manufacturer of ginger beer was held to owe a duty to the ultimate consumer, who found a decomposing snail in the empty bottle. The consumer could not sue in contract because the ginger beer had been purchased by a friend, and in any event the default was that of the manufacturer, not the seller. In a celebrated judgment Lord Atkin held:

> "The rule that you are to love your neighbour becomes in law, you must not injure your neighbour; and the lawyer's question who is my neighbour? receives a restricted reply. You must take reasonable care to avoid acts or omissions which you can reasonably foresee would be likely to injure your neighbour. Who, then, in law is my neighbour? The answer seems to be—persons who are so closely and directly affected by my act that I ought reasonably to have them in contemplation as being so affected when I am directing my mind to the acts or omissions which are called in question."

The law of tort remained comparatively dormant for some time after Donoghue's case and it was not until *Dutton v. Bognor Regis UDC*[2] that the courts began to grapple with the questions which subsequently came to dominate this area of law, namely the circumstances in which a duty would arise and the inter-relationship between such duties and the type of damage suffered.

The expansion period

Whenever a change in the law is argued for, the "floodgates" argument is raised, *i.e.* if this claim is allowed, it will be followed by a flood of other claims. *Donoghue v. Stevenson* was not followed by any flood but *Dutton's* case was quite the opposite. That case produced not merely a flood of litigation but also an almost unprecedented upheaval in the common law, which ultimately led to the House of Lords deciding to disapprove its own previous decisions. A brief account of this piece of legal history is necessary, not least because there will undoubtedly be more development

---

[1] [1932] A.C. 562.
[2] [1972] 1 Q.B. 373.

before this branch of the law can be said to be settled. The episode also illustrates the perils of expansion and contraction of the common law: during the expansion period many claims will be brought, which, if still pending during the contraction, will be found unsustainable with disastrous consequences for the unfortunate plaintiff, who may have to withdraw his claim and pay costs. Thus, although in theory the common law does not "change" but is restated, the practical effect is not only that the law does change, but that it changes retrospectively, which would be regarded as constitutionally improper for changes in statute law.

In *Dutton's* case the plaintiff, a second purchaser of a house built on a rubbish tip, was unable to sue in contract. Proceedings were brought in tort against the original builder and against the local authority who had approved the plans and inspected the work on site. The case against the builder was settled but Mrs Dutton proceeded against the local authority. In the Court of Appeal it was argued that the council owed no duty and that there was no physical damage to found an action in tort. It was further pointed out that the council should not be held liable because the builder was not liable in tort. Lord Denning M.R. rejected all these arguments holding not only that the council was liable but, incidentally, the builder would also have been liable in tort.

Dutton's case had two important consequences. First, claims in tort were potentially available against parties not hitherto thought liable; and secondly, claims in tort could be brought as an alternative to claims in contract. The importance of the second point is that claims in contract become statute barred within (usually) six years of the date of the breach, whereas claims in tort do not arise until damage is suffered. In addition to more claims being available, *Dutton* created the possibility of bringing claims which would hitherto have been long barred by limitation. Thus, the flood produced by *Dutton* was two-fold. A rush of cases was brought in tort against local authorities, contractors and professionals of all sorts; and compounding this flood, many of the claims were extremely stale. Plaintiffs alleged that, although a house was built many years before, damage had occurred only within the period of six years before issuing the writ. This development led to yet another series of cases on the question when the cause of action in tort arose. In regard to latent defects in buildings, the doctrine was developed that the cause of action arose when the plaintiff ought reasonably to have become aware of the existence of the damage. The courts had, therefore, to consider many stale claims.

In *Anns v. LB Merton*[3] the House of Lords, for the first time, considered the principles arising from *Dutton's* case. *Anns'* case concerned allegations of negligence against the local authority's building inspectors, the primary issue being whether the claim was statute-barred. The House of Lords approved, with some modification, the decision in *Dutton* so that the flood of litigation continued. The judgment of the House of Lords in *Anns* restated the circumstances in which a duty of care would arise. Lord Wilberforce set out a two-stage test:

> "First, one has to ask whether as between the alleged wrongdoer and the person who has suffered damage there is a sufficient relationship of proximity or neighbourhood such that, in the reasonable contemplation of the former, carelessness on his part may be likely to cause damage to the latter-in which case a prima facie duty of care arises. Secondly, if the first question is answered affirmatively, it is necessary to consider whether there are any considerations which ought to negative or reduce or limit the scope of the duty or the class of person to whom it is owed or the damages to which a breach of it may give rise."

On limitation it was held that the cause of action arose only when the state of the building was such that there was present or imminent danger to the health or safety of occupiers. This point was subsequently developed in a number of other cases.

The next step which represented the zenith in the development of the law of tort was the case of *Junior Books v. Veitchi.*[4] This was a Scottish case (the Scots law of delict was for the purpose of the appeal regarded as identical with the English law of tort), in which the plaintiff owners of a factory brought a claim in tort for economic loss against a nominated sub-contractor who was alleged negligently to have installed flooring, the loss representing the financial consequences of having to replace the defective floor. The House of Lords allowed the claim, holding that there was no good reason to restrict the loss recoverable to the cost of making good physical damage. Lord Brandon delivered a strong dissenting judgment which has subsequently achieved greater currency than the case itself.

The retrenchment period

Within months of *Junior Books*, the House of Lords had to consider again the effect on limitation of the law of tort. In the

---

[3] [1978] A.C. 728.
[4] [1983] A.C. 520.

leading case of *Pirelli v. Oscar Faber*,[5] the House held that the cause of action against consulting engineers for negligent design which had resulted in cracks in a chimney arose when that damage came into existence and not at the later date when the cracks were or should with reasonable diligence have been discovered. The case was complicated because the Law Reform Commission proposals which were to lead to the Latent Damage Act 1986 indicated that if the House decided that the claim was statute-barred the law would then be changed back by the new Act.

The next opportunity to reconsider the law came in a number of cases dealing with local authorities. In *Peabody v. Parkinson and Lambeth*[6] the Court of Appeal reversed the findings of an Official Referee that the local authority had been negligent in permitting the installation of drainage not complying with the relevant byelaws. The House of Lords, in dismissing the appeal, held that the local authority owed no duty to the plaintiff owners, and that the appropriate test as to whether there was such a duty, given the existence of proximity, was whether it was just and reasonable. Thereafter, in *Investors in Industry v. South Beds D.C.*[7] the Court of Appeal, following *Peabody*, held that it would not normally be just and reasonable to impose a duty on a local authority where the building owner relied on other professional advisers, and that the duty owed by the local authority would ordinarily be limited to a duty to subsequent owners. The flood gates had thus been closed, at least against local authorities. This left, however, the enlargement in the law of tort against other parties. There followed a series of cases in which *Junior Books* was heavily criticised by the Court of Appeal. The case is now regarded as one where there was a special relationship between the parties. Other cases held that the right to sue for damages in tort was available only where the plaintiff had a proprietary or possessory right in the property damaged: it was not sufficient for the plaintiff to have a mere contractual right.[8]

The most far-reaching decision in this period of retrenchment was the case of *D. & F. Estates v. Church Commissioners*.[9] Here, the owners of a flat brought a claim in tort, *inter alia*, against Wates, who were main contractors when the block of flats was constructed, in respect of alleged negligence by their plastering sub-contractor

---

[5] [1983] 2 A.C. 1.
[6] [1985] A.C. 210.
[7] [1986] Q.B. 1034.
[8] *Candlewood v. Mitsui* [1986] A.C. 1.
[9] [1989] A.C. 177.

(who was not a party to the action). The plaster was found to be cracked and unsound, and the plaintiffs claimed the cost of renewing it. The House of Lords held that Wates were not under the duty alleged, but went on to consider the claim that would have been available had they been under such a duty. The House referred to and approved the dissenting judgment of Lord Brandon in *Junior Books* in which he said:

> "It is, however, of fundamental importance to observe that the duty of care laid down in *Donoghue v. Stevenson* was based on the existence of a danger of physical injury to persons or their property . . . the relevant property for the purpose of the wider principle on which the decision . . . was based was property other than the very property which gave rise to the danger of physical damage concerned.
>
> It has always been either stated expressly or taken for granted that an essential ingredient in the cause of action relied on was the existence of danger or the threat of danger, of physical damage to persons or their property excluding for this purpose the very piece of property from the defective condition of which such danger, or threat of danger, arises."

Lord Bridge went on to say that if the principle of *Donoghue v. Stevenson* is applied to the liability of the builder of a structure which is dangerously defective:

> "Liability can only arise if the defect remains hidden until the defective structure causes personal injury or damage to property other than the structure itself. If the defect is discovered before any damage is done, the loss sustained by the owner of the structure, who has to repair or demolish it to avoid a potential source of danger to third parties, would seem to be purely economic. Thus, if I acquire a property with a dangerously defective garden wall which is attributable to the workmanship of the original builder, it is difficult to see any basis in principle on which I can sustain an action in tort against the builder for the cost of either repairing or demolishing the wall."

The question of the existence of a duty of care, considered in relation to local authorities in the *Peabody* and *Investors* cases, arose again in the case of *Caparo Industries v. Dickman*.[10] The plaintiff had purchased shares in a public company. After publication of statutory accounts prepared by the defendant, the plaintiff, relying on the report, decided to purchase more shares and eventually launched a successful takeover of the company. Subsequently errors were found in the accounts, and the plaintiff claimed damages in tort from the accountants. *Hedley Byrne v. Heller*[11]

---

[10] [1990] 2 A.C. 605.
[11] [1964] A.C. 465.

established that such an action was maintainable, but the particular issue in the *Caparo* case was whether a duty was owed by accountants to shareholders and to investors. The Court of Appeal held that a duty was owed to shareholders only. The House of Lords allowed the defendant's appeal, holding that no duty at all was owed. Lord Bridge described the test as follows:

> "What emerges is that, in addition to the foreseeability of damage, necessary ingredients in any situation giving rise to a duty of care are that there should exist between the party owing the duty and the party to whom it is owed a relationship characterised by the law as one of 'proximity' or 'neighbourhood' and that the situation should be one in which the court considers it fair, just and reasonable that the law should impose a duty of a given scope upon the one party for the benefit of the other."

The latest and decisive stage in the retrenchment of the law of tort in regard to buildings occurred in the case of *Murphy v. Brentwood D.C.*[12] In that case, the Court of Appeal had upheld the decision of the Official Referee, allowing recovery by a plaintiff houseowner who discovered cracks which were attributable to the negligent passing of plans by the local authority. Instead of carrying out repairs, Mr Murphy sold his defective house and sued for the drop in value. The House of Lords now had the opportunity to reconsider *Anns*, and did so in an unusual court of seven Law Lords, including the Lord Chancellor. The decision, which is of far-reaching importance, established, or restated the law of negligence in the following terms:

(1) A builder owed duty of care within the principle of *Donoghue v. Stevenson*, to persons likely to suffer injury as a result of his negligence.

(2) This extended, however, only to injury caused by latent, *i.e.* undiscovered defects in the building.

(3) Where a defect came to light, whether through the existence of cracks or through a survey, expenditure on remedial work was to be regarded as pure economic loss, not recoverable in tort.

(4) Contrary to *Dutton* and *Anns*, cracks representing the manifestation of underlying defects were not to be regarded as material damage.

---

[12] [1991] 1 A.C. 398.

(5) The question whether a local authority exercising powers to secure compliance with building regulations owed any duty to owners or occupiers of the relevant building was left open.

Ordinarily, such liability in tort will be limited to injury to persons or other property, excluding the property which gave rise to the injury (*i.e.* the building itself). Lord Bridge, in the *Murphy* case, recognised two possible exceptions. The first was expressed in the following passage, which also summarises the general extent of the builder's liability:

> "If a builder erects a structure containing a latent defect which renders it dangerous to persons or property, he will be liable in tort for injury to persons or damage to property resulting from that dangerous defect. But if the defect becomes apparent before any injury or damage has been caused, the loss sustained by the building owner is purely economic. If the defect can be repaired at economic cost, that is the measure of the loss. If the building cannot be repaired, it may have to be abandoned as unfit for occupation and therefore valueless. These economic losses are recoverable if they flow from breach of a relevant contractual duty, but, here again, in the absence of a special relationship of proximity they are not recoverable in tort. The only qualification I would make to this is that, if a building stands so close to the boundary of the building owner's land that after discovery of the dangerous defect it remains a potential source of injury to persons or property on neighbouring land or on the highway, the building owner ought in principle to be entitled to recover in tort from the negligent builder the cost of obviating the danger, whether by repair or demolition, so far as that cost is necessarily incurred in order to protect himself from potential liability to third parties."

The second exception arises from the possibility of damage to "other property." What is the position if the other property is part of the defective building itself? This was considered in the *D. & F.* case and dubbed the "complex structure theory." Despite wide and searching academic criticism, the House of Lords in *Murphy* appear to regard the theory as still sound. Lord Bridge re-stated his view of the theory as follows:

> "A critical distinction must be drawn here between some part of a complex structure which is said to be a 'danger' only because it does not perform its proper function in sustaining the other parts and some distinct item incorporated in the structure which positively malfunctions so as to inflict positive damage on the structure in which it is incorporated. Thus, if a defective central heating boiler explodes and damages a house or a defective electrical installation malfunctions and

sets the house on fire, I see no reason to doubt that the owner of the house, if he can prove that the damage was due to the negligence of the boiler manufacturer in the one case, or the electrical contractor in the other, can recover damages in tort on the *Donaghue v. Stevenson* principles but the position in law is entirely different where, by reason of the inadequacy of the foundations of the building to support the weight of the superstructure, differential settlement and consequent cracking occurs. Here, once the first cracks appear, the structure as a whole is seen to be defective, and the nature of the defect is known."

The above represents the position of English law as it currently stands. The developments in terms of liability have been accompanied by major changes to the law of limitation, dealt with below, which have to a large extent been left in the air as a result of the retrenchment. Nevertheless, there remain many issues still to be clarified and little in this area can be regarded as finally settled. Of particular note is three recent decisions from the Commonwealth in which the *Murphy* decision has not been followed, demonstrating that much of the earlier law may still have relevance.[13] It is also to be noted that the Defective Premises Act 1972 (see Chapter 7) still affords to residential owners rights equivalent to those held in the *Dutton* case. A continuing debate in the construction industry is whether the Act should be extended to cover commercial buildings.

Losses recoverable in negligence

The law of negligence is now to be regarded as concerned with actual damage in the form, usually, of physical injury to persons or property, necessarily caused by latent defects. Ordinarily, the property must be something distinct from that which the negligent defendant has supplied or constructed. The question whether a local authority will ever be liable for the negligent exercise of its powers in accordance with these principles, remains to be decided under English law. It is to be noted that in all the local authority cases, none involved loss or injury of a type that would now be regarded as recoverable. The *Murphy* case has also settled, for the present, the question of when damage to a building is to be regarded as "purely economic loss" and therefore unrecoverable. In both *Dutton* and *Anns*, it was held that the cracks to the building themselves constituted physical damage. This analysis was decisively rejected in *Murphy*. In the result, expenditure on repairing damaged or defective parts of buildings will not normally be recoverable, unless falling within one of the exceptions referred to above.

---

[13] See I.N. Duncan Wallace Q.C., "*Murphy* Rejected." (1995) 11 Const. L.J. 249.

In some negligence cases, the question arises which losses are properly to be regarded as economic, and where the line is to be drawn. In *Spartan Steel v. Martin*,[14] the Court of Appeal had to consider issues arising from the negligent cutting of an electricity cable to a factory. This caused physical "damage" to certain products which were being manufactured in the plaintiff's machinery at the time of the power cut. The plaintiff recovered the value of the material damaged and also loss of profit on this material as consequential (economic) loss. The Court of Appeal refused to allow recovery in respect of other material which the plaintiffs were unable to process during the power cut, regarding this as purely economic. The case has been much discussed, particularly as to whether these distinctions were matters of law or policy. Lord Oliver, in his judgment in the *Murphy* case commented on *Spartan Steel* as follows:

> "The solution to such borderline cases has so far been achieved pragmatically, not by the application of logic but by the perceived necessity as a matter of policy to place some limits—perhaps arbitrary limits—to what would otherwise be an endless cumulative causative chain bounded only by theoretical foreseeability."

Debate will continue as to when and on what legal basis certain losses are irrecoverable. However, the general rule that losses in tort must arise from physical damage is now firmly established.

Breach and damage

If a duty of care exists, it is necessary to establish a breach of that duty. The standard of care required is that of a "reasonable man." This is a legal abstraction which represents a person who weighs up the circumstances, considers the characteristics of the persons endangered, takes greater care when there is greater danger, and never loses his temper. He is sometimes epitomised as the man on the Clapham omnibus. The duty is to guard against probabilities, not bare possibilities. But where the risk is greater, such as where children are involved, reasonable possibilities must be guarded against. The required standard of care thus depends on the circumstances, but in any particular case there is one appropriate standard below which a person is legally negligent. The term "gross negligence" is sometimes used in contracts, but the adjective has no legal significance in the law of tort. There is only one

---

[14] [1973] Q.B. 27.

appropriate standard of care, any breach of which, gross or slight, incurs liability in law.

Finally, the injury to the plaintiff must have been caused by the defendant's act, and the damage must not be too remote (see below). If the plaintiff succeeds in his action in negligence but the loss was caused partly by his own default, the court may reduce the damages recovered under the Law Reform (Contributory Negligence) Act 1945. This provides by section 1:

> "Where any person suffers damage as a result partly of his own fault and partly of the fault of any other person or persons, a claim in respect of that damage shall not be defeated by reason of the fault of the person suffering the damage, but the damages recoverable in respect thereof shall be reduced to such extent as the Court thinks just and equitable having regard to the claimant's share in the responsibility for the damage."

Contributory negligence applies also to breaches of statutory duty. The application to a claim for breach of contract is covered in Chapter 6.

Negligent mis-statement

Liability for statements has developed, historically, along different lines from liability for acts or omissions. The leading case is *Hedley Byrne v. Heller*,[15] where a bank gave a gratuitous reference for a customer in respect of a company with whom they proposed to do business. The reference was favourable but was given "without responsibility." The reference was given negligently, and the customer lost money. The House of Lords held that the bank owed a duty of care and would have been liable, but was protected by the express disclaimer of responsibility. In giving judgment, the House considered the circumstances in which liability might arise for statements. Lord Devlin expressed the matter thus:

> "Wherever there is a relationship equivalent to contract, there is a duty of care. Such a relationship may be either general or particular. Examples of a general relationship are those of solicitor and client and of banker and customer. . . . Where, as in the present case, what is relied on is a particular relationship created ad hoc, it will be necessary to examine the particular facts to see whether there is an express or implied undertaking of responsibility."

Thus, engineers, architects and other professionals must be cautious when making statements to their clients, even concerning

---

[15] [1964] A.C. 465.

matters in which they are not directly instructed. And a duty of care may equally arise when giving gratuitous advice to strangers if the circumstances are such that there is an implied undertaking of responsibility.

This branch of the law was further developed in *Esso Petroleum v. Mardon.*[16] The defendant had taken a lease of a garage after representations were made by the plaintiffs as to the likely throughput of petrol. Their figures were based on the original design for the garage, but planning permission was refused and the garage was in fact built fronting away from the main road. In the result, the volume of trade was considerably reduced and the defendant lost money. The question arose whether the plaintiff's statements could be relied on when they were followed by a contract. Lord Denning held that the defendant was entitled to succeed in his counterclaim:

> "If a man who has or professes to have special knowledge or skill makes a representation by virtue thereof to another (be it advice, information or opinion) with the intention of inducing him to enter into a contract with him, he is under a duty to use reasonable care to see that the representation is correct, and that the advice, information or opinion is reliable. If he negligently gives unsound advice or misleading information or expresses any erroneous opinion, and thereby induces the other side to enter into a contract with him, he is liable in damages."

In both of these cases, the defendant had made a relatively simple statement upon which the plaintiff had relied, in the second case the statement being followed by a contract. What is less clear, is the precise ambit of this principle. Before the recent developments in the law of tort this question was of limited importance, but since the retrenchment elsewhere, the *Hedley Byrne* principle may be the only avenue by which pure economic loss can be recovered in tort. It is therefore to be anticipated that there will be further attempts to bring cases within this principle. In *I.B.A. v. E.M.I.*[17] the designing sub-contractor responded to a request from the employer concerning the performance of the television mast, stating "we are well satisfied that the structure will not oscillate dangerously." The Court of Appeal treated this as contractually binding, but the House of Lords regarded it as falling within the principles of *Hedley Byrne*. It is difficult to see any logical dividing line between the provision of a design, upon which the building

---

[16] [1976] Q.B. 801.
[17] (1980) 14 B.L.R. 1; and see Chap. 9.

owner will invariably place reliance, and the simple provision of information, particularly if a statement that "our design is adequate" is regarded as sufficient.

A further, unsuccessful, attempt to apply the *Hedley Byrne* principle occurred in *Pacific Associates v. Baxter*.[18] In this case the defendant was appointed engineer under a FIDIC contract. As a result of rejecting claims brought by the contractor, arbitration proceedings had to be pursued against the employer, which resulted in a partially successful settlement. The plaintiff contractor then attempted to sue the engineer in tort to recover the remainder of his loss. The engineer applied to strike out the claim as unsustainable. The application succeeded both before the official referee and the Court of Appeal, the latter holding that there was no voluntary assumption of responsibility. The case was complicated by argument based on a clause in the main contract providing that the engineer was not to be liable, but the case is relevant in defining the limits of the *Hedley Byrne* principle.

OTHER ASPECTS OF NEGLIGENCE

Liability of occupiers

Under the Occupiers' Liability Act 1957 an occupier of premises owes a duty of care to all visitors lawfully on the premises. Unless the occupier can and does modify or exclude his obligations by agreement, he owes to any visitor a duty to take reasonable care so that the visitor will be reasonably safe in using the premises for the purposes for which he is permitted to be there. The occupier may escape liability by giving adequate warning of existing dangers, and he may also expect persons such as workmen entering to carry out a job to guard against special dangers of their trade. In *Roles v. Nathan*[19] two chimney sweeps had been warned of the danger of fumes, which they disregarded. They were asphyxiated but the occupier was held not liable. An occupier is not liable for the faulty work of an independent contractor (see below) unless he is himself to blame for the defects.[20] An occupier may avoid such liability if he has taken reasonable steps to ensure that the contractor was competent and the work properly done.

---

[18] [1990] 1 Q.B. 993.
[19] [1963] 1 W.L.R. 1117.
[20] See also *D & F Estates* above.

The Occupiers' Liability Act applies to those who occupy land and buildings (including building and construction sites) and any fixed or movable structure such as a vehicle, vessel, lift or scaffolding. An "occupier" need not be the owner of the premises but is merely a person having some degree of control. There may therefore be more than one occupier of the same premises. In *A.M.F. v. Magnet Bowling*[21] both the general contractor and the employer were held to be occupiers with respect to a specialist direct contractor. A sub-contractor may also be an occupier of the whole or part of the site.

The Act does not apply to persons who are not visitors, whether they come onto the land lawfully, such as for the purposes of using a right of way, or unlawfully as trespassers. In the past, this has created great difficulty, particularly in the case of children who may stray into dangerous areas. Construction sites are a particular case where children and others may be at risk. The common law extended the principle of the Occupiers' Liability Act 1957 to provide that an occupier might, in such circumstances, owe a limited duty. This development is superseded by the Occupiers' Liability Act 1984, which deals expressly with these difficulties. The 1984 Act provides that an occupier owes a duty to a person who is not a visitor if he is aware (or has reasonable grounds to believe) that a danger exists and also knows (or has reasonable grounds to believe) that the other person is or may come into the vicinity of the danger. The risk must be one against which the occupier may be expected to offer some protection. Where such a duty exists, the occupier must take reasonable care to avoid injury, but the duty may be discharged by giving appropriate warning or discouraging persons from incurring the risk.

The Unfair Contract Terms Act 1977 provides, in respect of a business occupier, that liability as an occupier cannot be excluded (by a contract term or notice), in respect of death or personal injury resulting from negligence. In respect of other loss, liability can be excluded or restricted only so far as it is fair and reasonable. The 1984 Act restricts the operation of this provision to persons who are granted access for the purpose of the business of the occupier, so that, for example, owners of quarries who allow access for rock climbing may now exclude liability for the dangerous state of the premises.

---

[21] [1968] 1 W.L.R. 1028.

Employers' liability

An employer may be liable for injury caused to his employee in three ways:

(1) if the injury is caused by the negligence of a fellow employee acting in the course of his employment (see below);

(2) if it is caused by the employer's breach of a statutory duty (see Chapter 16); or

(3) if it is caused by the employer's negligence. The third possibility is discussed here.

There is no doubt that an employer owes his employees a duty of care. The problem for the common law has been to trace its extent. It has been defined as a three-fold duty: the provision of a competent staff of men, adequate material, and a proper system and effective supervision. However, the duty can be viewed as a single duty to take reasonable care for the safety of employees in all the circumstances. Thus the duty is not absolute, and an employer is only liable for injury caused by his failure to take sufficient care.

The standard of care that is required varies with the circumstances so that where potentially dangerous plant is being used the employer may have to provide safety devices or protective equipment. The employer remains liable for breach of his duty even though he may delegate its performance. In *McDermid v. Nash Dredging*,[22] the plaintiff had been employed by the defendant but was required to work on board a tug owned by a Dutch company and under the control of their captain. The plaintiff was injured by an accident caused by the captain's negligence and the issue arose whether the employer remained liable. Lord Brandon in the House of Lords restated the relevant principles of the law as follows:

"First, an employer owes to his employee a duty to exercise reasonable care to ensure that the system of work provided for him is a safe one. Secondly, the provision of a safe system of work has two aspects: (a) the devising of such a system and (b) the operation of it. Thirdly, the duty concerned has been described alternatively as either personal or non-delegable. The meaning of these expressions is not self-evident and needs explaining. The essential characteristic of the duty is that, if it is not performed, it is no defence for the employer to

---

[22] [1987] 1 A.C. 906.

show that he delegated its performance to a person, whether his servant or not his servant, whom he reasonably believed to be competent to perform it. Despite such delegation the employer is liable for the non-performance of the duty."

However, where an employee is experienced and the danger apparent, the duty upon the employer may be a limited one, particularly where the employer is not in control of the premises. In the case *Wilson v. Tyneside Window Cleaning*[23] the plaintiff, an experienced window cleaner, was injured when a handle came away from a window, causing him to lose his balance. He had never received any instructions regarding safety, except that if he found a window which presented unusual difficulty or risk he was to report for further instructions. It was held that there was no breach of duty by the employer. Pearce L.J. observed:

"The master's own premises are under his control: if they are dangerously in need of repair he can and must rectify the fault at once if he is to escape the censure of negligence. But if a master sends a plumber to mend a leak in a respectable private house, no one could hold him negligent for not visiting the house himself to see if the carpet in the hall creates a trap. Between these extremes are countless possible examples in which the court may have to decide the question of fact: did the master take reasonable care so to carry out his operations as not to subject those employed by him to unnecessary risk. Precautions dictated by reasonable care when the servant works on the master's premises may be wholly prevented or greatly circumscribed by the fact that the place of work is under the control of a stranger. Additional safeguards intended to reinforce the man's own knowledge and skill in surmounting difficulties or dangers may be reasonable in the former case but impracticable and unreasonable in the latter."

The employee must show regard for his own safety, and if he is injured as a result of his own negligence this may reduce or even extinguish the employer's liability. The employee also owes the employer a duty to exercise reasonable skill and care at his work and may be liable to his employer for causing injury in breach of this duty.

These general common law principles apply equally to work on construction sites. However, in many industries, particularly construction, there exist detailed regulations and this is particularly so following recent directives from the European Union (see Chapter 16).

---

[23] [1958] 2 Q.B. 110.

Strict liability

When a person keeps some potentially dangerous object on his land or carries on a dangerous operation there, the ordinary law of negligence may not afford adequate protection. Instead of extending the duty of care, the common law has set apart certain things for which liability is strict, without regard to lack of care. A person who deals in such things does so at his peril. This special type of liability is known by the name of the case in which it was first formulated. In *Rylands v. Fletcher*[24] the defendant built a reservoir on his land using reputable engineers and with the necessary permission. But when the reservoir was filled, the water escaped down through a disused mine shaft and flooded the plaintiff's coal mines on adjoining land. Although the defendant's actions were without fault, he was held liable for the plaintiff's loss. In giving judgment Lord Cairns held:

"On the other hand if the defendants, not stopping at the natural use of their close, had desired to use it for any purpose which I may term a non-natural use, for the purpose of introducing into the close that which in its natural condition was not in or upon it, for the purpose of introducing water either above or below ground in quantities and in a manner not the result of any work or operation upon or under the land—and if in consequence of their doing so, or in consequence of any imperfection in the mode of their doing so, the water came to escape and to pass off into the close of the plaintiff, then it appears to me that that which the defendants were doing they were doing at their own peril; and if in the course of their doing it, the evil arose to which I have referred, the evil, namely the escape of water and its passing away into the close of the plaintiff and injuring the plaintiff, then for the consequence of that, in my opinion, the defendant would be liable."

In addition to reservoirs, strict liability has been attached to colliery spoil heaps and inflammable goods. In *Hoare v. McAlpine*[25] a contractor was held liable for the escape of vibrations from pile driving operations which caused damage to an old house. Although vibrations may also constitute a nuisance (see below) it may be a defence to nuisance that the property damaged was unusually frail. Liability in *Rylands v. Fletcher* is, however, strict.

To incur liability the object or operation must be non-natural, so that while there is strict liability for a reservoir, the owner of a natural lake can be liable only under the ordinary principles of tort,

---

[24] (1868) L.R. 3 H.L. 330.
[25] [1923] 1 Ch. 167.

such as in negligence or nuisance. To establish strict liability there must also be an "escape" from the plaintiff's land which causes the damage, such as a slide of material from a spoil heap. In *Read v. Lyons*[26] an explosion in a munitions factory which injured persons on the premises did not incur strict liability since there had not been an escape from the defendant's land. In this case, Lord Macmillan observed:

> "The two prerequisites of the doctrine [of *Rylands v. Fletcher*] are that there must be the escape of something from one man's close to another man's close and that that which escapes must have been brought upon the land from which it escapes in consequence of some non-natural use of that land, whatever precisely that may mean. Neither of these features exists in the present case. I have already pointed out that nothing escaped from the defendant's premises and were it necessary to decide the point I should hesitate to rule that in these days and in an industrial community, it was an non-natural use of land to build a factory on it and conduct there the manufacture of explosives."

The principle of strict liability arose in a stark form in the recent case of *Cambridge Water v. Eastern Counties Leather*[27] where the defendants, leather manufacturers, used a chemical solvent. Small quantities seeped through the floor of their premises and into the soil below, eventually entering the ground water from which the plaintiff drew supplies. The escape rendered the water unfit for human consumption and involved substantial losses on the part of the plaintiff without any question of negligence on the part of the defendant. The House of Lords rejected the claim, taking the view that instead of extending the concept of "natural" use of land, the principle should be restricted by the need to establish foreseeability of harm. The House also recognised the role of legislation in such a sensitive area. Lord Goff said:

> "I incline to the opinion that, as a general rule, it is more appropriate for strict liability in respect of operations of high risk to be imposed by parliament than by the courts. If such liability is imposed by statute, the relevant activities can be identified, and those concerned can know where they stand. Furthermore, the state can, where appropriate, lay down precise criteria establishing the incidence and scope of such liability."

The result of this case can be seen as an attempt to subsume this area of strict liability into the ordinary law of negligence, as has

---

[26] [1947] A.C. 156.
[27] [1994] 2 A.C. 264.

already occurred in Australia.[28] The result would be the emergence of a duty in respect of particular types of harm, rather than a duty arising in particular circumstances, the extent of which leaves room for doubt.

Product liability

Claims based on negligence or on the Defective Premises Act (see Chapter 7) require proof of fault. An alternative claim may now be available under the Consumer Protection Act 1987, which is based on the European Community Directive on Product Liability. Subject to very limited defences, the Act imposes strict liability for personal injury and also for damage to property, other than the defective product itself. The Act provides special limitation periods of three years from discoverability of the damage with a longstop period of 10 years from the time of supply of the product rather than 15 years from the negligent act as under the Latent Damage Act 1986 (see below).

NUISANCE

Private nuisance may be defined as an unlawful interference with the use or enjoyment of another person's land. The interference may result in damage to property, such as by flooding or vibrations, or it may be only an annoyance, such as excessive noise or dust. There must be a substantial interference. A nuisance is often a continuing state of affairs, although an isolated happening may support an action in nuisance. Neighbours must exercise give and take, but deliberate acts intended to annoy neighbours can create an actionable nuisance. Persons who live in noisy or industrial neighbourhoods must usually put up with the attendant discomforts, although actual damage to property will be actionable.

Usually the only person who can sue for nuisance is the occupier of the land, although other persons may be able to sue on the same facts, for instance in negligence. The person liable is usually the occupier of the land or premises where the nuisance exists, but the person who created the nuisance may be liable. Thus, prima facie a building contractor will be liable for interference with adjoining land caused by the construction operations, but the employer may also be liable (see below).

---

[28] *Burnie v. General Jones* (1994) 120 A.L.R. 42.

Unlike negligence, liability for nuisance does not depend primarily on the standard of conduct of the defendant. Thus, it is not necessarily a defence to nuisance that reasonable care was taken to avoid it. But in the context of building and construction operations, those carrying out such work are under a duty to take proper precautions to see that nuisance is reduced to a minimum. Thus, in *Andreae v. Selfridge*[29] where a demolition contractor took no steps to minimise noise and dust near to the plaintiff's hotel, an actionable nuisance was created for which the employer was liable. Sir Wilfred Green M.R. held:

> "Those who say that their interference with the comfort of their neighbours is justified because their operations are normal and usual and conducted with proper care and skill are under a specific duty, if they wish to make good that defence, to use that reasonable and proper care and skill. It is not a correct attitude to take to say: 'We will go on and do what we like until someone complains.' That is not their duty to their neighbours. Their duty is to take proper precautions, to see that the nuisance is reduced to a minimum. It is no answer for them to say: 'But this would mean that we should have to do the work more slowly than we would like to do it or it would involve putting us to some extra expense.' All those questions are matters of common sense and degree and quite clearly it would be unreasonable to expect people to conduct their work so slowly or so expensively, for the purpose of preventing a transient inconvenience, that the cost and trouble would be prohibitive. It is all a question of fact and degree and must necessarily be so."

A nuisance may also be controlled by the local authority under the Environmental Protection Act 1990. Sections 79 and 80 empower the authority to serve an abatement notice in respect of a "Statutory Nuisance" as defined by the Act. This includes:

> "(a) any premises in such a state as to be prejudicial to health or a nuisance;
>   . . . . . .
> (d) any dust or effluvia caused by any trade, business, manufacture or process and being prejudicial to the health of or a nuisance to the inhabitants of the neighbourhood;
> (e) any accumulation or deposit which is prejudicial to health or a nuisance."

Where a statutory nuisance is not abated, the authority may acquire powers to carry out necessary work. Under the Control of Pollution Act 1974 a local authority has powers to control noise on construction sites. They may serve a notice restricting the use of

---

[29] [1938] Ch. 1.

specified plant, restricting hours of work and limiting the level of noise. The act also permits the contractor to obtain the prior consent of the authority to the methods of work proposed.[30]

Rights of support

A nuisance may be committed by interference with a right of support of land. There is a natural right of support for unweighted land, and a nuisance is committed if subsidence is caused either by removing the lateral support by excavation, or by undermining. Generally it is unimportant how the withdrawal of support occurs, but as an important exception there is no liability for causing subsidence by withdrawal of subterranean percolating water. Thus, in *Langbrook v. Surrey C.C.*[31] pumping carried out to keep excavations dry resulted in lowering of the water table and settlement of buildings on adjacent land. The adjoining owner was held to have no redress. Plowman J. after reviewing the authorities, concluded that a landowner was entitled to abstract water under his land which percolates in undefined channels, notwithstanding that this may cause neighbouring land to subside. The judge went on to consider whether there could, in such circumstances, be liability for nuisance or negligence, and held:

> "Since it is not actionable to cause damage by the abstraction of underground water, even where this is done maliciously, it would seem illogical that it should be actionable if it were done carelessly. Where there is no duty not to injure for the sake of inflicting injury, there cannot, in my judgement, be a duty to take care not to inflict the same injury."

The right of support for a building, as opposed to the land on which it stands, is not a natural right and must be acquired as an easement by grant or by usage (see Chapter 15). Once acquired, the right is usually a right of support both from adjacent land and from adjoining buildings. The right, however, is one which must be exercised with caution.

In *Redland Bricks v. Morris*[32] a landowner, the stability of whose property was threatened by excavation in an adjoining quarry, obtained a mandatory injunction against the quarry owner compelling him to carry out work to restore stability. But, in *Midland Bank v. Bardgrove*[33] the plaintiff failed to recover damages for the

---

[30] ss. 60, 61; and see Chap. 15.
[31] [1970] 1 W.L.R. 161.
[32] [1970] A.C. 652.
[33] (1992) 60 B.L.R. 1, CA.

cost of work which he had carried out to protect his property against threatened instability from excavation on the defendant's adjoining site. The defendant had constructed an inadequate restraining wall which was likely to cause damage at some time in the future, but the claim failed on the basis the plaintiff's cause of action in damages arose only when physical damage was suffered. This conclusion was based on a series of nineteenth-century mining cases which held that where property was damaged by undermining (the same principle applied to lateral withdrawal of support) a fresh cause of action arose each time damage was suffered.[34] Consequently, although the plaintiff whose land is threatened can recover compensation for every occurrence of damage, without the right to claim becoming statute barred, it follows also that a claim cannot be maintained until the loss has occurred. In respect of coal mining, the rights to compensation is now governed by statute.[35] The common law rule remains and has major implications for claims as between adjacent owners where one wishes to carry out construction works involving excavation. A sensible solution is to enter into an express agreement defining the measures to be taken to preserve stability so that any breach or non-observance will give a right to claim compensation. Where major civil engineering undertakings require special parliamentary authority, the enabling Acts invariably contain express provisions governing compensation claims. These do not, however, generally oust common law rights nor do they preclude the possibility of express agreement where the work may involve a particular risk of damage.

Other rights which may give rise to an action in nuisance for interference include rights of light and air, water rights and rights of way. Where an interference is caused to a wider group of persons than the occupier of neighbouring land, such conduct may constitute a public nuisance. This is primarily a crime, but a civil action may be brought in respect of a public nuisance by an individual who has suffered special damage different from that suffered by the public at large. Common examples of public nuisance are obstruction of the highway, and creating dangers upon or near the highway.

Tree roots and forseeability

Interference with a right of support is generally actionable irrespective of whether the consequences were foreseeable by the defen-

---

[34] *Darley Main Colliery v. Mitchell* (1886) 11 App. Cas. 127 and see Chap. 2 above.
[35] Coal Mining (Subsidence) Act 1957, amended and re-enacted 1991.

dant. However, in some cases of nuisance the defendant's conduct will not be actionable unless the consequences were foreseeable. The leading case, as regards damage to land, is *Leakey v. National Trust*[36] where two houseowners lived near the foot of a historic mound known as Burrow Mump in Somerset. Over the years soil and rubble had fallen from the mound and the defendant, who was responsible for its upkeep, knew of potential instability that might result in larger falls. After a substantial fall proceedings were brought. An interlocutory injunction was granted ordering the defendant to carry out protective works and at trial modest damages were awarded. The Court of Appeal upheld the judgment stating that, while the action was properly brought in nuisance rather than negligence, the distinction was of no practical importance. The defendant's duty was limited to doing what was reasonable.

The incursion of tree roots from neighbouring land has long been recognised as a species of nuisance, resulting in settlement of buildings through desiccation of clay soil. Some older cases were decided on the basis that liability was absolute. However, it is now clear that forseeability of damage is an essential element of liability. This was held in *Paterson v. Humberside CC*[37] by Roger Toulson Q.C. (now Toulson J.) who referred to the actions of the defendant as Highway Authority and held:

> "The Highway Authority was therefore clearly aware of the potential for tree roots to cause problems for buildings on the shrinkable subsoils in the Hull area. Hull City Council also published advice to householders and developers on trees in relation to buildings. The same publication recommended a safe planting distance in Hull for lime trees . . . I am therefore satisfied that the risk of damage by tree roots was foreseeable. It may not have been thought likely unless there were drought conditions; but that is not the test and, in any event, periods of drought do occur."

An alternative cause of action on the same facts may exist in negligence, as was found in Paterson. As between neighbours, however, an action in nuisance would be more likely to succeed, provided that the damage can be shown to have been foreseeable.

VICARIOUS LIABILITY

Vicarious liability in this context means liability for the torts of others. In the construction industry this may arise in two ways.

---

[36] [1980] Q.B. 485.
[37] (1995) 12 Const. L.J., 64.

First, the employer may be liable for the torts of the contractor or the contractor, for his sub-contractors; secondly, any of the parties involved in the work may be liable for the torts of their own individual employees.

As to the first type of vicarious liability, as a general rule a person is not liable for torts committed by his independent contractors. However, there are substantial exceptions to the general rule, whereby the employer may be liable. These include:

(1) where the liability is strict, such as under the rule in *Rylands v. Fletcher*;

(2) where work involves danger on or near a highway;

(3) where work will involve danger to other property unless proper care is taken (see above).

Even where the employer is not prima facie responsible he may still be liable for his own negligence in employing an incompetent contractor, or for failing to give adequate directions to avoid damage to another. The employer will also be liable under the law of agency if he authorises or ratifies his contractor's wrongful act.

Master and servant

A master is liable for a tort which he authorises or ratifies and is also vicariously liable, in general, for the unauthorised tort of his servant if it is committed within the course of his employment. The terms "master" and "servant" have acquired a special meaning in law which is rather wider than employer and employee. A servant may be said to be a person employed to carry out work other than as an independent contractor. The work of the servant is an integral part of the master's business, while an independent contractor undertakes only to produce a given result. Persons are frequently found on construction sites who are technically self-employed, but who may be difficult to categorise as servants or independent contractors. The master is generally not liable for the latter.

In addition to deciding who is a servant, it may be necessary to decide who is the master, for instance, where a servant is hired by his employer to another employer. It is presumed that the original employer remains liable as the master, unless the right to control the way in which the work was done passes to the temporary employer. Thus, in *Mersey Docks v. Coggins & Griffiths*[38] a contrac-

---

[38] [1947] A.C. 1.

tor hired a crane together with its driver to carry out unloading work. The hirer supervised the work but not the management of the crane. The original employer was held to be responsible for the driver's negligence. In this case Lord Porter said:

> "Amongst the many tests suggested I think the most satisfactory, by which to ascertain who is the employer at any particular time, is to ask who was entitled to tell the employee the way in which he is to do the work upon which he is engaged. If someone other than his general employer is authorised to do this he will, as a rule, be the person liable for the employee's negligence. But it is not enough that the task to be performed should be under his control, he must also control the method of performing it."

The master will be liable for the tort of his servant only if the tort is committed during the course of his employment; that is, it must be a wrongful way of doing that which he is employed to do. Thus, an employer will be liable if the employee carries out his duties negligently or fraudulently. The employer may even be liable if the employee does something which he has expressly been forbidden to do, provided it is within the scope of his employment. But in *Conway v. Wimpey*[39] a driver, employed on a building site to carry only fellow-employees, carried an employee of another firm on the site. This was held to be an act which he was not employed to perform. The employer was therefore not liable for the driver's negligence. In giving judgment in the Court of Appeal, Asquith L.J. held:

> "I should hold that taking men not employed by the defendants on to the vehicle was not merely a wrongful mode of performing the act of the class this driver was employed to perform, but was the performance of an act of a class which he was not employed to perform at all. In other words, the act was outside the scope of his employment."

Whether or not the master is liable, the servant is generally liable for his own tort and may be sued jointly with the master, or separately. Similarly where an employer is liable for his independent contractor, the contractor may also be sued jointly or separately. The practical importance to a plaintiff of vicarious liability is that it affords more likelihood of finding a defendant who is either solvent or who has the benefit of insurance cover. Where more than one defendant is held liable, the principles of contribution come into play (see below).

---

[39] [1951] 1 All E.R. 56.

Remedies

The remedy claimed in most tort actions is damages. The successful plaintiff in an action for damages will generally be awarded a sum which is intended to compensate for the real loss suffered. The sum awarded must take into account future loss since usually only one action may be brought. There is an exception in the case of withdrawal of support (see above). Damages may be proportionally reduced if contributory negligence is found against the plaintiff. In addition to damages, the plaintiff may apply for an injunction (see below) and this may in some cases be the only substantial remedy required (see the *Redland Bricks* case above).

Remoteness and causation

Once liability is established, the question may arise whether the damage claimed is too remote to be recoverable. The general test is that compensation may be recovered for damage which is of a reasonably foreseeable kind. If this is so it does not matter if the damage occurred in an unforeseeable manner or to an unforeseeable extent; the defendant will be liable for the whole loss. The rules of remoteness in contract and in tort are not identical. The tortfeasor is liable for loss which is foreseeable as the possible result of his conduct and therefore may bear a heavier burden than the contract breaker, who is liable only for the probable result of his actions (see Chapter 6).

The test of foreseeability was discussed in the leading case *The Waggon Mound*,[40] in which the Privy Council had to consider the following facts. A large quantity of furnace oil was discharged through the negligence of the defendants from their ship while moored. The oil spread to a wharf belonging to the plaintiffs who were engaged while refitting work, including welding. Believing there to be no danger, the plaintiffs continued the welding. The oil ignited and caused serious damage. The Privy Council held the defendants not liable. Lord Simonds, after reviewing the authorities said:

> "The essential factor in determining liability is whether the damage is of such a kind as the reasonable man should have foreseen. This accords with the general view thus stated by Lord Atkin in *Donoghue v. Stevenson*: 'the liability for negligence . . . is no doubt based on a general public sentiment of moral wrong-doing for which the

---

[40] *Overseas Tankship v. Morts Dock & Engineering* [1961] A.C. 388.

offender must pay.' It is not a departure from this sovereign principle if liability is made to depend solely on the damage being the direct or natural consequence of the precedent act. Who knows or can be assumed to know all the processes of nature? But if it would be wrong that a man should be held liable for damage unpredictable by a reasonable man because it was direct or natural, equally it would be wrong that he should escape liability however indirect the damage, if he foresaw or could reasonably foresee the intervening events which lead to it being done. . . . Thus forseeability becomes the effective test."

The question of causation, rejected in *The Waggon Mound* may still be of relevance. In *Barnett v. Chelsea Hospital Committee*[41] a nightwatchman presented himself at the hospital casualty department complaining of vomiting after drinking tea. He was told to go home to bed, where he later died of arsenic poisoning. In an action by the widow for negligence, it was held that the plaintiff had failed to establish that the defendant's negligence had caused the death. Conversely in *Baker v. Willoughby*[42] the plaintiff sued for injury to his leg caused by the defendant's negligent driving. But before the claim was heard, he was involved in an armed robbery in which the injured leg was shot and had to be amputated. The House of Lords declined to reduce the damages on this account, holding the second injury a mere concurrent cause. The questions of remoteness and causation, both in tort and contract, will continue to produce many complex legal problems.

Contribution

Where there are two or more defendants responsible for the same loss the plaintiff is entitled to recover judgment against each to the full amount of their individual liability. The plaintiff is then entitled to enforce the judgment obtained against either, provided that he does not recover more than the total damages proved. It would be a matter for the plaintiff to decide against which defendant first to enforce. To mitigate this, the court has power under the Civil Liability (Contribution) Act 1978[43] to apportion liability between defendants. The apportionment is dealt with by the judge after deciding upon the liability of the defendants to the plaintiff. The effect is to give any defendant entitled to contribution the right to recover from another defendant the amount of that defendant's liability. If, therefore, one or more of the defen-

---

[41] [1969] 1 Q.B. 428.
[42] [1970] A.C. 467.
[43] Previously the Law Reform (Married Women and Tortfeasors) Act 1935.

dants is without the means to pay, the remaining defendants must bear the whole loss. In many construction cases, this has operated to the disadvantage of professionals who have indemnity insurance available, when contractors or sub-contractors may be insolvent. The case of *Eckersley v. Binnie*[44] is an example of the harshness of the application of these rules. The facts are set out in Chapter 6.

The rules as to contribution apply whether or not the other persons liable have been sued by the plaintiff. A defendant may therefore bring into an action as a third party anyone whom he considers should contribute to the plaintiff's loss as a joint tortfeasor. Or the defendant may bring separate proceedings claiming contribution after the plaintiff has obtained judgment.

Injunction

An alternative remedy to damages, which may be appropriate particularly in cases of nuisance, is an injunction, either to restrain the defendant from doing some act or to compel the performance of an act. An injunction is an equitable remedy and therefore lies in the court's discretion. It will usually be refused where damages would be an adequate remedy, or where to grant it would be in vain. A valuable feature of this remedy is the power of the court to give an interlocutory injunction. Temporary relief may be obtained within days or even hours of a cause of complaint arising. The power to grant an injunction is applicable particularly where the defendant is threatening an act which is arguably unlawful and likely to cause damage. An interlocutory injunction may be issued *ex parte* (in the absence of the defendant) in the first place, and then reconsidered after a short period when the defendant can be heard. The court will then consider whether to continue the injunction for a further period or even up to trial. The test upon which the court decides whether to give an interlocutory injunction has been restated by the House of Lords.[45] The plaintiff is required to establish that he has a good arguable claim to the right which he seeks to protect. The court then considers the balance of convenience in granting or refusing the injunction.

Limitation

An important consideration in any action must be the period during which the action may be brought. The Limitation Act 1980

---

[44] (1987) 18 Con. L.R. 1 at 44.
[45] *American Cyanamid v. Ethicon* [1975] A.C. 396.

provides that an action founded on tort shall not be brought after the expiration of six years from the date on which the cause of action accrued (section 2). There is a further limitation where the claim is for damages in respect of personal injuries arising from negligence, nuisance or breach of duty. Here, the action must normally be brought within three years from the date on which the cause of action accrued (section 11) subject to certain extensions (see below).

The main difference between limitation in a contract action and in tort is that in the former case, the cause of action accrues at the date of the breach. In tort, however, the cause of action accrues only when damage is suffered, so that the cause of action may not arise until long after the relevant act or omission occurred. Thus, if a builder negligently erects a chimney stack, a cause of action in contract arises in favour of the employer when the work is done, or when the builder purports to finish it. If the work remains in place and no complaint is made, the right of action in contract will be lost after six years. If, after 10 years, the chimney falls to injure a passer-by, a right of action in tort then immediately vests in the injured person.

Latent damage

There has been great development in the law of limitation relating to defects in building work in recent years. The impetus came from those cases (see above) which held that a right of action in tort was available, and this led to a series of cases in which local authorities, builders, designers and other professionals were held liable long after the relevant work was carried out. In a number of cases it was held that the cause of action accrued only when the building owner became aware, or ought reasonably to have become aware, of the existence of the defect. A further period of six years was then available to issue proceedings. However, these decisions were contrary to the law of limitation as applied in personal injury cases. Here, the House of Lords held in *Cartledge v. Jopling*[46] that the plaintiff's action was statute-barred before the damage (contraction of pneumoconiosis) could reasonably have been known. As a result of this case, an amendment to the Limitation Act was passed,[47] allowing a further three years to bring an action from the date when the plaintiff knew or ought to have known that he had a cause of action for personal injury. The question of limitation in

---

[46] [1963] A.C. 758.
[47] Limitation Act 1980, ss. 11 and 14.

relation to claims for damage to buildings was referred to the Law Reform Commission, but the issue came directly before the courts in *Pirelli v. Oscar Faber*.[48] This concerned the design of a chimney, built in 1969. Not later than 1970, cracks developed near the top, but they were not discovered until 1977. A writ was issued in 1978. The House of Lords, while agreeing that the law was unsatisfactory, held that the cause of action arose in 1970 and was therefore statute-barred. *Cartledge v. Jopling* applied equally to damage to property and the previous contrary decisions of the Court of Appeal were overruled.

The effect of this case was changed by the Latent Damage Act 1986 which introduced new sections into the Limitation Act 1980. The material effect of these provisions is, first, that an action for damages for negligence may be brought within three years of the date upon which a reasonable person would conclude that proceedings could and should be instituted (section 14A). Secondly, a long–stop period of 15 years is applied to the bringing of any such action, this period running from the date of the last alleged act of negligence (section 14B). The act also introduces a change to the law where a person buys a house which has a defect such that a cause of action has already arisen. In the case of *Perry v. Tendring District Council*[49] the official referee held, in such circumstances, that the cause of action vested in the original owner and could not ordinarily be transferred to a purchaser. Section 3 of the Latent Damage Act 1986 reverses this decision by providing that a fresh cause of action is to accrue to the purchaser on the date that he acquires an interest in the property, that cause of action being treated as having accrued on the same date as the original cause of action of the previous owner (so that, the defendant is not to be placed in any worse position). These cases must all be reconsidered in the light of the *Murphy* case (see above).

It is ironic that, while the Latent Damage Act was going through Parliament, the courts were embarking on the series of tort cases already referred to which, by 1990, finally established that tortious liability would normally be limited to cases of actual physical injury. Thus, the resolution of doubt over the law of limitation in tort has coincided with emasculation of the rights of action.

Defective premises

A further development, overshadowed at the time by changes in the law of tort, was the passing of the Defective Premises Act 1972

---

[48] [1983] 2 A.C. 1.
[49] (1984) 30 B.L.R. 118.

(see Chapter 7). This Act creates a general duty on persons to see that work is done in a workmanlike or professional manner, with proper materials and so that the dwelling will be fit for habitation (section 1). For the purpose of limitation, the Act provides that any cause of action in respect of a breach of these duties:

> "shall be deemed . . . to have accrued at the time that the dwelling was completed, but if after that time a person who has done work for or in connection with the provision of the dwelling does further work to rectify the work he has already done, any such cause of action in respect of that further work shall be deemed for those purposes to have accrued at the time when the further work was finished."

The Act, therefore, creates a statutory duty similar to (although rather wider than) that which the courts had sought to impose under the law of tort, and also provides what may be regarded as a fair limitation rule, which has no need of a long–stop provision. Since the *D. & F. and Murphy* cases the Defective Premises Act represents the principal remedy outside contract in respect of damaged buildings.

Concealment

Section 32 of the Limitation Act 1980 postpones the limitation period in a case where facts relevant to the plaintiff's right of action have been deliberately concealed from him by the defendant. The section is amended by the Latent Damage Act 1986 with the effect that sections 14A and 14B (see above) do not apply in the case of deliberate concealment, where the normal period of limitation will apply after the plaintiff has discovered or could reasonably have discovered the concealment. Normal limitation provisions will also be overridden where a claim for contribution is brought under the Civil Liability (Contribution) Act 1978. Here, by section 10 of the Limitation Act 1980, an action to recover contributions may be brought within two years of the judgment, award or settlement which establishes the liability of the person claiming contribution.

CHAPTER 15

LAND, PLANNING AND ENVIRONMENT LAW

In England the law relating to land has always been different and distinct from law relating to other property. There are many reasons for this. Perhaps the most obvious is that a piece of land is indestructible and unique. No other land is quite the same. On a practical level, it is common for two or more persons to hold simultaneously different interests in the same land, and this is one of the reasons why a sale of land is more complicated and lengthy than a sale of other property.

Although the expression "land owner" is often encountered it requires some qualification in legal terms. It is not possible to "own" land in the absolute way that other property (such as a motor car) may be owned. Instead, the law speaks of owning an estate or interest in land, which gives the owner certain rights over that land. The largest estate which may be owned is called a fee simple absolute in possession. This is what is commonly known as "freehold" and for convenience it is so called in this section, although in legal terms an estate lasting only for the life of the holder may be a type of freehold. Interests in land are termed "real" property as opposed to "personal", which covers other forms of property.

Another type of estate in land is a tenancy. This may be created out of the estate of the freeholder or of a superior tenant. The word "tenancy" refers to a right to possession of land for a limited period. This may be for a fixed term of years (when the tenancy must normally be created by a lease) but also includes periodic tenancies such as a weekly or quarterly tenancy. In practice a building owner is likely to be concerned only with long tenancies created by lease.

Interests in land generally indicate something less than an estate; they may be of many kinds. Two of the most important are easements and restrictive covenants, and these are mentioned further below. Another very common interest in land is a mortgage, where the land is used as a security. Two further provisions may illustrate the special legal status of land. First, an infant (*i.e.* a person under 18 years), cannot own an estate in land. Secondly, a contract for the sale of land or any interest in land is unenforceable unless in writing and signed by or on behalf of each party, or in the case of contracts which are exchanged, unless each counterpart is signed.[1]

Throughout its history land law has been profoundly affected by the principles of equity (see Chapter 1). The result is that interests in land are for some purposes classified as being either legal or equitable. The practical importance of this distinction is that a legal interest attaches to the land itself and is enforceable against any person. An equitable interest binds only certain persons, and is not enforceable against a bona fide purchaser of the land who has no notice (actual or constructive) of such interest. An equitable interest is, therefore, less secure. An example of an equitable interest is an agreement for a lease. Practically, this is as good as an actual lease but if another person purchases the land without notice of the agreement, it becomes unenforceable against that person.

RIGHTS OF THE OWNER OR OCCUPIER OF LAND

When a person wishes to build on land there are many factors to be considered. He must obtain a sufficient interest to give him a right of possession. He must consider what restrictions there are as to what may be done with the land. He will also wish to know who is entitled to ownership or use of things on or in the land. These points are considered below. Another factor, which may be of great importance, is the question of rights which other persons hold concurrently over the land. Such rights are considered in the following section.

Leases

The most common legal device for obtaining possession of premises for business use is by a lease. There is no fixed definition of the term, but it is usually taken to mean a formal tenancy

---

[1] Law of Property (Miscellaneous Provisions) Act 1989, s. 2.

granted by deed. For most purposes a simple written tenancy agreement will have the same legal effect. The process is also colloquially referred to as "letting." Whatever term is used, the process has two separate elements. First, there is conveyance of the property as defined for the period stated (which may be periodic and renewable); secondly, there are terms (sometimes called covenants) which operate as a contract enforceable between the parties. However, because the transaction relates to land, the terms or covenants may be enforceable directly by or against other parties who take over the interest of either landlord or tenant (see below), and there are also statutory restrictions on enforcement.

Leases and tenancy agreements are in many ways more convenient than buying a freehold interest. Commercial buildings are erected as investments, and the letting, subletting, and assignment of such premises are part of the commercial activity of any business community. It is now comparatively rare for commercial companies, at least in inner city areas, to own the freehold of their own premises. The mobility afforded by letting arrangements has allowed the substantial re-development of commercial areas which has been carried out over the last decades. Business leases are subject to security of tenure under the Landlord and Tenant Act 1954. At the end of a lease, the tenant may apply to the court for a new tenancy. However, the need to demolish and reconstruct the premises or to carry out substantial reconstruction work is a ground upon which the court can refuse a new lease, the tenant receiving compensation instead; and it is possible to contract out of the provisions for security in the Act.

The detailed operation of leases and their covenants is not dealt with. But mention should be made of two classes of construction dispute which frequently arise out of lettings. Leases invariably contain repairing covenants of some variety. These require the tenant to carry out to defined parts of a building certain works, usually defined in terms of the intended result, such as to keep the premises in "good and tenantable repair." Leases usually require the premises to be yielded up in a repaired state, and when the end of the tenancy arrives, there may be a dispute as to its state. This is referred to as a "dilapidations" claim, which often consists of long schedules and counter schedules settled by surveyors for each party. Such a dispute is concerned only with the assessment and valuation of wants of repair in accordance with the covenants. Sometimes disputes of this sort arise during the term of a tenancy. Here, the position is more complex because, if there are a number of years of tenancy still to run, the landlord usually suffers no damage. Accordingly, where there are three years or more of the term remaining, the landlord must obtain leave of the court before proceeding with any action for breach of a repairing covenant, and the court will grant such leave only on exceptional grounds, such as where it is shown that immediate remedying of the breach is necessary.[2]

In addition to specific covenants covering repairs, payment of rent and other matters, leases usually contain a "forfeiture" clause permitting the landlord to take back the lease, or "re-enter" for breach of covenant. For the protection of tenants, the law has evolved a series of protective measures. First, no such right of forfeiture may be enforced (except in the case of non-payment of rent) unless the landlord has first served a notice specifying the breach and requiring remedy and compensation.[3] When such a notice has been served, and not complied with (perhaps because the tenant disputes that repairs are necessary), the landlord must seek an order from the court for forfeiture, and the tenant can then apply for the equitable remedy of relief (now embodied also in statute) which may be granted on such terms as the court thinks fit. The result of the dispute is, in effect, to decide the rights of the parties under the contractual terms or covenants of the lease, and to enforce them. Only in exceptional circumstances will leases be declared forfeit. Indeed, there is little purpose in the tenant allowing the lease to be forfeited, because he will remain liable for the financial consequences of previous breaches of covenant.

Another type of construction dispute that may arise out of a lease is one relating to the nature of the repairs that the tenant is responsible for. What is the tenant's position under a normal repairing covenant if the premises become defective due to design defects in the original construction? In the case of *Ravenseft Properties v. Davstone*[4] a tenant undertook covenants which included an obligation "well and sufficiently to repair, renew, rebuild, uphold, support, sustain, maintain" the premises. The building had been constructed in concrete with external stone cladding, but with no expansion joints to allow differential movement between the stone and concrete. The cladding bowed and needed substantial repair as a result of this inherent defect. It was held that it was a question of degree whether the remedying of an inherent defect was work of repair, and where (as in this case) the work of inserting expansion joints was a comparatively trivial part of the whole building, so as not to involve giving back to the

[2] Leasehold Property (Repairs) Act 1938.
[3] Law of Property Act 1925, s. 146.
[4] [1980] Q.B. 12.

landlord a wholly different building, the tenant was liable for the repairs. In answer to a further argument that the tenant should not be liable for repair necessary to remedy an inherent defect Forbes J. said:

> "It was proper engineering practice to see that such expansion joints were included, and it would have been dangerous not to include them. In no realistic sense, therefore, could it be said that there was any other possible way of reinstating the cladding than by providing the expansion joints which were in fact provided."

Thus, the tenant was held liable for the whole cost of repair, in the only way that it could realistically be done. The effect of this, and other cases to like effect, has been to cause tenants to bring claims for their own loss (including loss of use or loss of profit) against contractors and designers. The fact that such claims are now not generally maintainable in tort (see Chapter 14) has meant that tenants in particular have sought direct contractual warranties from those responsible for the work (see Chapter 9).

Licences

A right to occupation of land which is not sufficient to create a true tenancy is said to create a licence. This is essentially a right to do some act which would otherwise be a trespass. Examples of the operation of licences are a person occupying a cinema seat or an hotel room; or a contractor in possession of a building site. Essentially a licence is a personal arrangement between grantor and grantee which does not bind third parties. There may, however, be circumstances where an interest in the land affecting third parties is created. In *Inwards v. Baker*[5] a man allowed his son to build a bungalow on his (the father's) land and then died leaving the land to others. It was held that the son should be allowed to remain in the bungalow as long as he desired. Lord Denning said:

> "All that is necessary is that the licensee should, at the request or with the encouragement of the landlord, have spent the money in the expectation of being allowed to stay there. If so, the court will not allow that expectation to be defeated where it would be inequitable so to do. In this case it is quite plain that the father allowed an expectation to be created in the son's mind that this bungalow was to be his home. It was to be his home for his life, or at all events, his home as long as he wished it to remain his home. It seems to me, in the light of that equity, that the father could not in 1932 have turned

[5] [1965] 1 Q.B. 29.

to his son and said: 'You are to go. It is my land and my house.' Nor could he at any time thereafter so long as the son wanted it as his home."

A gratuitous licence merely to enter land may be revoked at any time; while a licence coupled with an interest in the property (such as a right to dig gravel) cannot be revoked. A more usual type of licence in business transactions is one given under a contract. Such a licence will be regarded as part of the contract creating it and its revocation in breach of contract may in some circumstances be resisted by injunction. In *Hounslow v. Twickenham Gardens*,[6] a contractor's employment under a building contract had been terminated by the employer. The contractor, contending the termination was invalid, refused to leave the site. The employer claimed an injunction to remove him, which was refused. The employer was held to be under an implied obligation not to revoke the contractor's licence except in accordance with the contract. In this case Megarry J. said:

> "Now in this case the contract is one for the execution of specified works on the site during a specified period which is still running. The contract confers on each party specified rights on specified events to determine the employment of the contractor under the contract. In those circumstances I think that there must be at least an implied negative obligation on the Borough not to revoke any licence (otherwise than in accordance with the contract) while the period is still running."

Having decided that the Borough had not conclusively established the validity of its determination notices, the judge went on to say:

> "I fully accept the importance to the Borough on social grounds as well as others of securing the due completion of the contract, and the unsatisfactory nature of damages as an alternative. But the contract was made, and the contractors are not to be stripped of their rights under it, however desirable that may be for the Borough. A contract remains a contract, even if (or perhaps especially if) it turns out badly."

This case, however, has been criticised and the result in other similar circumstances may not be the same.[7]

Rights over the land

The question of what the owner or occupier is entitled to do with the land depends upon many factors. It is subject to numerous

---

[6] [1971] Ch. 233.
[7] See *Chermar v. Pretest* (1992) 8 Const. L.J. 44.

statutory provisions, such as the Housing Acts and Public Health Acts. Building work itself is closely controlled by regulations and byelaws (see Chapter 16). The occupier of land may become liable to his neighbours under the law of nuisance (see Chapter 14). Perhaps the most fundamental restriction upon the user of land arises through statutory planning controls (see below). Rather than to say that land is owned subject to restrictions, it is probably more accurate to say that land is held for the benefit both of the owner or occupier and of the community.

As to ownership of things on or in the land (as opposed to "ownership" of the space occupied by the land), it is presumed that the owner of the freehold owns everything upon or below the land. He is generally entitled to everything which is attached to the land. Such items are commonly called fixtures (as opposed to mere fittings) and they will belong to the freeholder as against a tenant. The question of what attachment is sufficient to make an object a fixture is a matter of degree and purpose. Thus in *Webb v. Bevis*[8] a corrugated iron building which was bolted to, but not embedded in, a concrete floor was held not to be sufficiently attached to be a fixture. Scott L.J. said:

> "That the concrete floor was so affixed to the ground as to become part of the soil is obvious. It was completely and permanently attached to the ground and, secondly, it could not be detached except by being broken up and ceasing to exist either as a concrete floor or as the cement and rubble of which it had been made. Does that fact of itself prevent the superstructure from being a tenant's fixture? I do not think so. If it had been erected on concrete blocks, one under each post, the top level with the surface of the ground and the attachment of post to block had been plainly removeable at ground level, "the object and purpose" of the attachment would have been obvious namely to erect a mere tenant's fixture. In my opinion it was equally so in the actual construction adopted for holding the posts in position on their concrete supports."

Building materials will become the property of the freeholder as soon as they are (like the concrete floor in the case) attached to the land or to the permanent works, whether paid for by the employer or not. As between landlord and tenant there are certain exceptions to the rule, which relate, particularly to trade and agricultural fixtures.

---

[8] [1940] 1 All E.R. 247.

RIGHTS OVER LAND OF OTHERS

Of the many types of interest over land belonging to or in the possession of other persons, the most important so far as a building developer is concerned are easements and restrictive covenants. In this context the building developer is seen as the "other person" over whose land rights exist which may affect its use or development. The short account given below describes the nature of these interests.

Easements

An easement is a right which allows the holder to use, or restrict the use of, the land of another person in some way. Common examples are private rights of way, rights of light and rights of support. An easement can exist only in relation to other land which is nearby, which is said to "benefit" from the easement. A "quasi-easement" (referred to in clause 22 of the ICE conditions) is a term used to describe an habitual right exercised by a person over a part of his own land, which would be an easement if the two parts were in different occupation, such as a right of support between adjoining buildings. A quasi-easement may become a real easement upon a sale of one or both parts of the land.

Rights of light often restrict development on sites adjoining the land which enjoys the right, called the "dominant tenement." An action to preserve a right of light lies in nuisance. But even where the right has been acquired by long usage, the holder cannot demand unlimited light. The courts have held that the right is to have enough light adequately to light a room "according to the ordinary notions of mankind".[9] It is not the function of a planning authority to guard or preserve rights of light or other easements. Thus, the fact that a right of light is enjoyed does not prevent a developer obtaining planning permission for a building which will block the right. But nor does the grant of planning permission guarantee that the developer will not be prevented from erecting the building in breach of the rights of neighbours. Building in apparent breach of a right of light is a good example of nuisance which could be restrained by interim injunction (see Chapter 14).

There are a number of other rights which are similar to, but which do not comprise easements. For example, a profit (or *profit à*

---

[9] *Colls v. Home and Colonial Stores* [1904] A.C. 179.

prendre) is a right to take something from the land of another person, such as grass or sand. A licence is a private right to go upon another's land (see above). Either of these rights may exist without the holder owning land which is benefited. An easement of support relates only to a building and is distinct from the natural right to have unweighted land supported by adjoining land (see Chapter 14).

Covenants

A contractual provision which seeks to constrain the way in which the holder of land may use it is termed a restrictive covenant. An example of the type of covenant which might be relevant to a building developer is one not to build on certain land. As between the original parties a covenant is binding, for instance, when the covenantee and covenantor are respectively landlord and tenant. It will also be binding on successors if it constitutes an easement (see above). Otherwise, only in limited circumstances will a restrictive covenant be enforceable by and against successors in title. A plaintiff who wishes to enforce a restrictive covenant must show that he has acquired the "benefit" of the covenant and that the defendant has acquired the "burden."

An example of the operation of restrictive covenants occurs where an estate is laid out in lots to be sold or leased for building and each purchaser or lessee agrees to similar restrictive covenants. It is necessary that the area be clearly defined and that restrictions are imposed by the common vendor or lessor which are consistent with the general scheme of the development. The covenants must be for the benefit of all the lots and the sale or leasing must be transacted with this intention. Provided these conditions are satisfied, the covenants will be enforceable by and against the owners or lessees for the time being of any plot on the estate. The covenants therefore constitute a local law for the estate.

Party walls

Problems frequently occur in heavily built-up areas over the rights possessed by adjoining owners in a party wall. This occurs where the boundary line between two premises is built on, or where one adjoining owner has acquired rights in a wall built up to the boundary on neighbouring land. The problem is frequently to define where the boundary lies, and this is often a major problem where there have been alterations and extensions over the years. Having identified the position of the boundary, it is often found that walls vary in line and in thickness, and foundations usually extend out beyond the wall itself, so that there is often no simple answer to the question of ownership.

Assuming that ownership can be established, there are rights at common law to carry out work to a party wall, notwithstanding that part of the wall is not owned by the party wishing to carry out the work. In London, where the problem is most acute, the rights were codified in the London Building Acts. These provisions worked well for London, and have now been extended with some amendments, to the whole country by the Party Wall, etc., Act 1996. The Act, for the first time, contains a definition of "party wall" which means:

> "(a) A wall which forms part of a building and stands on lands of different owners to a greater extent than the projections of any artificially formed support on which the wall rests; and
> (b) so much of a wall not being a wall referred to in paragraph (a) above as separates buildings belonging to different owners."

This definition covers both a wall which straddles a boundary and a wall which is built up to a boundary. Section 1 of the Act makes provision for building new party walls. Section 2 covers the general problem of rights in an existing party wall where one neighbour wishes to build. These rights include underpinning, raising, demolishing and rebuilding or carrying out various work to the wall in question. The person wishing to carry out work must normally serve a "Party Structure Notice" at least two months before the date for beginning the proposed work. No notice is required if the adjoining owner consents in writing, nor if the building owner is required to comply with some statutory obligation, such as complying with a Dangerous Structure Notice. The adjoining owner may serve a "counter notice" and any dispute which is not settled by agreement is to be referred to surveyors under section 10 of the Act. The old London Building Acts required each party to appoint a surveyor. The new Act, consistent with the Civil Procedure Rules (see Chapter 2) provides for the appointment of an "agreed surveyor" or two surveyors if there is no agreement. The two surveyors settle the dispute by a "Party Wall Award". If they cannot agree they must appoint a third surveyor to determine the matter. The award is to determine:

(a) the right to execute any work;

(b) the time and manner of executing any work; and

(c) any other matter arising out of or incidental to the dispute including costs of making the award.

Section 10 of the Act further provides that either party may appeal the award to the county court within 14 days; otherwise the award is to be conclusive.

Planning Law

Planning law is substantially a creature of the twentieth century. The first attempts at systematic town planning were introduced in 1909, and since then the scope and complexity of planning law has widened enormously. Very largely unrestricted urban sprawls in the 1930s, together with the destruction brought about by the Second World War with its consequent opportunities for replanning, has been responsible for much of this development. The principal enactment is now the Town and Country Planning Act 1990, which consolidates the previous statute law relating to planning. There are also other important Acts together with regulations, rules and orders made under statutory powers. Parallel legislation has shown increasing interest and emphasis on the conservation of ancient monuments and historic buildings and urban areas.[10]

The practical effect of modern planning law is that the owner's rights to use his land are to a large extent subordinated to the good of the community. In the historic 1947 Act, land use planning was made of general application; and both development rights and development values were effectively nationalised (the latter being returned to landowners by later legislation). In general, a land owner has no right to use his land for any purpose other than its present use unless he obtains permission. Further, he may be dispossessed of even its present use by authorities exercising powers of compulsory acquisition. The economic importance of planning law to the individual is demonstrated by the direct effect which planning consent has upon the value of land and buildings.

In origin, planning law is entirely statutory, that is to say, it is all contained in Acts of Parliament and the delegated legislation made under the Acts. Most statutes are periodically considered and interpreted by the courts and this usually makes them easier to understand. But planning law is brought before the courts on comparatively rare occasions. The usual procedure is that planning decisions are made initially in a purely administrative way, by the

---

[10] Planning (Listed Buildings and Conservation Areas) Act 1990.

local authority or government department. An appeal is usually allowed to the Minister, and his function is sometimes described as quasi-judicial. But in either case the decision is not of legal rights (as in the case of a decision in the courts) but of the proper administration of planning policy. Thus, in a case which arose under the New Towns Act 1946, the Minister stated publicly that Stevenage would be the first new town. He made a draft order to this effect and an inquiry was held into objections. The Minister then confirmed his order. The court held that there was no judicial duty upon the Minister, but only a duty to consider the objections: *Franklin v. Minister of Town and Country Planning* (1948). In this case, Lord Thankerton, in the House of Lords, held:

> "In my opinion, no judicial or quasi-judicial duty was imposed on the (Minister) and any reference to judicial duty, or bias, is irrelevant to the present case. . . . The (Minister) was required to satisfy himself that it was a sound scheme before he took the serious step of issuing a draft order. It seems clear also, that the purpose of inviting objections and, where they are not withdrawn, having a public inquiry, to be held by someone other than the (Minister) to whom that person reports, was for the further information of the (Minister). . . . I am of the opinion that no judicial duty is laid on the (Minister) in discharge of those statutory duties and that the only question is whether he has complied with the statutory directions to appoint a person to hold the public inquiry and to consider that person's report."

There may be an application to the High Court for judicial review if a point of law is disclosed in the manner or form of the action taken by a local planning authority or by the Minister. It is only through such decisions of the court that case law is created. Otherwise, decisions made by planning authorities or by the minister are administrative decisions only which do not create any precedent. While such decisions may serve as a guide to enable applicants to assess the likelihood of planning consent being granted, each decision is a matter of discretion, limited by the application of relevant local and national planning policies and guidance, for example, Planning Policy Guidance (Notes) (PPGs) or Minerals Planning Guidance (Notes) (MPGs) issued periodically and updated by the Secretaries of the Environment and Transport.

Planning authorities

The administration of planning law and planning control at local levels is carried out by the local planning authority. This body will be either a county council or a district council or, in London, the relevant Borough Council. The authority responsible for central administration is referred to in the Acts as the Secretary of State (for the Environment, or for Wales or Scotland). In addition to these authorities, there are other bodies which have specific functions in relation to planning administration. In particular, the Lands Tribunal, among its various functions, has powers to settle disputes over the valuation of land arising out of planning decisions.

Development plan

A primary creative duty of the planning authority is regularly to produce and review plans for future development. The development plan has two parts. First, the structure plan formulates policy and general proposals for development and use of land. This will take account of Regional Planning Guidance (RPG) issued for the standard planning regions of England (*e.g.* West Midlands) which will include the conservation of natural beauty, the improvement of the physical environment and the management of traffic. The structure plan is prepared by the county planning authority, where this still exists and by unitary authorities elsewhere. Secondly, local plans may be prepared by district and unitary planning authorities to show development proposals in any part of the area. Local plans must be provided for action areas, but different plans may cover the same area for different purposes. Both the structure and local plans must be given adequate publicity and inquiries may be held if there are objections as to what they do or do not contain. The structure plan must be submitted to the Secretary of State. These and other plans are adopted by the local planning authority, but the Secretary of State may direct that his approval is required. A local authority, when considering a planning application is required to "have regard" to the provisions of the development plan. Recent amendments to the 1990 Act additionally require any determination under the planning Acts to be "in accordance with the (development) plan, unless material considerations indicate otherwise".[11]

The ultimate objective of a planning authority is thus to see that its development plan is carried out. Where this can be achieved by planning restrictions the problem is reduced to one of imposing limits or conditions on consents or of enforcement. But positive

---

[11] Planning and Compensation Act 1991.

development may require more than mere control. One of the most important powers of local authorities in this respect is the power to acquire land compulsorily for planning and related purposes, with the authorisation of the Secretary of State. The measure of compensation which is payable upon compulsory acquisition is basically the open market value, but subject to some statutory modifications, for example, where land or buildings are being acquired for a public or non-profitable use. There are also provisions for additional compensation for such matters as distur-bance and severance of lands. Local authorities may compulsorily acquire land within their own areas and also in other areas. They may also acquire land by agreement, when the consent of the Secretary of State is generally not required.

Where land has been acquired or appropriated by a local authority for planning purposes, the local authority itself may carry out building or work upon the land; or instead of carrying out development itself the authority may make arrangements with an authorised association to carry out such operations. Alternatively the local authority may dispose of land to others (such as building developers) so as to secure the use or development of the land needed for the proper planning of the area. This method has been used to secure the re-development of many inner city areas. Where land which was acquired compulsorily is to be disposed of, the previous occupants must, so far as is practicable, be afforded an opportunity to return to their land.

Requirement of planning consent

In general any development of land requires a formal application for planning consent to be made to the local planning authority and the development may not be carried out unless such consent is granted. "Development" is defined by the Town and Country Planning Act 1990 as meaning the carrying out of building, engineering, mining or other operations in, on, over or under land, or the making of any material change in the use of any buildings or other land (section 55). The Act contains further definitions of many of its terms, such as "building," "land" and "use" (section 336). There are, however, some classes of exceptions under which things may be done to or with land without the necessity of obtaining planning consent.

The first exception is that no planning permission is required if the project is not within the meaning of "development." The Act states that certain operations or uses of land are not to be taken to involve development. These include most works which do not materially affect the external appearance of a building, maintenance works or works within the public highway. But the division of a dwelling-house into two or more separate units requires planning permission, as does also the extension of a refuse tip, in area or in height, so as to exceed the level of adjoining land (section 55).

Where land is being used for a classified purpose[12] the change to another use of the same class does not constitute development. Thus, the change from a grocery shop to a tobacconist does not require planning permission; but a change to a fried-fish shop or to an office not within the exemption and requires permission. If there is uncertainty as to whether any project constitutes a development within the meaning of the Act, it may be resolved by determination of the local planning authority (section 64) with an appeal to the Secretary of State and a further appeal to the High Court (section 290).

Secondly, the Secretary of State has power to make development orders which permit either particular development in a given place or area or some class of general development. An order may itself grant planning permission or provide for permission to be granted by the local planning authority. An order may be limited in its area of application and may be subject to conditions. The current order sets out the types of operation and change of use which the order itself sanctions.[13]

The third exception is that planning permission is not required in some specified cases relating to the resumption of a former use of land, such as after the expiry of planning permission granted for a limited period (section 57). A fourth exception applies to local authorities and statutory undertakers. Where authorisation is required from a government department for a development, the department may itself grant deemed planning permission (section 90).

Applications for planning consent

If a project requires planning permission to be obtained then the usual course is to make application to the local planning authority. An application must be made in such a manner as is prescribed by regulations,[14] and must be accompanied by plans and other particulars of the project. In some cases the applicant is required to certify

---

[12] Town and Country Planning (Use Classes) Order 1987.
[13] Town & Country Planning (General Permitted Development) Order 1995.
[14] Town & Country Palnning (General Development Procedure) Order 1995.

that he has given certain notices. For example, where the applicant is not the freeholder or leaseholder of all the land in question, he must take steps to notify the owner of the land of the application. Also, applications relating to certain classes of development must arrange for the local advertisement of the application (sections 65–69). Planning applications require payment of a fee, which is periodically reviewed.

Application may be made for "outline permission," in accordance with the provision of a development order (section 92). Such permission will be subject to subsequent approval of "reserved matters" not particularised in the application. Local authorities will generally resist the granting of outline planning permission if the site lies within a conservation area.[15] Outline planning permission will only be granted subject to application for approval of reserved matters, and commencement of the development, being made within specified periods. The procedure for outline planning applications may be useful for a prospective developer who does not own the land in question. The procedure does not apply to "change of use" development.

The local planning authority must generally give its decision within two months of an application for planning permission. In coming to a decision the authority must always have regard to the adopted or emergent development plan, and to all other material matters. There are provisions requiring various persons and bodies to be consulted. Where applications require local advertisement, the planning authority must take into account representations received. Where the applicant is not the freeholder or leaseholder the representations of owners of the land must be considered (section 71). In particular, works of demolition, alteration or extension of buildings listed as being of historic or architectural importance are given special consideration, and require a separate or additional application.

Effect of planning consent

Planning permission may be granted or refused or granted subject to conditions, and unless it is granted unconditionally reasons must be given for the decision. The conditions imposed may be permanent or of limited duration, although they must relate to the development. For example, a condition that payment should be made to the local authority does not relate to the

---

[15] Planning (Listed Building and Conservation Areas) Act 1990.

development, and is invalid. The planning permission itself is granted for a specific period (section 72). As an alternative to planning or listed building consent applications being determined by the local planning authority, the Secretary of State may "call in" for determination particular applications or classes of application. This is particularly common in applications to demolish listed buildings or for controversial or unusual forms of development.

Once planning permission is obtained it attaches to the land for the benefit of subsequent owners. However, there is generally a condition that the development will be begun within five years unless other express time limits are laid down (section 91). The local planning authority is also given powers to promote timely completion of developments. Thus, the authority may serve a "completion notice" (to be confirmed by the Secretary of State), whereby planning permission will cease to have effect after a specified period in respect of uncompleted work (section 94). The authority also has powers to modify or revoke planning permission, with the confirmation of the Secretary of State (section 97).

Appeals

An appeal against the local planning authority's unfavourable decision or failure to decide may be made to the Secretary of State, generally within six months. He may reconsider the whole application as if made to him in the first place, so that an appeal against the conditions imposed with a consent may result in a refusal of consent. The appeal may be considered on the basis of written representations or by way of a hearing or public local enquiry if either party so requests or the Secretary of State so decides (section 78). There is a final appeal from the Minister's decision to the High Court on a point of law. As an alternative, there is provision for an appeal from a decision of the local planning authority relating to the design or external appearance of a building to be heard by an independent tribunal but this is very rare. If the appellant is a central government department or a statutory undertaker or if there are considerations of regional or national importance involved, a Planning Inquiry Commission may hear the appeal.

Powers of control

If a development is carried out without planning consent, or contrary to conditions imposed with such consent, the local planning authority may enforce their control by serving an enforcement notice or listed building enforcement notice (section 172). This must specify the breach of planning or listed building control, the precise steps required for their rectification and the period for compliance. The steps required to be taken may include demolition or alteration of buildings or works. Such a notice must be served generally within four years of the offence, or 10 years of the breach of conditions. A copy must be served on the owner and on the occupier of the land and on any other persons having a sufficient interest. The period for compliance must be reasonable, but development may be stopped on three days' notice by an interim "stop notice" (section 183). An appeal against a notice may be made to the Secretary of State and the appeal is deemed also to be an application for planning permission in respect of the offending building or works (section 174). There is a further appeal to the High Court on a point of law. If the enforcement notice takes effect and the required steps are not taken then the local planning authority may itself carry out the work and recover the cost from the land owner. The owner is also liable to a fine. The wrongful exercise of enforcement controls or stop orders will entitle the injured developer to compensation.

Compensation

Since permission to develop land has such a direct influence on its value it is obviously right that a measure of redress should be provided for those who suffer loss through planning restrictions. Redress may be available in one of two ways. First, if planning restrictions prevent or hamper development or cause depreciation in the value of the land, compensation may be payable in a limited number of cases. Such restrictions may take the form of a refusal of planning consent or may arise from the exercise of powers, such as those to revoke or modify an existing consent or to order removal of an existing building (sections 97, 102). Compensation may also be payable where land value drops as a result of the future possibility of compulsory purchase.

Secondly, when planning permission is refused or granted subject to conditions, the owner may in some cases require his land to be purchased by the local authority. But this is possible only if the land is incapable of reasonably beneficial use (*i.e.* is virtually worthless) in its existing state. The discussion above has been concerned principally with building and construction works. The 1990 Act also contains provisions relating to trees, waste land and advertisements. Buildings of special architectural or historic interest are covered by the Planning (Listed Buildings and Conservation Areas) Act 1990.

New town development

The concept of building new towns dates from the end of the Second World War when the policy of rebuilding devastated city centres and slum clearance produced an excess of population which needed to be rehoused. New towns were seen as the best alternative to continued urban sprawl. The main Act currently in force is the New Towns Act 1981.

The first step in creating a new town is the designation of the area by the Minister. In some cases this has been virgin land but more usually an existing small town is chosen as a nucleus. Before designating a site the Minister will consult local authorities in the area, and if any person or body objects to the site chosen a public local inquiry must be held. However, after considering the objections, the Minister may properly override them and make his decision on the grounds of policy: see *Franklin's* case above.

The development of a new town will be undertaken by an ad hoc authority called a development corporation. This is a corporate body consisting of a number of persons selected by the Minister, and having statutory objects and powers. The purpose of the corporation is to prepare proposals for the development of the new town and to secure their implementation. The corporation does not displace the local planning authority, but is absolved from the normal requirements for obtaining planning consents.

To achieve their purposes, development corporations are invested with wide powers. These include acquisition of land (by compulsory purchase if necessary), carrying out building or other operations, provision of services, and carrying on any business or undertaking for the purposes of the new town. However, the Minister retains a general power of control over any of the powers of the development corporation. Further, a corporation may not undertake, *inter alia*, the supply of water, electricity or gas without first obtaining specific powers to do so.

A new town development corporation is intended to have a limited life. When the laying out and development of a new town has been substantially achieved the corporation will be wound up and its property transferred to the Commission for New Towns, although since 1976 the Secretary of State has had powers to transfer dwellings and associated property to the district council for the area.

Environment Law

This term applies to a group of topics formerly existing under titles such as public health. In some areas, the law exists solely as private rights between individuals, for example, under the law of nuisance. There has been a growing trend towards the creation of wider powers and controls over the use of land and the environment, and increasingly this has taken the form of administrative powers exercisable by public authorities. This has naturally relegated the role of the individual to that of requesting public authorities to exercise their powers, and in appropriate cases of seeking public law remedies where the individual is affected (see Chapter 1).

Much of the modern structure of environment law goes back to the Public Health Acts enacted over the last hundred years and particularly to the massive Act of 1936. This Act dealt with matters such as domestic water supply and sewerage, collection of refuse, provision of recreation grounds and many other topics. Substantial amendments were made in the Public Health Act 1961, which contains provisions dealing with sanitation and trade effluent. A major environmental impact resulted from the Clean Air Act 1956, which controlled the omission of smoke and other effluent from both industrial and domestic premises. This had a dramatic effect on pollution in all urban communities. The current legislation is the Clean Air Act 1993 which contains further controls over emissions of dark smoke and other effluent from chimneys. The other major piece of historical legislation on the topic is the Control of Pollution Act 1974.

In the general sphere of environment law and practice, the Environmental Protection Act 1990 makes far-reaching changes to the whole field by establishing a new integrated approach to pollution control, to be operated by the Inspectorate of Pollution. In regard to less serious pollution, new functions are placed on local authorities; and in regard to waste disposal, local authorities are given new powers and duties. In some areas, including Greater London, new waste authorities are created. The Act amends and replaces Part I of the Control of Pollution Act 1974, with extended controls over waste. There are other provisions for dealing with statutory nuisances and new measures to deal with litter.

Environment law and practice has introduced some new acronyms into the language. Recognising that environment protection must involve questions of priority, controls are often expressed in terms of the principle of Best Available Techniques Not Entailing Excessive Cost (BATNEEC) or the simpler concept of Best Practical Means (BPM). Because environment law is a comparatively new concept, there is, as yet, little case law. The following section concentrates on three topics: pollution of land; the control of noise on building sites; and use and abuse of water. In each area, the courts have made some progress in considering the relevant legislation.

Pollution of land

Statutory controls, now under the Environmental Protection Act 1990, regulate the disposal of waste which is generally categorised as "controlled" waste or "special" waste. The former covers most types of waste, including that arising from construction, demolition or excavation. Materials do not cease to be waste even though they have use or value. The 1990 Act makes it an offence to deposit controlled waste on any land or to keep it without a waste management licence. There are specific duties to avoid causing pollution to the environment and a general duty of care with regards to waste. Licences may be revoked or suspended.[16] No one can now store or deal in waste without complying with the new Act and its regulations.

Special waste is material, designated by regulations, which is considered to be so dangerous or difficult to treat, keep or dispose of that special provision is required for dealing with it (section 62). Regulations are also to make provision for treating, keeping or disposal of special waste. Where special waste is discovered on land, for example in the course of carrying out construction work, regulations require the owner to make provision either for its safe retention on the site or for its lawful removal.

The 1990 Act makes general provisions for a public register of information concerning waste, for inspectors with powers of entry and taking records, samples, etc., for requiring information to be furnished and for powers to deal with imminent danger of serious pollution (sections 64 to 71).

Control of noise

The Control of Pollution Act 1974 has specific application to noise on construction sites. Under section 60 of the Act the relevant local authority is empowered to serve a notice imposing requirements as to the way in which the works are to be carried out. The Act makes the following provisions about notice:

"60(3) The notice may in particular—

    (a) specify the plant or machinery which is or is not to be used;
    (b) specify the hours during which the works may be carried out;
    (c) specify the level of noise which may be omitted from the premises in question or at any specified point on those

---

[16] Environment Protection Act 1990, Pt II.

premises or which may be so omitted during specified
hours; and

(d) provide for any change of circumstances."

Section 60(5) requires the notice to be served on the person who
appears to be carrying out or going to carry out the works and on
others who appear to be responsible for or have control over the
works. The recipient may appeal the notice to a magistrates' court;
otherwise, it becomes an offence to contravene any requirement of
the notice "without reasonable excuse." Section 61 of the Act
makes provision for obtaining prior consent of the local authority
to work which is to be carried out. This is a useful device if the
contractor needs to know whether he will be permitted to use
particular plant and machinery, and if so, during what hours.
Although the Act uses only the term "noise," this is defined as
including vibration.

Enforcement proceedings

Where contravention of a notice occurs, the local authority may
prosecute the contractor who, in the event of conviction, will be
liable to a fine. If the offences are serious or repeated, consideration may need to be given to other action. This is particularly so
where the contractor is under financial or other pressure to
continue with the work in contravention of the notice. In such a
case, the local authority may invoke the civil jurisdiction of the
court to grant an injunction to restrain further contravention of the
notice. If granted, any further contravention of the injunction
would be punishable by all the means available to the court,
including imprisonment. The circumstances in which the local
authority can take this step is, therefore, of general interest,
because it applies in many other areas where a local authority is
charged with the proper administration of the law. One example is
the proper administration of planning law, which has given rise to a
number of cases where there have been flagrant breaches of the
law relating to caravan sites, tree preservation orders, and the like.
If the relevant enforcement notices have been issued and are being
ignored, despite criminal proceedings and the imposition of
(modest) fines, the local authority may need to take sterner action.
Another example (now historical) is Sunday trading, where fines
which could be imposed even for persistent and intentional
breaches of the law, were trivial compared to the profits made from
illegal trading.

The legal basis for action in these circumstances by the local
authority is section 222 of the Local Government Act 1972, which

empowers the local authority to take action in their own name where this is expedient "for the promotion or protection of the interests of the inhabitants of their area." In resolving to take such action, the local authority is exercising a public law power, and as such is liable to control through judicial review if it acts improperly (see Chapter 1). Further, in deciding whether or not to grant an injunction, the court has to consider whether it is appropriate to grant this civil remedy in aid of enforcement of the criminal law, which provides its own remedy. Thus, over a number of cases, the courts have evolved principles upon which an injunction will be granted. These matters were considered by the Court of Appeal in *City of London v. Bovis*,[17] where the City were seeking an injunction to restrain further breaches of a notice served under section 60 of the Control of Pollution Act. Bingham L.J. stated that the guiding principles were:

(1) that the jurisdiction was to be invoked and exercised exceptionally and with great caution;

(2) there must be something more than mere infringement of the criminal law; and

(3) the court must conclude that the defendants' unlawful operations would continue unless restrained by injunction.

He also summarised the alternative courses of action that might have been open as follows:

"Any individual resident of Petticoat Square could have sued in private nuisance. The Attorney General could have sued in public nuisance either ex officio or on the relation of the local authority or a resident of Petticoat Square. The local authority could have sued in their own name for a public nuisance by virtue on section 222 of the Local Government Act 1972 if they considered it expedient for the protection of the interests of the inhabitants of their area. . . . As it was none of these procedures was invoked. Instead, the local authority decided . . . to issue summonses under section 60(8) alleging contraventions of the section 60 notice.

A point argued in the *Bovis* case was whether a construction manager under a form of contract providing for work to be carried out by trade contractors, should be regarded as the appropriate person for service and enforcement of a notice under section 60. Sub-section (5) provides:

---

[17] (1988) 49 B.L.R. 1.

"A notice under this section shall be served on the person who appears to the local authority to be carrying out or going to carry out the works and on such other persons appearing to the local authority to be responsible for or to have control over the carrying out of the works as the local authority thinks fit."

The Court of Appeal held that a construction manager came within the section and was properly to be held accountable for breaches of the notice. The categories under section 60(5) were not mutually exclusive, and it was sufficient that *Bovis* under their contract (and under parallel provisions in the trade contracts) were given control of the building operations. Where there is any doubt about the person responsible, the local authority has power under section 93 of the Act to serve notice requiring information.

This may allow the authority, for example, to obtain copies of relevant commercial documents which will establish who is in control for the purpose of notice abatement.

Use of water

Water-borne pollution may be seen as one of the most serious problems which environment law has to tackle. Off-shore and deep-sea pollution are matters for international law and treaty; but inland and coastal waters are subject to close statutory control of their use as well as to historical common law principles.

A land owner who borders a watercourse is known as a "riparian owner". He normally owns the land up to the middle of the stream. At common law the riparian owner is entitled to the use of flowing water for ordinary purposes such as domestic use. He may also use the water for an extraordinary purpose such as manufacturing or irrigation, provided the water is returned in the same volume and character to the stream. However, there are very stringent statutory controls on the right to abstract water from any source of supply, whether it be a watercourse or a well or bore-hole, for other than domestic purposes, contained now in the Water Resources Act 1991.[18]

Most matters concerning use of inland waterways and resources come under the statutory jurisdiction of the Environment Agency.[19] The question of what person are entitled to put into water is strictly controlled by the Water Resources Act 1991, Part III, re-enacting earlier statutes. The Act makes it an offence to cause or

---

[18] Pt II, Chap. II.
[19] Until April 1, 1996 the National Rivers Authority: see Environment Act 1995 and Water Resources Act 1991, ss. 1–23.

knowingly permit any poisonous, noxious or polluting matter or any solid waste matter to enter any controlled waters (section 85(1)). In respect of existing discharge of trade effluent, sewage effluent, etc., the Act provides for prohibition notices which may limit the amount or type of discharge or impose conditions, and for any discharge outside these requirements to be treated as an offence (section 86). "Controlled waters" includes rivers, lakes and ponds, ground water, coastal waters and territorial waters (section 104).

Where the "person" causing or permitting pollution is a corporate body, an officer of that body (such as a director or manager) may also be liable. A manufacturer was held liable for causing pollution of a river, even though he had taken elaborate precautions to prevent it and accidental spillage occurred only because of a defect in the apparatus: *Alphacell Ltd v. Woodward*.[20] In this case, the House of Lords considered whether an offence under the Act was committed by a person who had no knowledge of the fact that pollution was occurring and had not been negligent. Lord Dilhorne held:

"Here, the acts done by the appellants were intentional. They were acts calculated to lead to the river being polluted if the acts done by the appellants, the installation and operation of the pumps, were ineffective to prevent it. Where a person intentionally does certain things which produce a certain result, then it can truly be said that he has caused that result, and here in my opinion the acts done intentionally by the appellants caused the pollution."

Environment law and Europe

The European Commission has been active in promoting environment protection measures for many years. The Fifth Action Programme covers the period 1993 to 2000, and indicates the general framework in which directives or regulations may be issued by the commission. Unlike the previous Action Programmes, the present one is backed by new provisions inserted in the Single European Act by which the Community has a mandate to "preserve protect and improve the quality of the environment . . . contribute towards protecting human health . . . [and] to ensure a prudent and rational utilisation of natural resources."[21]

Specifically in relation to construction a Green Paper was issued in 1993. The paper addresses the questions of civil liability of

---

[20] [1972] A.C. 824.
[21] Art. 130, r. (1).

individuals for pollution, and the remedying of damage not met by civil liability principles, through some joint compensation system. The broad principle adopted is that "the polluter pays."[22] It is to be noted that there are other massive developments on foot elsewhere in the world, some of which are in advance of European developments, for example, the United States "superfund". Among the major problems forced by all such schemes is the need for rapid response and prevention measures, as to which ordinary legal processes offer little help; and the international nature of much water borne and air borne pollution.

---

[22] See generally *Construction Law and the Environment*, (Uff, Garthwrite and Barber ed., 1994, King's College, CCLM).

CHAPTER 16

CONSTRUCTION STATUTES AND SAFETY

CONTRACTS under which work is carried out in the construction industry are affected directly by few statutory provisions. However, matters such as the design of the works and the mode of carrying out operations are likely to be subject to many statutory controls. These may impose obligations on one or more of the parties involved. A breach of such an obligation may give rise to statutory penal sanctions and also to consequences at civil law, in tort or for breach of contract (see JCT, clause 6 and ICE, clause 26).

The more directly relevant statutes which affect the construction industry include the Public Health Acts 1936 and 1961, the Health and Safety at Work Act 1974, the various planning Acts and the Building Act 1984. There are also a number of statutes which apply only to London, which include the three Acts known as the London Building Acts. Many of the Acts referred to contain powers for the creation of further delegated legislation in the form of regulations or byelaws, which lay down detailed provisions.

In this chapter three types of statutory provision are discussed, which are of importance to construction work. First, the Building Regulations, which govern the design and construction of building works. Secondly, a short account is given of the law relating to highways, which is primarily governed by statute. Thirdly, statutory provisions relating to health and safety are described.

BUILDING REGULATIONS

Statutory provisions relating to building regulations are now consolidated into the Building Act 1984, which applies throughout England and Wales, including the Inner London Area. Before 1965 the design and construction of buildings was regulated by byelaws made under the Public Health Acts by individual local authorities. There were variations between different authorities, but latterly provisions became based on model byelaws issued by central government. In 1965 the first set of Building Regulations applying throughout England and Wales was issued. These have been amended periodically. The provisions now in force are the Building Regulations 1991.

The current regulations are shorter than previous Building Regulations, and are so expressed that rigid enforcement is discouraged. The intentions are that the new regulations will be interpreted so as to achieve reasonable standards of health or safety. A particular innovation in the Building Act 1984 is the use of "approved documents" under section 6. The effect of such documents is defined by section 7 as follows:

> "(1) A failure on the part of a person to comply with an approved document does not of itself render him liable to any civil or criminal proceedings; but if, in any proceedings whether civil or criminal, it is alleged that a person has at any time contravened a provision of building regulations—
>
> > (a) a failure to comply with a document that at that time was approved for the purposes of that provision may be relied upon as tending to establish liability, and
> > (b) proof of compliance with such a document may be relied on as tending to negative liability."

Powers and duties of the local authority with regard to the passing or rejection of plans are set out in section 16 of the Building Act 1984. In addition, the new regulations contain detailed provisions as to notices and forms. There are, however, provisions by which Building Regulation compliance or approval may be secured other than by a decision of the local authority. These are:

(1) Under section 17 of the Building Act 1984, "approved persons" may be designated to give a certificate of compliance with Building Regulations. Such a person must also provide evidence of insurance arrangements complying with regulations; alternatively there may be an "approved scheme" to provide insurance.

(2) Under sections 12 and 13 of the Building Act 1984, there may be approvals of any type of "building matter," *i.e.* any matter to which the Building Regulations are applicable. Such approvals may be made with or without specific application.

Content and application of regulations

As well as being comparatively short, the regulations are arranged differently from previous Building Regulations, and are as follows:

(A) Structure;

(B) Fire safety;

(C) Site preparation and resistance to moisture;

(D) Toxic substances;

(E) Resistance to the passage of sound;

(F) Ventilation;

(G) Hygiene;

(H) Drainage and waste disposal;

(J) Heat producing appliances;

(K) Stairways, ramps and guards;

(L) Conservation of fuel and power;

(M) Access and facilities for disabled people;

(N) Glazing-materials and protection.

These matters are set out In Schedule 1 to the regulations and are referred to as "Requirements." They are preceded by numbered regulations which set out the application of the technical provisions. Thus, regulation 4 requires that the building work should be carried out so that it complies with the relevant requirements of the schedules. As regards extension to or alteration of an existing building, regulation 4(2) requires that it shall not be adversely affected in relation to compliance with Schedule 1 (that is, there is no obligation to bring the whole building up to current regulations).

The regulations apply, *inter alia*, to the erection or extension of a building or to a material alteration (regulation 3). Control is also imposed on a material change of use, defined by regulation 5 as including changing to use as a dwelling, the creation of a flat within a building or changing a building for use as an hotel or public building. Regulation 6 sets out requirements that apply to a material change of use.

In addition to the technical requirements, regulation 7 makes a general requirement that any building work shall be carried out "with proper materials and in a workmanlike manner," thus reflecting similar obligations which would arise by implied terms in a construction contract, or by statute, for example, the Defective Premises Act 1972, or the Supply of Goods and Services Act 1982. However, there is a general limitation on application of the technical requirements in that regulation 8 provides that no obligation imposed by Schedule 1 or by regulation 7 is to require anything to be done "beyond what is necessary to secure reasonable standards of health and safety for persons in or about the building or others who may be affected. . . ."

The regulations do not apply in a number of circumstances. Some buildings and works are exempt from the regulations, including buildings not frequented by people, temporary buildings, greenhouses and agricultural buildings. The technical requirements set out in Schedule 1 are accompanied by notes stating the limits beyond which the requirements are not applicable.

Notices and compliance with regulations

A person who intends to carry out building work must submit to the local authority either a "building notice" containing the particulars specified in regulation 12, or he may at his option submit "full plans" complying with regulation 13. Alternatively, there may be compliance instead with the "approved persons" procedure (see above). Full plans are required where the work concerns means of escape, and the Fire Precautions Act 1971 applies (regulation 11(2)). The authority must pass or reject the submitted plans within a period of five weeks, which may be extended by agreement to two months.[1] Where plans are defective or show a contravention the authority may, as an alternative to rejection, pass the plans subject to conditions requiring modification and further deposit of plans.

In addition to obtaining approval under the regulations, a person intending to carry out building work must give notice at various stages of the work. These include notice 48 hours before commencement of the works, 24 hours before covering up a foundation or damp-proof course, etc., 24 hours before covering a drain and seven days after completing drain work. Notice must also be given within seven days of completion of building work. During the course of work, the local authority may exercise a right to make tests on drains, or to take samples of materials, to establish compliance with the requirements of the regulations.[2]

---

[1] Building Act 1984, s. 16.
[2] See regs 14(1), 14(5), 15, 16.

Building Regulations in London

Historically, London has had its own statutes and building control system, with features differing materially from those applying to the rest of England and Wales. For this purpose, London means the 12 inner London boroughs. Powers to introduce a uniform system throughout England and Wales, including London, were contained in the Building Act 1984. Shortly after their issue, the Building Regulations were applied by statutory instrument[3] to the inner London area, with comparatively minor provisions where the application in London differs. The former building control system in London was contained partly in three statutes known as the London Building Acts, being the Acts of 1930 and 1935 and the London Building Acts (Amendment) Act 1939. Parts of these Acts still remain in force, applying only to the inner London area. These parts relate primarily to precautions against fire, dangerous structures and party walls. There are proposals to extend the London statutory party wall procedure to the remainder of the country.

Civil liability

The Public Health Acts, the Building Act and Building Regulations are enforced primarily through the criminal law. But the regulations are frequently incorporated into building contracts so that failure to comply will be a breach by the builder. The JCT and ICE forms of contract contain general obligations to comply with statutory provisions[4] which include the regulations. A provision for imposing general civil liability for breach of a duty imposed by Building Regulations is contained in section 38 of the Building Act 1984.[5] This has not been brought into effect.

During the 1970s and 1980s there were many cases before the courts where local authorities and builders were held liable in negligence for failing to ensure compliance with Building Regulations. These decisions were based on the case of *Anns v. L.B. Merton*,[6] now overruled by *Murphy v. Brentwood D.C.*[7] (see Chapter 14). The present law is that, where a building is found to be defective but there is no injury to person or property, money spent in repairing the building must be regarded as a purely economic

---

[3] Building (Inner London) Regulations 1985.
[4] See JCT, cl. 6, ICE, cl. 26.
[5] Formerly Health and Safety at Work Act 1974, s. 71.
[6] [1978] A.C. 728.
[7] [1991] 1 A.C. 398.

loss, not normally recoverable under the law of tort. Effectively a whole chapter of litigation has been closed by the *Murphy* decision. The principal right of redress outside the field of contract, is now under the Defective Premises Act 1972 (see Chapter 7).

HIGHWAYS

This section illustrates the operation of statute law in a particular area, which is of importance both to those involved in highway construction and in other fields of construction, such as housing development. The principal governing legislation is now the Highways Act 1980, which consolidates and amends a number of previous Acts. The Act does not attempt to define the term "highway," this being a term of some antiquity and of importance under the common law, for example in relation to the acquisition of rights of passage. Generally, a highway is a defined way over land which is exercisable by the public in general, although the right of use may be limited to particular classes of traffic. Highways thus range from a country footpath or a suburban cul-de-sac to motorways. The 1980 Act provides that a bridge or tunnel over or through which a highway passes is to be taken as part of the highway (section 328(2)).

Rights and duties

Questions that arise in relation to construction projects are: who owns or is responsible for a highway? In particular, who is responsible for its maintenance? What rights and duties exist in regard to services or apparatus under a highway? The first question raises interesting issues about rights and easements (see Chapter 15) as opposed to legal ownership. There are many possibilities which depend on the way in which the highway was created. There is a general presumption that the owner of land adjoining a highway also owns the soil, and all other rights that go with ownership, up to the middle line of the road. A highway built over land is, in law, no more than a right of passage, together with such rights and obligations as are necessary for maintenance and operation as a highway. These rights and obligations are vested by statute in a "highway authority" who will usually be either a county council or central government, acting through the relevant Minister of state. Some minor highway functions are vested in district councils and there are other exceptions.

A highway authority is generally responsible for the construction, maintenance and improvement of highways. A local highway authority may undertake work on behalf of the Minister, either pursuant to an agency agreement or by the provision of services. Different local highway authorities may make agreements between themselves for the carrying out of their functions.

The creation and designation of major roads falls into two categories. Trunk roads, as defined under section 10 of the Highways Act 1980, are intended to form a network of routes for through traffic, relieved of local traffic. This was the main national road network up to the 1950s. Section 16 of the Act deals with a category known as Special Roads, which includes motorways. The creation of a Special Road requires a scheme to be drawn up by the proposed authority, being either the Minister or the local highway authority, and the Act lays down detailed provisions for publication and enquiry before confirmation of the scheme.

Adoption of highways

Builders and developers of housing and industrial estates often construct road systems to serve the new buildings. When the buildings have been sold or leased off, the question arises, who is to undertake responsibility for continuing maintenance of the new roads? This may go beyond mere resurfacing, as the new highways may include earthworks and structures, including bridges. The same question arises in relation to new sewers constructed to serve an estate.

There is no reason why the developer or the purchasers of the buildings should not keep the roads in their own private ownership and at their own expense. However, a convenient alternative is provided by the Highways Act 1980. Section 38 provides that the highway authority may enter into an agreement with the person responsible for maintenance. The terms of the agreement (colloquially known as a "section 38 agreement") usually provide for the road to be constructed and completed to the highway authority's standard specification or satisfaction, and for the highway then to become a public highway maintained at the expense of the authority. Similar provisions apply to new sewers, which may become vested in the relevant water undertaker.[8]

Street works

Many highways, particularly in urban areas, carry services belonging to public utility companies, such as gas, electricity, water,

---

[8] Water Industry Act 1991, Pt IV.

sewerage, telephones and others. This gives rise to many potential problems when the services are to be installed or need repair. Excavations lead to questions of reinstatement and repair of the carriageway, as well as the possibility of damage to other services. These matters are now regulated by the New Roads and Street Works Act 1991 which has introduced an entirely new code to replace the former Streetworks Code.[9]

In the terminology of the new Act, streetworks require a streetworks licence, which is to be granted by the street authority.[10] The term "undertaker" is retained from the old Act and means the person by whom the relevant statutory rights is exercisable or the licensee under the streetworks licence. A licence is required to place, to retain and thereafter to inspect, maintain, etc., "apparatus" in the street, which includes a sewer, drain or tunnel.[11]

The Act requires the undertaker to give not less than seven days' notice prior to the start of works to any other person whose apparatus is likely to be affected. There are special provisions for emergency works. Provision is made for regulations requiring other notices. Notice lapses after seven days if the work is not begun.[12] The Act places many obligations on the undertaker including a duty to carry on and complete works with despatch, a duty to reinstate and a general obligation to co-operate with the street authority and with other undertakers.[13] Regulations may provide for charges to be levied against the undertaker where work is delayed.[14] The licensee is required to indemnify the street authority against third party claims.[15]

The new Act is dealt with expressly under the ICE conditions, sixth edition, clause 27. This provides that, for the purpose of obtaining any licence required for the permanent works, the undertaker is to be the employer, who is also to be the licensee. For all other purposes the undertaker is the contractor. The employer is to obtain any licence or consent required for the permanent works but the contractor is responsible for giving notices. Many of the provisions previously contained in the contract are now found in the Act itself.

---

[9] Under the Public Utilities Streetworks Act 1950.
[10] See ss. 48, 49, 50.
[11] s. 89(3).
[12] See ss. 54, 55, 57.
[13] See ss. 60, 66, 70.
[14] s. 74(1).
[15] Sched. 3, para. 8.

Health and Safety

This section deals with the design and implementation of construction projects, and the safety of those engaged on the work. Construction is notoriously dangerous, not through the callous attitude of employers, but usually because workmen take unnecessary risks. This may be seen as a failure by those in control to impose safety requirements on those working on sites. Statistics show that in 1988, 157 workers were killed in construction accidents, and over 20,000 injured.

As between employer and employee, there is a clear duty of care (see Chapter 14), and it is of little importance whether this is regarded as arising in tort or in contract, since the damages for personal injury will be recoverable under either head. However, historically there has been considerable statutory intervention aimed at providing positive duties for the protection of workers, formerly through the Factories Acts, subsequently through the Health and Safety at Work Act 1974 and most recently from the European Community. The 1974 Act created the new Health and Safety Executive (HSE), whose inspectors administer the Act and its regulations. These provisions apply throughout industry, but there are particular regulations applying to construction work. The new CDM Regulations also take effect under the 1974 Act and depend upon the HSE for enforcement.

Health and Safety at Work Act 1974

The HSE operates under the direction of the Health and Safety Commission, which is itself charged with the overall duty of achieving the purposes of Part I of the Act, which are to secure the health, safety and welfare of persons at work (section 1(1)). The Commission may order investigations or enquiries into accidents, which may be carried out by the HSE itself or by others. Enforcement of the Act and its regulations is carried out through inspectors.

The Act creates general duties on employers to ensure, so far as is reasonably practicable, the health, safety and welfare at work of employees. A similar duty is placed on employers and self-employed persons to ensure, so far as it is reasonably practicable, that other persons are not exposed to risks to health or safety.[16]

---

[16] See ss. 2 and 3.

Persons having control of premises must ensure, so far as is reasonably practicable, that the premises are safe and without risk to health.[17]

Powers available to inspectors include the right to enter premises to make examination and investigation, to take samples, and to require the production of documents and information. Where there is a contravention, the inspector may serve an "Improvement Notice"; and where the contravention involves the risks of serious personal injury, he may give a "Prohibition Notice", which may have immediate effect.[18] The Act provides penalties for breach of its provisions and for other offences.

Detailed regulation governing day-to-day activities on construction sites have been issued and amended periodically. Regulations presently in force were issued before the Health and Safety at Work Act, although they now take effect under it. These are the Construction (General Provisions) Regulations and the Construction (Lifting Operations) Regulations, both of 1961; and the Construction (Working Places) Regulations and the Construction (Health and Welfare) Regulations, both of 1966, also the Management of Health and Safety at Work Regulations 1992, issued subsequently.

CDM Regulations

Since their issue in 1994 and coming into force on March 31, 1995, the Construction (Design and Management) Regulations 1994 (CDM) have had a major impact on the planning of all major construction work in the United Kingdom. The Regulations are made under section 15 of the Health and Safety at Work Act 1994 but derive from a European directive.[19] The Regulations embody new concepts in operational safety planning, as opposed to the previous statutory approach of merely attempting to avoid unsafe situations. The key to the new regulations is the appointment and placing of functions upon two individuals, the planning supervisor and the principal contractor.

By regulation 6 the client (or the developer) must make these two appointments. The planning supervisor is given duties to (1) notify the HSE about the project; (2) fulfil specific requirements as to design; and (3) ensure that the health and safety plan is prepared to comply with the requirements.[20]

---

[17] s. 4.
[18] See ss. 20, 21, 22.
[19] Temporary or Mobile Construction Sites Directive.
[20] See regs 14 and 15.

The principal contractor, who must be a contractor but may be the same person as the planning supervisor, is given specific duties to co-ordinate all contractors on the project, to ensure that they comply with the health and safety plan, to restrict access to authorised persons, to display notices and to ensure that information is provided to the planning supervisor. Other contractors are required to co-operate with the principal contractor, to provide him with information and to comply with his directions.[21]

The design for the project is required to give adequate regard to the need to avoid foreseeable risks to health and safety of any person carrying out construction or cleaning work or of any other person who may be affected by the work. Adequate regard must also be given to the need to combat at source risks to health and safety of any such person and to give priority to measures which protect such persons. It is the duty of the planning supervisory to ensure so far as is reasonably practicable that the design includes the above matters. The Health and Safety plan must contain a general description of the work including its timing and details of risks to the health and safety of persons carrying out the construction work and other information.[22]

The regulations draw particular attention to the need to ensure competence of individuals involved, adequate resourcing and availability of information. The regulations require the planning supervisor to prepare a health and safety file. This must contain adequate information about any aspects of the project, structure or materials which might affect the health or safety of persons working on or cleaning the structure or subsequently being affected by it. The client must ensure that the file is kept available for inspection and deliver it to any person who subsequently acquires an interest in the building or structure.[23]

The regulations are enforceable at the suit of the HSE through penalties and other sanctions under the Health and Safety at Work Act. The importance of the new regulations is that, for the first time, they place duties enforceable under the criminal law, on persons other than the contractor, notably the designer. Limitations apply where the project is carried out "in house"[24] and a project is notifiable under the regulations only if the construction phase will exceed 30 days or involve more than 500 person days of work. Subject to these limits, the client must appoint a planning

---

[21] See regs 16 and 19.

[22] See regs 13(2), 14(a) and 15(3).

[23] See regs 12 and 14(d).

[24] See regs 2(3) and 13(1).

supervisor who thereby becomes liable to ensure compliance with the design requirements. Application of the regulations is assisted by an Approved Code of Practice issued by the HSE.

Civil liability

The CDM regulations do not confer any general right of action in civil proceedings (regulation 21), and nor does the Health and Safety at Work Act itself (section 47(1)). However, a breach of safety regulations may give rise to civil liability, and there are many cases in which the courts have held employers liable on this basis, either in addition to or as an alternative to liability in tort (see Chapter 14).

To succeed in a civil action for breach of a regulation in which there is no provision creating civil liability, it must be shown that:

(1) the regulation was intended to protect a class of which the plaintiff was a member;

(2) the regulation was broken;

(3) he has suffered damage of a kind against which the regulation was intended to protect; and

(4) the damage was caused by the breach.

These requirements may appear self-evident, but their effect in practice may be less than obvious. In a number of cases it has been held that self-employed workmen were not entitled to the protection of the Construction Regulations. Instead of having a right of action in respect of their injuries, self-employed men may themselves be liable to prosecution for breach of the regulations. However, the Court of Appeal has now laid down that the question whether a man is employed or self-employed is to be approached broadly. The fact that he pays his own income tax is not decisive. In *Ferguson v. John Dawson & Partners*[25] a roofing worker was employed expressly as a "self-employed labour only subcontractor." Despite this, the court found that the reality of the relationship created was that of employer and employee. Megaw L.J. said:

> "The parties cannot transfer a statute-imposed duty of care for safety of workmen from an employer to the workman himself merely

---

[25] [1976] 1 W.L.R. 346.

because the parties agree, in effect, that the workman shall be deemed to be self-employed, where the true essence of the contract is, otherwise, of a contract of service."

The kinds. of damage against which the Construction Regulations are intended to protect include personal injuries, but not, for example, loss of earnings due to unsafe scaffolding preventing work. The damage must also be caused by the breach. In *Cummings v. Arrol*[26] a workman was killed in a fall through not wearing a safety belt. It was found that the contractor, in breach of regulations, failed to provide belts for the men. But the contractor showed that the deceased would not have worn a belt even if it had been provided, and consequently was held not liable in civil law for damages. Viscount Kilmuir said:

"The necessity, in actions by employees against their employers on the grounds of negligence, of establishing not only the breach of duty but also the causal connection between the breach and the injury complained of is, in my view, part of the law of both England and Scotland."

Safety on construction sites will continue to be an important and developing area of Construction Law.

[26] [1962] 1 W.L.R. 295.

GLOSSARY OF LEGAL TERMS

ADJOURN: To put off the hearing of a case or matter to a later date.

AFFIDAVIT: A written statement sworn on oath which may be used in certain cases as evidence.

AGENT: A person with authority to act for another (the principal).

APPELLANT: The party who brings an appeal to a higher court.

ARBITRATION: Proceedings before a private tribunal to which the parties agree to submit disputes.

ASSETS: Property which is available for paying debts.

ASSIGNMENT: The transfer by agreement of a right or interest to another person.

AWARD: The decision given by an arbitrator.

BAILMENT: Delivery of goods into the possession of a person who is not their owner.

BANKRUPT: A person who cannot pay his debts and who is adjudicated a *bankrupt*.

BAR: The profession of barristers from which most judges are recruited.

BREACH: Non-fulfilment of some contractual (or other) obligation.

BYELAWS: Rules, usually made under statutory authority and having full force of law.

CASE STATED: A statement of facts prepared by a lower court (or tribunal) for the decision of a higher court on a point of law.

CAVEAT EMPTOR: "Let the buyer beware"; a maxim indicating that any risk is upon the buyer and not the seller.

CHANCERY: One of the three divisions of the *High Court*.

CHARGE: An interest, usually over land, given as security.

CHATTELS: Personal property.

CHOSE IN ACTION: An intangible right which can be enforced by action; such as a debt.

CLAIM FORM: A docoument which initiates a civil claim (formerly a writ).

CLAIMANT: The party who initiates civil proceedings.

COMMERCIAL COURT: A special court in the *Queen's Bench* Division for dealing with commercial actions.

COMMON LAW: Law embodied in case precedent as opposed to *Statute* law or *Equity*.

CONDITION: An important term of a *contract*.

CONSIDERATION: The bargain or inducement provided by a party to a *contract*.

CONSOLIDATION: Joining of two separate actions so that they may be tried together.

CONTRACT: An agreement which is binding in law.

CONVEYANCE: A written instrument which transfers property (especially land) from one person to another.

COUNTERCLAIM: A cross action brought by a *defendant* or *respondent* against the *claimant*.

COUNTY COURTS: Local courts which deal with smaller civil claims.

COURT OF APPEAL: The court which hears appeals from (*inter alia*) the *High Court* and *county courts*.

COVENANT: An undertaking contained in a document, especially in a *deed* or *lease*.

CROWN COURT: The branch of the *Supreme Court* which deals with criminal trials (and also some civil cases).

DAMAGES: The money award made to a successful party in a civil action.

DEED: A formal written instrument, signed and witnessed.

DEFENCE: A pleading from the *defendant* in answer to the *claim form*.

DEFENDANT: The person sued in an ordinary civil action.

DETINUE: An action in *tort* for recovery of a specific *chattel.*

DISCLOSURE: The production before trial of documents relating to a case.

DOMICILE: The country or state with which a person or company is most closely connected.

EASEMENT: A right enjoyed over land belonging to another person.

EQUITY: Law based upon discretion and conscience, derived from the old Court of Chancery.

ESTATE: General term for an interest in land.

EX-PARTE: By one side only, for example an *ex parte* application for an *injunction*.

EXECUTION: Methods of enforcement of a *judgment* in an action.

EXHIBIT: A document used in evidence, especially when annexed to an *affidavit*.

FORCE MAJEURE: Irresistible compulsion, such as war or Act of God.

FORFEITURE: A provision, especially in a *contract* or *lease*, enabling one party to strip the other of his whole interest in certain events; for example determination of a building contract.

FRUSTRATION: Determination of a *contract* by some intervening event, such as destruction of the subject matter.

GARNISHEE: A person against whom a *judgment* debt is enforced by ordering him to pay a debt owed to the debtor to the judgment creditor instead.

GENERAL DAMAGES: Unascertained *damages*, to be assessed by the judge.

HEARSAY: Testimony by a witness as to a matter not within his personal knowledge.

HIGH COURT: The principal court in which civil actions are heard at first instance.

HOUSE OF LORDS: The highest court of England and Scotland.

INCORPOREAL: Rights and interests which are intangible are said to be incorporeal, such as debts or shares in a company.

INDORSEMENT: Something written on the back of a document.

INJUNCTION: An order of the court which commands a person to do or refrain from doing some act.

INTERIM: Provisional, until further direction, for example, an *interim* order.

INTERLOCUTORY: A matter dealt with before the trial of an action; such as an *interlocutory injunction*.

JOINT: Where two or more persons share some right or obligation such that their interest is not severed, each having an interest in the whole, for example a *joint* tenancy or account.

JUDGMENT: The order given by a court after hearing a case.

JURISDICTION: Authority of a court or arbitrator to hear and determine causes, thus, the *county court jurisdiction* is limited financially and geographically.

KING'S BENCH: A division of the *High Court*, called *Queen's Bench* when the sovereign is female.

LAW REPORTS: Authenticated reports of decided cases in the superior courts.

LEASE: A letting or demise of land; also the instrument containing the demise and its covenants.

LEGAL AID: A system for providing free or assisted legal advice or representation, for persons of limited means.

LICENCE: An authority, especially to enter land without having exclusive possession.

LIEN: A right to retain possession of some article until a claim by the holder is satisfied.

LIMITATION: Statutory periods within which actions must be commenced.

LIQUIDATED DAMAGES: Ascertained or calculated monetary loss claimed in an action. *Also* a sum provided by a *contract* as payable in the event of breach.

LIQUIDATION: Winding up of a company.

LORD JUSTICE OF APPEAL: Title of a judge of the *Court of Appeal*.

LORD OF APPEAL IN ORDINARY: Title of a judge in the *House of Lords* or *Privy Council*; commonly called a "Law Lord."

MASTER: An official of the *High Court* who decides many *interlocutory* matters.

MITIGATION: Abatement of loss or damage.

MORTGAGE: A *conveyance, assignment* or *lease* of property as security for a loan.

NEGLIGENCE: Conduct falling short of the duty of care owed to persons generally.

NUISANCE: Unlawful interference with the use or enjoyment of another person's land.

OBITER DICTUM: Statement of a judge on a point not directly relevant to his decision and therefore not strictly of authority.

OFFICIAL REFEREE: Former title of a judge who tries technical cases in the TCC.

OFFICIAL SOLICITOR: An officer of the *Supreme Court* who acts for persons under a disability.

PARTICULARS: Details of some allegation pleaded in an action.

PARTNERSHIP: An unincorporated association of persons in business with a view to profit.

PENALTY: A sum provided by a *contract* as payable in the event of breach, which is not deemed to be *liquidated damages*.

PLEADING: A written statement of a party's case in a civil action.

POINTS OF CLAIM, Defence: The title of pleadings in an *arbitration*.

PRESCRIPTION: A claim to some right, based upon long user.

PRIVY COUNCIL: The judicial committee of the *Privy Council* is the final court of appeal for some Commonwealth countries.

PROFIT À PRENDRE: A right to take something from another person's land.

PUISNE JUDGE: A judge of one of the divisions of the *High Court*, who should be referred to as Mr Justice ----------.

QUEEN'S BENCH: One of the three divisions of the *High Court*.

QUEEN'S COUNSEL or Q.C.: A senior barrister.

QUANTUM MERUIT: An action claiming a reasonable price for work or goods.

RATIFICATION: Confirmation, for example of a *contract*, so as to make it binding.

RATIO DECIDENDI: The relevant part of a judge's decision in a case, which is authoritative.

REAL PROPERTY: Certain interests and rights in land; as opposed to personal property.

RECTIFICATION: Correction by the court of a document so as to express the parties' true intention.

REPLY: A *pleading* from a *claimant* in answer to the *defendant's defence*.

REPUDIATION: An express or implied refusal by one party to perform his obligation under a *contract*.

RESPONDENT: The *defendant* in certain types of action, including an *arbitration*.

RIPARIAN OWNER: An owner of land bordering on a watercourse.

SEQUESTRATION: Order of the *High Court* to seize goods and lands of a *defendant* who is in contempt of court.

SET-OFF: Diminution or extinction of the *plaintiff's* claim in an action by deducting the *defendant's counterclaim*.

SEVERAL: Where two or more persons share an obligation so that it may be enforced in full against any one of them, independently.

SHERIFF: A local office of great antiquity. The present-day duties of a *Sheriff* in civil cases are principally the *execution* of *judgments*.

SIMPLE CONTRACT: A *contract* not by *deed*, whether oral or in writing.

SPECIAL DAMAGES: Ascertained or calculated monetary loss; as opposed to unascertained or *general damages*.

SPECIFIC PERFORMANCE: An equitable remedy whereby a person may be compelled to perform his obligation under a *contract*.

STATUTE: An Act of Parliament.

STATUTORY INSTRUMENT OR S.I.: A form of delegated legislation, which has full force of law.

SUBPOENA: An order requiring a person to appear in court and give evidence or produce documents.

SUBROGATION: The right to bring an action in the name of another person.

SUE: To take legal proceedings for a civil remedy.

SUMMONS: An order to appear before a judge or magistrate. Some civil actions are begun by an originating *summons*.

SUPREME COURT: The *High Court*, the *Court of Appeal* and the *Crown Court* are collectively known as the *Supreme Court* of Judicature.

TECHNOLOGY & CONSTRUCTION COURT or TCC: The courts in which such cases, formerly known as Official Referee's business, are tried.

TORT: A civil wrong, independent of *contract* or breach of *trust*.

TRESPASS: A tortious injury to the person or goods of another, or an unauthorised entry upon his land.

TRUST: A disposition of property to be held by trustees for the benefit of beneficiaries.

UBERRIMA FIDES: Utmost good faith, required in certain transactions, such as insurance contracts.

ULTRA VIRES: Beyond their powers; especially of a limited company or statutory body. The opposite of *intra vires*.

UNLIQUIDATED DAMAGES: Damages which cannot be calculated as a monetary loss, and which are assessed by the judge; such as damages for personal injury.

WARRANTY: A term of a *contract* which is not a *condition*; especially a statement by the vendor as to the quality of goods.

WITHOUT PREJUDICE: Correspondence in connection with a dispute, thus headed, is privileged and cannot be taken as implying any admission by the writer.

INDEX

481